UNACCOMPANIED CHILDREN

UNACCOMPANIED CHILDREN

Care and Protection in Wars, Natural Disasters, and Refugee Movements

Everett M. Ressler, Neil Boothby,
and Daniel J. Steinbock

New York   Oxford
OXFORD UNIVERSITY PRESS
1988

Oxford University Press

Oxford   New York   Toronto
Delhi   Bombay   Calcutta   Madras   Karachi
Petaling Jaya   Singapore   Hong Kong   Tokyo
Nairobi   Dar es Salaam   Cape Town
Melbourne   Auckland

and associated companies in
Beirut   Berlin   Ibadan   Nicosia

Copyright © 1988 by Oxford University Press, Inc.

Published by Oxford University Press, Inc.,
200 Madison Avenue, New York, New York 10016

Oxford is a registered trademark of Oxford University Press

Library of Congress Cataloging-in-Publication Data
Ressler, Everett.
Unaccompanied children.
Bibliography: p.
Includes index.
1. Abandoned children.   2. Child welfare.
3. Abandoned children—Law and legislation.
I. Boothby, Neil.
II. Steinbock, Daniel.   III. Title.
HV873.R54   1988   362.7'044   86-31289
ISBN 0-19-504091-0
ISBN 0-19-504937-3 (pbk.)

1 3 5 7 9 8 6 4 2

Printed in the United States of America

Acknowledgments

This book is a product of a study carried out between June 1982 and March 1985 entitled "Unaccompanied Children in Emergencies: Considerations of Placement Options and Legal Implications for Unaccompanied Children Displaced by War, Natural Disasters, and Refugee Situations." The study, which was an independent research project, was carried out by the three authors of this book, who are responsible for the analysis and conclusions. It drew, nevertheless, on the participation of people in many countries, to whom the authors are indebted for their information, ideas, and criticism.

This project was made possible by grants from the Ford Foundation, Norwegian Save the Children (Redd Barna), the United Nations High Commissioner for Refugees (UNHCR), Norwegian Ministry of Foreign Affairs, the United Nations Children's Fund (UNICEF), Save the Children Federation, Diakonisches Werk, and the International Union of Child Welfare (IUCW). It was through their support that this project was carried out.

The Norwegian Save the Children organization (Redd Barna) in Oslo played a major role in the development of the study by encouraging the study from its inception, by providing administrative and financial support, and by coordinating the input of other agencies involved in the work. The project could not have been implemented without their assistance. The Chairman of the Board of Directors, Mr. Hans Olav Moen; the Secretary General, the late Mr. Sigmond Groven; and the present Secretary General, Mr. Hans Christian Bugge, actively encouraged and supported the study. It was a pleasure to work with the staff of Redd Barna.

Several staff members of the Ford Foundation were instrumental in the development of this study. It was because of the initial encouragement of Peter Geithner, then Representative for Southeast Asia, and Mr. Frank Sutton, then Executive Vice-President, that the project evolved from an idea to a reality. In particular, we wish to thank Diana Morris, the project officer, for her support and many positive suggestions.

Many persons within the Office of the United Nations High Commissioner for Refugees (UNHCR) in Geneva contributed significantly to the study and this book. We thank M. L. Zollner and Alan Simmance in the Assistance Division; Nicholas Morris and Mark Mallock Brown (formerly with UNHCR) from the Emergency Unit. In particular, we thank John Williamson from the Social Welfare Section and Pierce Gerety from the Legal Section for their helpful insights and comments on the manuscript.

From the United Nations Children's Fund (UNICEF) in New York, we thank Richard Jolly, Nyi Nyi, Jacques Beaumont, Vincent O'Reilly, and Ron Ockwell.

We thank the international Council of Voluntary Agencies (ICVA) in Geneva and its Executive Director, Anthony Kozlowski, for their support. The ICVA Sub-Group on Unaccompanied Children and Adolescents served as an adivsory board to the study and gave support through the individual assistance of its members.

In particular, we thank Audrey Moser, Secretary General of the International Social Services in Geneva, who was throughout the study a most valuable advisor. She encouraged the study, chaired meetings on its behalf, provided comments and suggestions, and read draft versions of the manuscript.

The coordinating office for the study was at the Henry Dunant Institute in Geneva, which provided excellent facilities and administrative support for the project. We wish to thank Jacques Meurant, Manfred Kill, and Ted Acherhielm (formerly with the Institute). In particular, we wish to thank Jiri Toman for his personal interest and assistance.

Thanks are due to both the Center for the Study of Human Rights at Columbia University and its director, Paul Martin, and to the University of Puget Sound School of Law and Dean Fredric C. Tausaund and Associate Dean Andrew Walkover, for providing congenial work environments for Daniel J. Steinbock during the study.

The International Union of Child Welfare in Geneva was particularly helpful by permitting access to its library and archives, which date back to the era of the Russian Revolution. They enabled us to photocopy thousands of pages of unpublished materials. We particularly appreciate the assistance and advice of Mr. Gaston Kung.

To address the many issues that are important in the care and protection of unaccompanied children, various background papers were commissioned from study funds. Jan Linowitz (Boston) reviewed the literature on transracial and transcultural adoption and provided an assessment of the resettlement of unaccompanied children. Evi Underhill (Geneva) reviewed the history of the rights of the child in international legislation and the history of the protection of children in armed conflicts. Birgitta Nylund (Uppsala) reviewed the legal situation of unaccompanied children in Norway, Sweden, and Denmark. Claire Rodier (Paris) reviewed the legal and historical situation of unaccompanied children in France, Belgium, and the Netherlands. John A. Paul (Harrisonburg) reviewed the admission of unaccompanied children into the United States. Marianne Kahnert (Geneva) reviewed the literature on the unaccompanied Tibetan children taken to Switzerland. Jan Williamson (Geneva) reviewed the care of unaccompanied children from Laos, Kampuchea, and Vietnam. David C. Chi (Seoul) reviewed the Korean literature about institutional care of children in Korea. The authors wish to thank these people for their contributions.

To learn about the care of unaccompanied children in resettlement countries, efforts were made to encourage independent national reviews within various countries. Howard Adelman (Toronto) led a research group at the University of Toronto in examining the Canadian experience. Helga Jockenhovel-Schiecke (Frankfurt), through the German branch of the International Social Service, reviewed the German experience. Joyce Pierce and Ron Baker (London), through the Ockenden Venture, provided a paper on past experience with unaccompanied minors in the United Kingdom. Susan Forbes (Washington D.C.), through the Refugee Policy Group, organized and carried out a review of the U.S. experience. Mr. Eric van der Houven (the Hague), from the Coordinatiecommissie Wetenschappelijk Onderzoek Kinderbescherming, shared his research on the experience of unaccompanied children in Holland. Most of these papers are being published independently.

Staff of the International Committee of the Red Cross (ICRC) and of the League of Red Cross Societies (LRCS), both in Geneva, assisted by providing historical information, as did the American and German branches of the International Social Service (ISS), the American Council of Voluntary Agencies (ACVA) in New York, the Lutheran Immigration and Refugee Services (LIRS) in New York, and the United States Catholic Conference (USCC) in New York.

Special thanks to Robert Coles for his valuable insights, and to Charles Willie and Charles Ducey, all from Harvard University. We also thank the Lyndhurst Foundation for its generous support of Neil Boothby's work over the past several years.

Other individuals who provided valuable comment and assistance include Atle Grahl Matson, Sue Peel-Morris, Marie de la Souderie, Pirrko Karoula, Pertti Kaven, Dorthy Lagaretta, Herman Stein, Mark Soler, Rhoda Berkowitz, and Henrik Beer. Many others assisted, gave interviews, and shared their views. Even though they are not named here, their contribution was essential.

We also thank those who provided direct services. Nevena Vukovic organized the hundreds of collected reference materials into a library, and Marlynn Geiger and Beth Davis provided secretarial support. Annie Wilson, Susan Abrams, Virginia Germino, and Linda Day Evans provided editorial comment for parts of the book. Claudia Huggins, Ira Gaugler, Evi Underhill, and Marianne Kahnert assisted in archive research. Wanida Hongyok assisted with the accounting and administration.

We also wish to thank Susan Rabiner, senior editor at Oxford University Press, for her interest in the project and her willingness to transform a manuscript into a book.

Finally, for all of their support and encouragement during the years of research and writing, we thank Phyllis Ressler, Martha Clark, and Laurie Jackson.

Bangkok, Thailand   E.M.R.
Durham, North Carolina   N.B.
Toledo, Ohio   D.J.S.
September 1986

Contents

UNACCOMPANIED CHILDREN

UNACCOMPANIED CHILDREN

Introduction

This is a study about unaccompanied children in emergencies: children who are separated from their families during wars, natural disasters, and refugee movements. Its primary purpose is to provide guidance for policy makers and program staff members in all phases of their dealings with unaccompanied children, from prevention through permanent placement. It aims to encourage the satisfaction of children's developmental needs and the protection of their rights.

The fate of unaccompanied children in emergencies is determined by their circumstances and the assistance provided or not provided them. Intervention on behalf of these children is often necessary, and experience, psychology, and law should guide these efforts. Therefore, by reviewing the history of unaccompanied children in selected past emergencies we first identify the lessons to be learned: common patterns and recurring problems. Unaccompanied children are then viewed from a psychological perspective with special attention to the factors which can increase or decrease their inherent vulnerability. Because the law provides the framework within which actions concerning unaccompanied children take place, we examine their situation in comparative and international law. These three analyses provide the basis for recommendations which we believe should guide future intervention.

Unaccompanied children have existed in virtually every past war, famine, refugee situation, and natural disaster. Unaccompanied children are also present in present-day emergencies. At the time of this writing, unaccompanied children from Vietnam, Kampuchea, and Laos are still to be found in refugee camps and holding centers throughout Southeast Asia. Unaccompanied children exist among the famine victims in Ethiopia, among Ethiopian refugees in the Sudan, and, undoubtedly, in the many other African countries affected by drought and famine. They are also among the multitudes affected by war in southern Africa. Children are separated from their families in Lebanon as a result of the conflict there; thousands of others have taken refuge in other countries. As a result of the conflicts in Central America, unaccompanied children are known to exist in El Salvador, Guatemala, Nicaragua, Honduras, and Costa Rica, as well as Mexico. Within the last several years, unaccompanied Haitian, Cuban, Guatemalan, and Salvadoran children have sought refuge in the United States and in Canada. These are only a few examples from current emergencies—there are surely many more. On the basis of past and present experience, it is certain that the future will produce its share of unaccompanied children as well.

For a child, being unaccompanied means living apart from the people who would

3

otherwise provide nurturance, care, and protection—essentials for healthy growth and development. In the milieu of war, refugee movements, and natural disasters, where there are added dangers and difficulties, such care is even more critical. Disconnected from their own families, some children are fortunate enough to be taken in by other caring adults; others are not. For these children, being unaccompanied means having to search for food, clothing, shelter, and other essentials on their own. It may also mean being passed from one adult to another. Unless special assistance is provided, unaccompanied children are dependent upon the chance charity of others, which can fall short of even minimal care and protection. For most unaccompanied children, what happens to them is not of their own choosing, but is forced upon them.

From an administrative or agency perspective, unaccompanied children pose special problems. Relief officials must decide whether to search for unaccompanied children or assume that they are being cared for. When an unaccompanied child has been located, his identity may be uncertain, the whereabouts of his family and their intentions at the time of separation may be unknown, and current responsibility for the child may be ambiguous. Those involved must determine whether the needs of the child are being met in his or her present situation or whether further assistance is required. In the latter case, administrators or agency staff must choose what care should be provided for the child and by whom, and whether the aid is needed on an immediate basis only, for an interim period, or over the long term.

In many past emergencies, however, policy and program staff have not been prepared to make these decisions and have been uncertain as to what actions should be taken, and, therefore, some unaccompanied children have received no help at all. They have been neglected, abused, abducted or exploited; some have become malnourished; some have died. Where there has been assistance, it has sometimes been inadequate or misdirected. Even when programs have satisfactorily met some of the needs, there has been little carryover of the lessons learned to subsequent emergencies. For all these reasons, this study seems needed and timely. While its focus is often on the past, it is directed to the benefit of those children who will be unaccompanied in future emergencies.

Although this book may also be of interest to psychologists, lawyers, sociologists, and historians, it is primarily intended for people directly involved with children. We have made every effort to limit the use of jargon and technical terms, as readers are likely to have diverse backgrounds. We have not attempted the exhaustive analysis common to some areas of historical, psychological, and legal research. Because of the many issues involved, they are not examined beyond the extent necessary to suggest policy. Rather, our aim is to provide an overview that will be useful to policy personnel, child care workers, relief workers, and resettlement staffs.

BACKGROUND

This book grew out of the experience of all three authors working with unaccompanied children in Thailand in 1980–82, during the influx of refugees from Cambodia, (Kampuchea). In this emergency more than 3,500 unaccompanied children were provided with special services. From the moment the unaccompanied children were first identified, relief workers and policy makers were faced with a number of pressing uncertainties including whether the parents of these children were alive and, if so, how to locate them; who was legally responsible for the children; what kind of care and placement would most effectively meet the psychosocial needs of the children; what would be the long term consequences of emergency actions undertaken.

Among the personnel involved, there were major differences of opinion as to what actions would best meet the needs of these unaccompanied Cambodian children. Some people advocated removing them immediately from the camps; others believed the children should not be moved prior to the completion of family tracing; still others defended the advantages of caring for children within their own community, even in refugee camps. There were differences of opinion about assessment, tracing, protection, kind of placement, services required, and the resettlement of the children to other countries.

Relief workers in Thailand had virtually no information which could help resolve these questions. A search for reference materials was undertaken. It was discovered that while the care of unaccompanied children in past emergencies had posed similar problems, there had been no attempt to compile lessons learned or suggest ways of dealing with the problem. From this context the study was conceived.

METHODOLOGY

Out of the Thailand experience, it became clear that the many problems faced by unaccompanied children in emergencies, as well as by those who would offer them assistance, needed to be investigated from historical, psychological, and legal perspectives. As a result, an independent interdisciplinary research team was established: Everett M. Ressler was the principal researcher for the historical and programmatic issues; Neil Boothby for the developmental and psychological concerns; and, Daniel J. Steinbock for the comparative national and international law sections. All three researchers assumed responsibility for integrating their various findings into recommendations to guide future intervention efforts, which compose the fourth and final part of this book. Everett Ressler served as the coordinator of the study.

The actual research began with an effort to identify all available information about the care and placement of unaccompanied children in natural disasters, wars, and refugee situations from the Spanish Civil War to the present. While library searches provided access to existing published materials, much of the most valuable information lay buried in the archives of various national and international relief organizations. This finding dictated that special attention be paid to the collection of unpublished materials. In a similar way, efforts were undertaken to stimulate a review of the national experience of various countries in which unaccompanied children have existed. In this manner, information was collected, background papers were commissioned, and independent research projects were undertaken by individuals and nongovernmental agencies in Australia, Canada, England, France, Holland, Korea, Nigeria, Sweden, the Federal Republic of Germany, and the United States.

Interviews were carried out with policy makers, program personnel, and social workers who had been major participants in past relief efforts. The purpose was to learn as much as possible about the unaccompanied children themselves, the various causes of separation, intervention strategies, and the particular assistance provided and not provided in these emergencies.

In order to better understand the contents within which emergencies occur, as well as the problems faced by countries offering refugees or displaced persons temporary and permanent asylum, visits to ongoing emergency areas and to current resettlement programs for unaccompanied children were a third component of this research project. Assessments of programs for unaccompanied children were undertaken in Thailand, Korea, Nigeria, and Lebanon. Site visitations and interviews with directors of resettlement programs for unaccompanied children from Southeast Asia were conducted in Canada, England, France, the Federal Republic of Germany, Holland, Switzerland, and the United States.

ORGANIZATION

This book has four parts. Part I provides an overview of the scope of the problem of unaccompanied children in emergencies. It then reviews the problems of unaccompanied children and gives examples of efforts that personnel have taken and problems they have encountered when providing services. Chapter 10 provides an overview of the sociological and program issues, indicating the scope of the problem. It examines the reasons why children and parents separate, describes the characteristics of unaccompanied children, and critiques the assistance provided in ten selected emergencies from the Spanish Civil War to the present. These were chosen from a larger number of emergencies on the basis of significance and available documentation. An analysis of the historical experience then follows.

Part II of the book deals with the psychological vantage point. It first provides a general overview of a child's psychological and social development as it unfolds within a family and is shaped by cultural influences. The discussion then deals with the importance of family attachments and community ties of children in wars, refugee situations, and natural disasters. It then focuses specifically on the unaccompanied child and looks at the effects of loss and separation, other trauma, and care and placement. Next, it addresses the psychological issues involved in reuniting families, and finally, the experience of children who have been moved to other countries through adoption and resettlement programs.

Part III of the book examines the major legal issues for unaccompanied children, beginning with the family and child welfare law framework relevant in most countries. After presenting the law of emergencies, particularly of armed conflict and refugee status, it goes on to discuss issues of jurisdiction and choice of law, and concludes with an analysis of the legal role of international and voluntary organizations.

Based on a distillation and integration of the most important findings of the first three sections, Part IV sets forth recommended principles to guide action in future emergencies. These recommendations stand on their own. People looking for practical implications can refer directly to this section.

The book was written by the three authors in collaboration. Everett M. Ressler was principal author of Part I, the historical review and statement of the problem. Neil Boothby was principal author of Part II on unaccompanied children from a psychological perspective. Daniel J. Steinbock was principal author of Part III, unaccompanied children in comparative and international law. The recommendations in Part IV were written by all three authors.

TERMINOLOGY

Some common understanding of certain terms is required for discussion. For the terms that follow we use the meaning given here unless the context indicates otherwise.

Unaccompanied child. *Person who is under the age of majority and not accompanied by a parent, guardian, or other person who by law or custom is responsible for him or her.* This definition focuses on the absence of any adult with firm legal or customary responsibility for the child. It excludes from consideration children who are accompanied by one parent or guardian. Historically, children without parents or guardians have been described by various terms, and even when the term *unaccompanied child* has been used, different meanings have been ascribed to it. The reference to the age of majority in this definition is intended to refer the issue of whether a person is a "child" to the relevant local law or custom. Every effort is made to use the term as it is defined here throughout the book, but at times in the historical chapters we use data based on contemporaneous, local definitions. Because there is no standard definition of unaccompanied child, one is proposed in Part VI.

Emergency. *A crisis such as a war, refugee movement, or natural disaster.* This definition aims to distinguish circumstances of crisis and social upheaval from more normal times, recognizing that there is often no clear line of demarcation between the two.

Parent. *A natural or adoptive mother or father.* When foster or other substitute parents are mentioned, they are so designated.

Family. *A group of people related by blood or marriage.* Unless otherwise noted, this term is used to denote the extended family as it is defined in the particular culture or cultures being discussed.

Refugee. *A person who has fled or has otherwise been displaced from his or her home.* The more limited definition of this term used in international law is reserved for and discussed in Part III.

I

THE PROBLEM OF
UNACCOMPANIED CHILDREN

Children have been separated from their families in virtually every war, refugee situation, famine, and natural disaster. (See Table I-1.) In a single emergency the number of children separated from parents can range from a few to hundreds of thousands. At the *end* of World War II, for example, there were at least 50,000 "homeless" children in most European countries and as many as 200,000 in some.[1] Estimates of the total number of orphaned and abandoned children during the entire war are as high as 13,000,000.[2] During the civil wars in Spain, Korea, and Nigeria, there were approximately 100,000 unaccompanied children in each country at any given time. Unaccompanied children comprised approximately three to five percent of the Hungarians who fled their homeland after the 1956 revolt, of Cuban refugees who went to the United States, and of Cambodians who entered Thailand after 1979. In situations where more of the refugees are predominately women and children, however, the percentage of unaccompanied children can be much higher.

Even the figures in these examples understate the magnitude of the problem. First, family separations have seldom been counted and cumulative totals are even rarer. Figures such as those above are usually estimates of the number of unaccompanied children at a particular time, for example, at some point during a war or at its end. Second, figures are usually based on select groups of unaccompanied children, generally the more obvious ones. Children living alone, children picked up by unrelated families, street children, and those abducted are often not included. Consequently, the number of unaccompanied children not counted in official tallies can be quite large. Among Kampuchean refugees, for example, the number of unaccompanied children who made informal family arrangements was estimated to be at least as high as that of those who sought official assistance. Third, statistics for unaccompanied children do not include those who died during the emergency. An estimated 1,800,000 children under the age of sixteen died in Poland during World War II. No one will ever know how many of them died as unaccompanied children. Yet it is known that children in general, and unaccompanied children more specifically, often have the highest mortality rates in emergencies. Fourth, statistics are often misleading because the terms used to describe unaccompanied children have varied and even the same terms have been used differently at different times. *Orphan*, for example, sometimes refers to children who have lost both parents, while at other times to those who

Table I-1. Selected Examples of Emergencies with Large Numbers of Unaccompanied Children

1915   Armenian Massacre. Some 132,000 Armenian children were rescued, including 63,000 who went to France with their families, 30,000 placed into orphanages in Russian Armenia, and 10,000 in orphanages in Greece.[1]

1919   Russian Famine and Revolution. The number of abandoned children was reported to be seven million, of whom 800,000 were provided institutional care.[2]

1936   Spanish Civil War. At one point during the war the number of orphaned and abandoned children was reported to be 90,000.[3] The total number of children separated from parents is unknown. More than 20,000 children were evacuated to other countries.[4]

1939   World War II. The number of orphans and abandoned children was reportedly as high as 13,000,000.[5] After the war national services provided for most unaccompanied children, and the United Nations organizations provided special services for more than 22,000 displaced, unaccompanied children.[6] The large evacuations of children during the war are noteworthy. Within Great Britain more than 730,000 children were moved from urban centers to rural communities.[7] In another evacuation more than 67,000 Finnish children were evacuated to Sweden.[8]

1948   Greek Civil War. More than 23,000 children were abducted to neighboring countries by guerrillas[9], and the Greek government relocated at least another 14,500 from war zones to safer places within Greece.[10]

1950   Korean War. At one period during the war the number of unaccompanied children was estimated to be as high as 100,000[11] of whom 10,000 were street children. The total number of unaccompanied children is unknown. At the end of the war, some 53,000 children were in orphanages.[12]

1954   Tibetan Refugees. Among the approximately 80,000 Tibetan people to take refuge in India were unaccompanied children. At least 2,000 children, some orphans, but some with absent working parents, needed and were provided residential care. At least 250 Tibetan children were subsequently resettled in Switzerland.[13]

1954   Vietnam War. Total number of unaccompanied children is unknown. In South Vietnam in 1973 there were 880,000 children reported to be "orphans" (this included true orphans and children with one surviving parent); 20,000 children were then in registered orphanages; and 5,000 were estimated to be living as street children, or in nonregistered orphanages.[14] At least 3,900 children were moved to other countries through adoption services including those moved in the babylift.

1956   Hungarian Revolt. Among the refugees during this revolt were at least 6,000 unaccompanied children.

1960   Cuban Revolution. Between 1960 and 1983, parents and Cuban authorities sent some 17,000 children from Cuba to the United States.

1970   Nigerian Civil War. During the war, the number of unaccompanied children is estimated to have been at least 100,000. The total number of unaccompanied children is unknown. An estimated 500,000 children died in one famine period during the war. Approximately 5,000 children were evacuated to neighboring countries. At the end of the war, 40,000 unaccompanied children lived in orphanages and relief centers.[15]

1970   Bangladesh Cyclone and Tidal Wave. Approximately 7,000 children were reportedly orphaned.[16]

1970   Bangladesh War of Independence. At the end of the war, estimates as to the number of orphans varied from 30,000 to 400,000.[17] Some 6,300 children were placed in institutions and over 4,000 were cared for in temporary reception centers.

1972   Famine in Ethiopia. Over 2,000 unaccompanied children were identified in the feeding centers and relief camps.[18]

Table I-1. (*Continued*)

1975   Vietnam Refugee Exodus. Between 1970 and 1984, some 22,000 unaccompanied children left Vietnam.[19]

1975   Laotian Refugees. Between 1975 and 1984 it is estimated that more than 2,000 unaccompanied Lao children who had taken refuge in Thailand were resettled in other countries.[20]

1979   Cambodian Crises. The total number of unaccompanied children is unknown. In 1980, approximately 6,000 children were living in institutions in Cambodia.[21] The number of unaccompanied Khmer children to take refuge in Thailand is estimated to have been at least 5,000.

1. Dorothy Legarreta, *The Guernica Generation* (Reno: University of Nevada Press, 1985), in Notes, quoting John Hope Simpson, *The Refugee Problem* (Oxford, 1937), pp. 30–38.

2. Save the Children Fund International Union, *Proceedings of the First General Congress on Child Welfare* (Geneva, 24 August to 28 August 1925), Part II, p. 43.

3. Patrick Murphy Malin, Report to the Committee on Spain and to the American Friends Service Committee, IUCW Archives, p. 19.

4. Legarreta, op. cit. This book documents the evacuation of the Basque children and their subsequent care.

5. Harbar et al., log. cit.

6. This figure refers to those unaccompanied children assisted by UNRRA and IRO through special programs between 1945 to 1951.

7. Gillian Wagner, *Children of the Empire,* (London: Weidenfeld and Nicholson, 1982), p. 248.

8. Pertti Kaven, "Evacuation of Finnish Children to Sweden During World War II," *Children and War,* edited by Marianne Kahnert, David Pitt, and Ikka Taipale, Proceedings of Symposium at Siuntio Baths, Finland, 24 March to 27 March 1983 (n.p., GIPRI, IPB, Peace Union of Finland), 1983, p. 76.

9. "Repatriation of Greek Children," quoting figures as reported by the League of Red Cross Societies. n. d.

10. United Nations Special Committee on the Balkans, "Report on Removal of Greek Children to Albania, Bulgaria, Yugoslavia, and other Northern Countries," Adopted at the 78th Meeting of the Special Committee on 21 May 1948, p. 6.

11. The American-Korean Foundation, "Report of the Rusk Mission to Korea," 1953, p. 1.

12. David C. Chi, "The Institutional Care of Children in Korea," study papers, 1984.

13. Marianne Kahnert, "Tibetan Children in Switzerland," a Review of the Emergency situation and a summary of the Comparative Study on Young Tibetans in CH (Switzerland), quoting Aeschimann, and Junge Tibeter, study papers, p. 5.

14. Jean and John Thomas, "Visit to the Republic of Vietnam," Report to USAID, November 1973.

15. Fredrick Forsyth, *The Making of an African Legend: The Biafra Story* (Middlesex: Penguin Books, 1977), p. 257.

16. Bette Sprung-Miller, "General Information Concerning Government Operated Orphanages," quoting the Directorate of Social Welfare, November 1972. It was also indicated that within one year all but 600 of the children had returned to their "parents or guardians."

17. Bette Sprung-Miller, "A Study of the UICW Programme in Bangladesh, with Recommendations," January 1973, pp. 4 and 6.

18. Mekuria Bulcha, "Final Report on Children Relief and Rehabilitation Activities in the Drought Affected Province of Wollo," Ethiopia (Dessie), April 1976, p. 3.

19. See Chapter 8, footnote no. 45.

20. Accurate statistics on the number of unaccompanied Lao children resettled from Thailand exist only for the last several years. The figure of 2,000 is based on a departure rate of between 27 and 47 children per month, estimated by the UNHCR.

21. Helga Jockenhovel-Schiecke, "The Unaccompanied Minor Refugee from Kampuchea in the Camps of Thailand," International Social Service–German Branch, November 1981.

have lost only one. Lastly, statistics about family separations are available for only a few of the wars, mass population movements, famines, and natural disasters since World War II.

The precise number of all children separated from their families in emergencies this century will never be known, but the tally would likely be in the millions. Such a large number seems plausible only when compared with the magnitude of the disruption and losses caused by the monumental crises that have plagued the world in the last eighty years. World War I, the Russian Revolution and famine, the Armenian massacres, the famines in India, and the Spanish Civil War were but a few of the earlier large-scale emergencies. At least fifty million people died in World War II alone, and in the forty years that followed some 145 smaller conflicts have claimed the lives of an additional twenty million people. One hundred million men, women, and children are estimated to have been forced to leave their homelands since 1900 because of war, political upheaval, and persecution, and even more people have been displaced within their own countries for the same reasons. Tens of millions have also died or been displaced by famines and by such natural disasters as drought, floods, cyclones, and earthquakes.

No attempt has been made to document the number of unaccompanied children in present-day emergencies, but on the basis of the scattered information available it is safe to say that the total is very likely in the range of hundreds of thousands of children. It should be kept in mind that at the time of this writing there are wars and internal conflicts occurring in some forty countries. An estimated ten million persons are living away from their homelands as displaced persons. Drought and famine in Africa threaten millions of people, and natural disasters continue to occur around the world almost daily. In all of these emergencies, children are separated from their families.

There is some correlation between the type of emergency and the relative number of unaccompanied children. More children are likely to be separated from their families in war, refugee, and famine situations than in cyclones and earthquakes. Children most often separate from families in which a death has occurred or the parents themselves have separated, or where there is a continuing threat to safety, abject poverty, or displacement. Such circumstances are more likely to follow in wars, famines, and refugee situations, than in natural disasters. While many lives may be lost and the hardships great after natural disasters, seldom is the extent of deprivation comparable to that created by man-made emergencies.

This then, is the broad picture of unaccompanied children in emergencies, but the plight of these children is best understood in the context of individual crises. What follows is the account of nine emergencies. Each presents a discrete story of children separated from their families and of the assistance subsequently provided. Together, these case studies provide a broader view of the nature and dimensions of the problems that unaccompanied children face in emergencies. Commonalities can be discerned and these will be described and examined in Chapter 10.

1

Spanish Civil War

Thousands of children separated from their families during the Spanish Civil War. To understand how and why such large numbers of children were unaccompanied in this emergency necessitates an examination of the actions of all involved—children, parents, families, organizations, and governments. It requires at least a brief review of the Spanish Civil War as social history, from the perspective of individual families. The evacuation of children here also provides a valuable case study of problems that can arise with the movement of children in time of crisis.

Civil unrest, rapid changes of government, and revolts preceded the war by several years. Actual fighting began on 17 July 1936 with a revolt by Spanish troops led by General Francisco Franco against the elected Republican government. The insurgents represented conservative values, defended the traditional feudal systems, and sought to implement an authoritarian regime. The elected liberal Republican government was composed of a coalition of parties (with the socialists and anarchists most prominent) and was dedicated to sweeping political and economic changes. Spain was quickly divided by provinces, cities, and in some cases within individual families for or against the insurgents. Helped by Moroccan troops and with military assistance from Mussolini in Italy and Hitler in Germany, Franco's forces fought Spanish troops loyal to the existing government and finally defeated them. The war ended in 1939.

Fighting in the Spanish Civil War was localized. Those who suffered most from the war lived in the areas controlled by the losing Republican government. About one million people were killed, and the war continues to be remembered for its brutality. Even young children were sometimes hanged. There was little discipline among the soldiers and hastily recruited civilians on both sides. Also, for the first time in history, airplanes were used for massive bombardment. The battle at Durango, early in the war, was described as "the most terrible bombardment of a civil population in the history of the world" (up to March 31, 1937). Over 250 people were killed instantly.[1]

In this chaotic situation, tens of thousands of children were separated from their parents. Separations occurred within Spain itself, during refugee movements into France and also as a result of organized evacuations of children to other countries. Some Spanish children who had been away from their parents temporarily before the war were unable to rejoin them because of the conflict. Some were orphaned or left unaccompanied by the absence of parents. Some children were placed with other families or with special children's programs because of difficult family circumstances. Some parents and children were separated accidentally, and some children were sent

13

away by parents for safety. The total number of children separated from their parents is unknown, but more than 20,000 children were involved in just one type of separation for which there are statistics, namely, that of organized evacuation out of Spain. A variety of child welfare programs were set up to deal with children unattended for different reasons; each had its own consequences for the children.

The sudden onset of fighting trapped at least 1,401 children in thirty different summer camps.[2] Because they could not cross the battle lines, these children were unable to rejoin their families immediately. As a result of negotiations for an exchange of women and children organized by the International Committee of the Red Cross, and with special transportation arrangements, at least 819 of the children were soon returned to their families. Many of the remaining children, however, could not reunite with their families until the war ended. During the war they were temporarily placed in local village homes or in institutions. International agencies helped local groups by providing material goods and helping the children to maintain communication with their parents.

After only one year of fighting there were estimated to be 90,000 orphans and abandoned children.[3] Orphanages were established throughout Spain, particularly for those children whose fathers had died while fighting. Contemporaneous press reports contained such headlines as "A model orphanage for children of Basque soldiers killed in combat was opened in Bilboa."[4]

Accidental separation of children from their families occurred during mass population movements within Spain. Family tracing services were established both for lost children and for adults who had lost contact with other family members.

Many children separated from their parents because of dire family circumstances such as destitution, loss of a parent, or displacement. Destitution was a common denominator. Poverty was either caused by or compounded by displacement, lack of opportunity to earn money, absence, or death or imprisonment of the father or mother. Finding food was a pressing problem. While Spain as a whole had been food-sufficient, the government-controlled areas were not agricultural regions. After the first few months of the war, finding food became increasingly difficult because production was disrupted and imports were significantly reduced. For those who could pay food was generally available, but many people lacked income. Relief efforts focused primarily on the acquisition and distribution of food. Feeding centers for children were set up in most cities and villages. By 1939, the number of children in a "pre-starvation" condition was estimated to be 100,000.[5] Basic commodities, such as clothing and blankets and even soap, were also needed by large numbers of refugees who had fled hurriedly, taking little with them. The harsh winters made their plight even more desperate.

Separation of parents from children was most likely in single-parent families. Many single parents had to work away from home which made caring for their children more difficult. In some cases the conflict between work and family responsibilities caused parents to place their children with other families or in institutions, believing this to be in the children's best interests.

Displacement caused much hardship. By the end of 1937, more than one million people were refugees.[6] Large numbers of people moved for various reasons. In contested areas, people fled because of immediate danger to their lives and through fear of the violence of invading armies. The people who flooded into neighboring villages and provinces, expecting only temporary displacement, were often beginning a trek of several years from one area of apparent safety to another. Some families were forced to flee three or four times in one year and with each move their difficulties increased.

> Soon, the highways became clogged with fleeing families, their household goods, and sometimes, their livestock. Many of the children walked or rode in carts or trucks first to Eibar, then on to Durango, Bemo, or Bilbao in September of 1936 were among those evacuated from Euzkadi the following spring and summer. They were urged along . . . by the news of the executions of the militia who had remained . . . and of the excesses of the Moorish Legionnaires . . . and the hope that safety could be found in Vicaya[7]

The general public was also displaced as a result of evacuations ordered by officials, in Madrid for example, to reduce the number of civilians in particular areas. The safety of women and children was presented as a major justification for the evacuation, and in this way, thousands of children were reported to have been evacuated from Madrid to children's colonies.[8]

Large numbers of parent–child separations occurred during mass movements of people across the northern border from Spain into France:

> Children were found marching along the roads, knowing nothing of the whereabouts of their parents. . . . Mothers lost their children as the throng pressed along the roadways, and streamed over the surrounding hillsides. . . . Many months were required before some semblance of order was brought into this shocking situation.[9]

Movement into France began with one of the first major battles, the battle of Irun, at the beginning of the war in 1936. Within six months, as many as 8,000 refugees had crossed the border, and over the next three years that number had increased to more than 500,000.[10]

Moreover, on arrival in France, family members were systematically separated by French authorities. Soldiers and able-bodied men were placed in prison camps— then called "concentration camps"—on the southern coast of France. By 1939 there were about 200,000 Spanish male refugees in three such camps.[11]

In response to this influx every province in France was asked to shelter refugees. Older men and the women and children were dispersed throughout France to more than 2,000 localities. They were sheltered in old factories, cellars, abandoned buildings, and hastily erected barracks, in groups as small as two or three people and as large as 1,000. As a rule, they lived without heat and with almost no medical attention.[12]

As concern mounted about imminent conflict between France and Germany, most men were released into work gangs, permitted posts in agriculture and industry, allowed to emigrate, or released for service in the French Foreign Legion. Only the elderly, the maimed, and the unemployable remained in the camps.[13] As the concentration camps for men were emptied, several were converted for women and children who were transferred to the camps from their temporary accommodations in northern France. Two such camps in 1939 contained over 9,000 women and children.

The response of the French public to the refugees was mixed, varying from open hostility to extraordinary benevolence, depending on political, economic, and religious identifications as well as humanitarian responses. Such factors were also reflected in the reaction to the unaccompanied children.

Nearly 20,000 children left their families in various evacuations to other countries during the Spanish Civil War.[14] While most were between the ages of five and twelve, some children were toddlers and some were over eighteen. More than 9,000 children went to France, 3,889 to England, 3,200 to Belgium, approximately 4,000 to the U.S.S.R., about 450 to Mexico, 245 to Switzerland, and 102 to Denmark. Groups in Sweden and Holland initially offered to take in children, but later decided to support programs in France instead. Efforts to evacuate children to the United States were blocked there and offers from Chile were declined.

Evacuations were originally planned as a type of vacation trip on the assumption—by parents, children, and receiving countries—that the war would be quickly won, and that the children would be away no more than a few months. The war dragged on for several years, however, with the Spanish Government, which the children's parents supported, losing in the end.

Once the children had been separated from their families, there were many obstacles to their return. First, the war continued longer than expected with increasingly difficult conditions in the government-controlled areas. Some children lost contact with parents as a result of population movements, or death or imprisonment of parents during their absence. Some parents did not wish the children to return, believing it better for them to remain in interim care. In some cases, children were so integrated into the foster family that neither the child nor the foster family wished to separate. Some children did not wish to return for other reasons, such as better educational opportunities in the host country, and some could not return because of logistical difficulties and political problems. A further important obstacle was the Decree of Responsibility imposed by Franco in 1938, which counted all exiles criminally responsible in the Civil War, except for children who had been less than fourteen years old in 1936. This decree was ended seven years later, in 1945, with a grant of amnesty for Spanish Republicans in exile. In addition, of course, World War II prevented movement from more distant countries such as Mexico and the U.S.S.R. During these years children grew to young adulthood, sometimes married, took jobs, and became integrated into the host society.

What everyone expected to be a sojourn of several months became an ordeal separating most children from their families for several years. Some children did return after only a few months. Some who went to France, England, and Belgium were away from home for at least two years. But many remained away much longer. In France, for example, only 4,400 of approximately 9,000 children had returned to Spain by 1940. Because Mexico and the U.S.S.R. refused to recognize the Franco government, children evacuated to the U.S.S.R. and Mexico could not return for over fifteen years. Most, in fact, never returned. Once they had lived in a country for an extended period and established new lives there, most chose to stay when the opportunity to return to the country of their birth was offered years later. In most of the host countries repatriation became a heated national debate, with some people demanding the return of the children and others asserting that repatriation was not in their best interests.

For large numbers of children, the experience of evacuation was significantly different from that intended by organizers or anticipated by parents. Many of the assumptions the organizers had made proved to be false, or even impossible, dreams. Some programs scrupulously maintained correspondence with parents, kept siblings together, provided caretakers of the same nationality, and amply provided for the children. Other programs, established with the best of intentions, were poorly managed by inadequate and inappropriate short-term staff, and were plagued by lack of funds. For many, the experience is remembered positively. Others remember it for its hardships, the loneliness experienced, and the difficulties encountered during their absence and on their return.

2

World War II

Unaccompanied children requiring special care existed in virtually every country affected by World War II. The total number is unknown. Both the International Committee of the Red Cross and UNESCO estimated there to have been thirteen million children who were described as "orphaned" or "abandoned." Over one million children were in institutions throughout Europe at the end of the war. Such large numbers seem plausible only when the statistics of the war itself are understood. The total number of casualties is calculated to have been between forty and fifty million. Of that number, thirty million people died in the European countries occupied by Germany, eighteen million of them as a result of the Nazi extermination policy. To the number of children affected must be added the children deported as slave laborers (over 700,000 were estimated to have been taken from Poland alone), children left alone as a result of the deportation of their parents, children abducted for a Nazi adoption program, illegitimate children of slave laborers and deportees who were not permitted to keep them, children who were accidentally separated, and children hidden or moved to safety by parents. To these numbers must be added the tens (or hundreds) of thousands who were among such displaced peoples as those who fled the Ukraine during the war, and those among the more than seven million ethnic Germans (Volksdeutsche) displaced from various parts of Europe after the war. Children were also separated from their families during the war in North Africa, China, the Far East, and in Pacific countries.

World War II began with the declaration of war on Germany by England and France on 3 September 1939, in response to Germany's invasion of Poland. The war was provoked principally by the aggressive acts of three countries: Germany, Japan, and Italy (then called the Axis powers.) World War II ended six years later with the unconditional surrender of Germany on 4 May 1945 and of Japan on 2 September 1945. Italy had surrendered on 8 September 1943. During this war, fifty-seven nations had been involved in the fighting.

No attempt is made in this study to give a comprehensive description of all unaccompanied children in World War II. We have chosen to focus principally on the separation of families in Europe before and during the war, on programs for unaccompanied children in Europe after the war, and on special programs for the evacuation of children in England and Finland.

FAMILY SEPARATIONS

Family separations during the war in Europe had their origins, to some extent, in prewar policies and ideologies. The fundamental principle of National Socialism was a simple one: racial supremacy of the Aryan peoples and of the Germans over all. Its aim was the rejuvenation of Germany and the "common enemy" was defined as the Jewish and Slavic peoples and anyone who questioned Nazi ideology and programs.

To a large extent the Nazi program began with and was based on the indoctrination of German children through the Hitler youth movement (Hitler Jugend), which was founded in 1926. Virtually all children from the age of ten were taught to idolize the Fuehrer, to show absolute obedience to authority, and to profess total devotion to the State, accepting the values and philosophy dictated by the Nazi party. The Nazi ideology advocated state control of children and the breakdown of traditional family structures. Before the war, many German parents sent their children to other countries to prevent their having to join the Hitler youth movement. Such plans were usually disguised as a way of gaining educational opportunities. Jewish youth were, of course, not eligible to join the Hitler Jugend and suffered greatly from its members' actions.

The harassment and persecution of non-Aryan people, particularly Jews, within Germany was part of the platform upon which the Nazi party was based. Concentration camps for non-Aryans and for any who opposed the Nazi regime, or who were even suspected of doing so, were built within the first year of Hitler's rule. The Nuremberg Laws in 1935 stripped the Jewish people of citizenship and institutionalized their oppression. This led to the confiscation of property and businesses, with the wrecking of shops, direct attacks on persons, and loss of the right to practice any profession. Destitution became the norm. It grew increasingly clear that survival for non-Aryan people depended on gaining asylum in another country; within this context various efforts were made to provide special help for children.

As the oppression in Germany and Austria became public knowledge, groups in other countries offered to care for the children. Many emigration and rescue schemes were devised, often based on the hope that families would leave as a unit. In practice, however, this proved almost impossible for many reasons and families had to separate. Initially, children left through ordinary emigration channels and visas were often available for children before they were available for adults. Relieved of responsibility for their children, parents were better able to push forward plans for their escape or emigration. Thus, for example, a Society of Friends program in Vienna concluded, "It therefore seemed wise for us to develop a special plan for the emigration of children, and the number of appeals from children for immediate help mounted to hundreds every week from November 1938 through January 1939."[1] At the outset the placement of children was regarded as a temporary arrangement, until parents and children could be reunited again under more favorable circumstances. But for many families this never happened.

The total number of children moved from Germany and occupied countries is difficult to determine. It is known, however, that from 1933 to 1939 at least 12,000 Jewish and non-Aryan Christian children were moved with parental consent to

England (10,000), Holland (1,500), Sweden (55), Belgium (1,026),[2] and the United States (590).[3] These countries, along with Switzerland, Norway, Denmark, Italy, Yugoslavia, Poland, and Lithuania were known as transit countries, where the reception and care of the children was organized by voluntary agencies. This was usually done on the understanding that the children were to remain there only temporarily and would not become public charges. Between 1933 and 1948, Switzerland provided temporary asylum for more than 10,000 children, of whom 1,278 were considered "completely unaccompanied." Of those children who went to Switzerland, all but forty were resettled to other countries.[4] Partly in recourse to the Nazi racial programs, an organization called Youth Aliyah (*aliyah* means "immigration" in Hebrew) was founded in 1933 to provide selected Jewish youth from central Europe with a two-year training program in what was then Palestine; it also promoted Jewish immigration to settlements there. Between 1933 and March 1938, 4,635 boys and girls, predominately between the ages of fifteen and seventeen, were moved to Palestine from Germany, Austria, Poland, Czechoslovakia, and Rumania, along with a small number of refugee children from Italy, England, Holland, and Denmark. These young people were incorporated into more than seventy different institutions and agricultural settlements and into work and educational programs established according to their organizers' social and religious philosophies. As the Nazis expanded their anti-Semitic designs, the Youth Aliyah program increasingly became a child rescue scheme. From 1939 to 1 October 1946, more than 15,000 children were assisted in entering Palestine, bringing the total number of children moved by Youth Aliyah to that date to over 20,000.[5]

But many more children at risk stayed in the occupied areas. Hundreds of thousands, if not millions, of children died in concentration camps and labor camps, or from hunger and disease as a result of the war. Yet many children without parents or families did survive. Some children who looked Aryan and whose true identity could be concealed were taken in by compassionate families at great risk to themselves. In Holland, for example, over 15,000 children remained hidden during the war. This became known ten days after Holland's liberation, when a public registry of their names was compiled. In many cities, children survived by depending entirely on their own skills, independently securing their own food, and living in cellars and destroyed buildings. In Italy, for example, at the beginning of 1945 there were about 180,000 "street urchins" in Rome, Milan, and Naples, many considered to be orphans and homeless.[6]

Unaccompanied children were part of a tragedy for non-Jews as well as Jews. As Germany expanded its territory, the Nazis expanded their plans for children. "The children of the annexed districts were to be Germanized; those of France and the Nordic democracies were to be reared as willing employees of the New Order; Greeks and Slavs were to be reduced to the condition of semiliterate labourers, while to Jewish and Gypsy and mentally defective children, a process of elimination was to be applied."[7] Programs were developed to "recover" ethnic Germanic elements for reincorporation into the Reich, to encourage illegitimate births among mothers of known Nordic extraction and German fathers, and to transfer selected children to Germany to "strengthen the Reich, and weaken the [occupied] country."[8]

The Lebensborn Society (Fountain of Life) was the central Nazi organization carrying out these programs. It was established by the SS (Staatspolizei [Secret Police]) in 1936. In addition to being a welfare agency for SS families and caring for Germany's war orphans, its goal was to secure "racially valuable" children who could be raised to become the future ruling generation. During its years of activity, the Lebensborn Society reportedly had about 92,000 children under its care, 12,000 of whom were born in its own facilities.[9] The Nazi machine took numerous measures to increase the German population. Illegitimate births were encouraged among young people, and social proclamations decreed that "every healthy girl of eighteen who is not already a mother is lacking in a sense of her social duty."[10] Special camps, or mating houses (Begattungsheime), were established within and outside Germany for procreation between German males and females chosen from various countries and forced to participate. The resulting infants were adopted by SS families, a service organized by the Lebensborn. Many children were also cared for in special institutions. In Germany, every SS man and every German woman acknowledged by a special authority to conform to the racial ideal of an Uebermensch ("super race") was obliged to give a child to the State. There were estimated to be approximately 60,000 such children who were reared in special homes, where they were indoctrinated with Nazi ideology, another service provided by the Lebensborn Society.[11]

The kidnapping of children was a secret Nazi program carried out in Poland, the U.S.S.R., and Czechoslovakia, and on a smaller scale in Yugoslavia, the Netherlands, Belgium, and France. The total number of children kidnapped for Germanization is unknown because records were destroyed in order to conceal the children's origins. After the war, the respective governments claimed that more than 550,000 children had been deported and kidnapped, including 200,000 from Poland, 3,000 from Czechoslovakia, 200,000 from France, 150,000 from the Soviet Union, and 500 from the Netherlands.[12] (The figure for France included children of French parents who had been slave laborers in countries other than France.) Children between the ages of two and twelve years were simply taken from the street, from schools, from homes and institutions and sent to clearing centers where "experts" carried out racial testing. Sometimes the populations of whole schools and institutions were moved. Children between the ages of six and twelve chosen to be "Germanized" were sent to special schools and then to approved homes. Germanization meant giving children German names, teaching them to speak German, indoctrinating them with Nazi teachings, and forcefully attempting to erase all elements of their pasts. Children between two and six years of age were given false birth certificates showing them to be German and then placed in institutions or adoptive homes by the Lebensborn Society. Even SS families believed them to be truly orphaned. Those children who did not qualify for institutions or homes were used in medical experiments, sent to labor camps, or exterminated.

Many children became unaccompanied when parents were deported as slave laborers. During the war, approximately twelve million people in Europe were internees and forced laborers.[13] They came from many different countries, but particularly from occupied Slavic countries, and the majority were moved to Germany. Many unaccompanied children and youth were also forcibly deported for the same purpose. After the battles of Smolensk, Stalingrad, Kiev, and battles in the Ukraine, thousands of young girls were abducted to work as forced laborers. Scattered throughout the forced labor group were two categories of unaccompanied children. First, there were boys and girls who had themselves been forcibly separated from their families and, second, there were illegitimate children later born to these youths. Young mothers in labor camps were not allowed to keep their children. They could either place them in special homes for infants or make private arrangements with a family.

Unaccompanied children were therefore found to be living in a great variety of circumstances; for example, in camps, homes, institutions, and on the streets. Some, called "mascot children," attached themselves to military units. Others returned alone from the concentration camps. As mentioned, children were found in adoptive homes and in institutions as a consequence of the Lebensborn kidnapping and adoption schemes. Others had been placed in foster and adoptive homes by German and foreign mothers—sometimes because they had been forced to do so, in other cases, they were placing the child for safekeeping or simply abandoning it. Many children were accidentally separated from parents during air raids and movements of mass populations. Some children had lost track of their families after having been sent by them to areas considered safe, and then there were the recently abandoned children and the Jewish children who had been moved from Europe to Palestine, called "infiltrees" because of the clandestine nature of the operation. Some children were cared for; others were not. One 1946 report, for example, noted that in almost all cases where unaccompanied children were living in Austrian foster homes, the foster parents wanted to keep the child. In one case, a Ukrainian woman, a slave laborer, had placed her two-year-old son with Austrian foster parents. The father was a German soldier who was not permitted to marry her; he was killed in 1944. Fearing difficulties in the U.S.S.R., she could not take the child with her when she returned in 1945, so she gave the foster parents permission to adopt him. Other surveys found children who had been ill-treated and kept for the work they could provide.

PROGRAMS FOR UNACCOMPANIED CHILDREN

In 1945 an estimated twelve million people around the world were living in countries other than their own because of the war. They included people who had fled their homes, been deported, been prisoners or soldiers, or been uprooted for a variety of other reasons. There were six-and-a-half million refugees and displaced persons in Europe alone.[14] In November 1943 the United Nations Relief and Rehabilitation Agency (UNRRA) was established by forty-four Allied nations to provide social, economic, and humanitarian assistance to this group, in collaboration with Allied military forces.[15] After the war, UNRRA and the Allied military focused initially on the repatriation of displaced people to their home countries, moving more than four million within the first several months. An additional 862,000 people were repatriated over the following two years,[16] still leaving more than 500,000 refugees and displaced persons who could not or chose not to return to their prewar homes. UNRRA camps and military camps for displaced persons existed in Albania, Austria, China, Czechoslovakia, Denmark, Egypt, France, Germany, Greece, Iran, Italy, and Norway, and were not closed until all residents had been either repatriated, settled locally, or resettled in third countries. In Europe, the largest number of displaced persons was in Germany and Austria, and the last displaced persons camp was not closed until 1967.

The children liberated from concentration camps at the end of the war, although a minority group among unaccompanied children, were among the first to be provided with special services by UNRRA and the military. The pathetic condition of these children made identification more immediate and they received assistance more quickly. For example, immediately at the war's end, children from the Buchenwald camp were provided care in France and Switzerland.

Toward the end of the war, UNRRA officials gave some thought to the existence of other unaccompanied children, but did not recognize the scope and complexity of the problem in the early months when massive numbers of people were repatriated. Policies and programs evolved as the needs became more apparent. Responsibility for displaced persons rested with the respective military commands in the French, British, American, and Russian zones of Germany and Austria, and consequently, the policies varied from zone to zone. Before any special programs were established the unaccompanied children in the displaced persons camps either took care of themselves or were helped by informal foster families, or in some cases, camp residents organized groups to care for them. UNRRA first addressed the issue of unaccompanied children when a small number of them were being moved out of the country by a quasi-voluntary agency, and also in response to offers from several countries to provide temporary asylum to orphans and other unaccompanied children.

In June 1945, UNRRA officials began collecting data to determine the total number of unaccompanied children in the displaced persons camps. As late as February 1947, a staff of forty to forty-five was still working on this project. Counting was an ongoing process because lost children continued to be found and separations continued to occur. The first registration of unaccompanied children in Austrian camps yielded only 130 names. (Most unaccompanied children, it was noted, were not registered because of resistance to repatriation by foster parents and staff.) Through interviews with children and requests for "lost" children by parents, the Lebensborn program of kidnapping, Germanization, and adoption was uncovered. It became evident that in addition to unaccompanied children living in displaced persons camps, there were probably large numbers of children living in institutions and homes throughout Germany and Austria who had been born in other countries or whose parents were foreigners. In September 1945, UNRRA, military authorities, and the governments agreed that these children should be located and identified. UNRRA staff, working in collaboration with voluntary agencies, notably the Belgian, Czechoslovak, Dutch, Polish, and Yugoslav Red Cross Societies,[17] began a child search program in the American and British zones of Germany and Austria. The search in the French section was carried out under independent French control. No search was known to have been carried out in the Russian zones, nor was UNRRA staff permitted to do one there. Italy allowed no searches for children until 1950.

The UNRRA child search program was originally designed for "location, documentation, and repatriation of Allied separated children."[18] The program was thus directed to identifying and assessing all children who were not living with a parent, who were foreign born, or who had been born in Germany and Austria of a foreign parent. Representing the various governments claiming abducted, lost, and deported children, the UNRRA program assumed that children who were separated from parents and from country belonged first to their families and second to the country of their nationality. UNRRA initially assumed that children should be repatriated to the country of origin, either to the care of their family, or if the whereabouts of the family was not known, to the care of a child welfare agency there. Provisions were made for nullifying certain adoptions of foreign children by German families during the war years on the grounds that the children had been taken illegally.

At first, in Austria, child search workers visited homes and institutions and, with the right of search and seizure, removed the children (who were often totally unsuspecting of their original origins) and placed them in temporary children's centers until they could be repatriated. The abrupt removal of the children from their foster or adoptive families, many of whom had cared for them for years, caused great anxiety in the children and was actively resisted by foster and adoptive parents. Even some participating social workers objected. Strong objections were raised as to whether such sudden removal was in the children's best interests, and in time, more consideration was given to their individual circumstances. For as long as UNRRA existed (until June 1947), repatriation remained the primary objective, but the child removal policy was changed to allow the children to stay in foster and adoptive homes unless their parents had been located and requested the children's return.

Children's centers, set up either as special centers or as special sections within camps, were established as transit centers to help repatriation of unaccompanied children. They also housed unaccompanied children who were found wandering in search of family; some who had been removed from homes; "wild" children who had lived in hiding, supporting themselves by begging, stealing, and black market activities; and children from orphan asylums. By 1947 there were thirty-five children's centers in the three German zones under western military control,[19] with more than 3,800 children in residence.[20] Although officials assumed that a child's stay in a center would be short, for some children, repatriation or resettlement took more than five years.

Jewish children were part of the postwar exodus of Jews from central and eastern Europe to Germany and Austria. By 15 March 1947, a total of 29,827 Jewish children under the age of eighteen had come under UNRRA's care. Of this number, 7,029 were classified as unaccompanied. These children usually arrived in the American zone of Germany, intending to start a new life in Palestine. The first group of 350 moved to Palestine in December 1945. Many of these children had either been in concentration camps, in hiding, or fighting with the partisans. They had returned to their home countries after the war and then joined the movement to Palestine. "They were organized into groups by Zionist leaders who had been working in the countries of origin locating orphans and buying children from non-Jewish families who had given them asylum. Some had been voluntarily 'signed over' by their parents to facilitate their emigration to Palestine".[21] Many parents expected to follow the children.

Authorities did not usually permit children to be resettled to countries outside of their countries of origin. As a result there existed an active underground movement of children from Europe to Palestine. However, in recognition of the horrors of the past and of individual dreams for the future, officials usually ignored these "illegal" movements of children, the false identification papers, and the illegal ships that frequently left Marseilles. On several occasions, however, UNRRA did block the movement of children until they had been properly documented. Unless the group was caught during their voyage (in which case the children and adults were placed in refugee camps in Cyprus), they were able to settle in Palestine.

Once UNRRA had identified the unaccompanied children, it changed the direction of its child welfare services in order to meet the demands of providing care, placement, and protection for this special group of children. Unfortunately, once the unaccompanied children became the focus of child care efforts in the camps, there was not sufficient staff left to provide social services to children still with their families.

UNRRA was faced with the responsibility for the immediate and interim care of the children and planning for their future, considering such factors as nationality, illegitimacy, guardianship, placement, repatriation, and resettlement. UNRRA assumed responsibility not only for children separated from their parents during the war, but also for new groups of unaccompanied children, since parent-child separations continued even after the end of hostilities in 1945. Some infants were abandoned in the displaced persons camps. Some children required temporary care because of parental death or illness, and other unaccompanied children continued to arrive in the camps, particularly Jewish children en route to Palestine.

UNRRA was disbanded in June 1947. Until then it had assisted 13,318 unaccompanied children and adolescents. They ranged from infants to eighteen-year-olds;[22] approximately sixty percent were Jewish, and the largest number from any one country were Polish. During the period 1945–1947 it had reunited 1,016 unaccompanied children with their families.[23] The emphasis had been on repatriation and 2,703 unaccompanied children had been officially returned to their country of origin with most returning to Poland and Yugoslavia. Resettlement in a third country had not been encouraged or permitted except for Jewish children, of whom a total of 961 were resettled to third countries, mostly to the United States and Palestine. These figures represent only the cases in which UNRRA was formally involved. The number of unaccompanied children repatriated through UNRRA is probably only a fraction of the total involved in the mass repatriation immediately after the war, or who returned with foster families or friends, or who stayed on their own. Resettlement figures are also misleading, for while 1,889 children were officially resettled in Palestine, an even greater number probably reached the country secretly through the underground movement.

UNRRA's refugee activities were then taken over by the Preparatory Commission of the International Refugee Organization, which existed for six months, until the International Refugee Organization (IRO) was established in 1948. Between 1948 and September 1950, IRO continued the UNRRA efforts for unaccompanied children and identified some 4,000 more "lost" children through its child search programs.

Initially, IRO followed the UNRRA policy of considering repatriation the normal solution for children of known nationality. As time went on, however, IRO and the military authorities in the American and British zones increasingly attempted to determine the "best interests" of the child and the child's own wishes. This involved recognizing and formalizing existing relationships between foreign children and foster parents and, eventually, also permitting resettlement to third countries.

In summary, from 1945 to 1951, 22,800 children were located through the UNRRA and IRO child search programs. Relatives were found for 5,800 of these children, 4,600 were repatriated to their country of origin, and 5,000 were resettled in third countries. When the IRO program ended in 1950 there were still 1,000 unaccompanied children in Germany, Italy, and Austria; most were placed under the supervision of a national organization. Again it must be stated that these figures indicate only the number of children assisted through UNRRA and IRO, and therefore represent only a fraction of the total number of unaccompanied children and of those repatriated or resettled in third countries. On 31 August 1950, 10,495 cases for child tracing were still on file. Attempts to locate family members continue to this day under the auspices of the German Red Cross and the International Committee of the Red Cross.

THE EVACUATION OF FINNISH CHILDREN

During World War II between 1939 and 1945 some 67,000 children were evacuated from Finland to Sweden through a program similar to those organized for the evacuation of children during the Spanish Civil War. Finnish parents sent children in response to an effectively organized program that included offers of temporary care from Swedish families. Parents and organizers initially believed that the children would be away no more than a few months. But the war continued for more than five years after the first children were evacuated. When it ended, contrary to initial expectations, many children did not return to their homes.

The war in Finland can be divided into two distinct periods. The first, called the "Winter War," occurred at the end of 1939 and early 1940, and the second, called the "continuation war," lasted from 1941 to 1944. The "Winter War" began on 30 November 1939 when Finland was invaded by Soviet Russia. The main battles occurred on the Karelian Isthmus, a Soviet-Finnish border area which stretches between Lake Lagano and the Baltic about thirty kilometers from Leningrad. The "Winter War" lasted for only 100 days, was fought in devastatingly cold weather, and ended in March 1940 when Finland lost the Karelia Isthmus and the area around Lake Lagano. For more than a year there was no fighting in Finland until 1941 when Germany moved a large army into Lapland in northern Finland. In 1941, under pressure from Germany, Finland became involved in Germany's conflict with Russia. While Russia was defending Stalingrad against Germany, Finland took back the isthmus. In 1944, with the defeat of Germany imminent, the Finnish Army pushed the occupying army out of Lapland, but as the German Army retreated it destroyed and burned the cities, devastated the infrastructure, and then mined the ruins. Also in 1944, when the German offensive collapsed, Russia again attacked the Karelia Isthmus and reclaimed what it had taken earlier. Peace was finally established with the signing of a provisional peace treaty on 19 September 1944 in Moscow.

While all of Finland was affected by the occupation and the war, most of the country suffered neither from the bombing nor actual fighting. The population of the Karelia region was the most severely affected, as most of the fighting took place within this limited region (approximately twelve percent of the land area), except for the bombing of Helsinki near the end of the war. Before the war this area had been rather densely populated by farmers. During the fighting the inhabitants (approximately twelve percent of the total Finnish population) withdrew, taking all equipment, cattle, and movable personal goods. This displaced population dispersed throughout the rest of Finland and was taken in by other farming families or housed wherever they could find space. The displaced Karelians stayed away for over a year, until Finland again gained control of the area in 1941 and families could return to begin reconstruction.

The Laplanders of the north were also displaced, but in this case by the German army during its occupation of the region. In 1944 the Karelians were permanently displaced by the second Russian offensive in the region, and the Laplanders suffered through the destruction caused by the retreating German army.

More than 500,000 people out of a population of 3,700,000 were displaced in Finland during the war. The war created some 47,000 orphans[24] who were cared for either by families or in one of 207 orphanages.[25] Before and during the war many people in Finland were rather poor, and the departure of the men for the military and other displacements increased family hardships.

Private Swedish initiatives led to the evacuation of Finnish children. Six days after the "Winter War" started, a philanthropic organization was established in Stockholm to assist Karelian refugees; it sent a letter to the Finnish Government offering to evacuate the children. The government at first declined the offer, suggesting that money should be provided instead. The Swedish committee (which included the wife of the foreign minister) continued its initiative, however, and with the help of the national press found more than 5,000 Swedish families willing to accept a Finnish child. The Finnish government was approached again, this time through selected public officials and prominent people sympathetic to the idea. The government reversed its earlier position, set up an evacuation committee, and established broad eligibility categories.

Finnish children were evacuated both during and after the "Winter War," and again between 1942 and 1945. During the first period, the displaced Karelians were the primary concern. Children between the ages of one and twelve were eligible and mothers could follow if one of the children sent was under the age of three. During the second period, after the German invasion, Karelian children whose parents had returned home to rebuild became eligible for evacuation. Other Finnish children affected by the war such as those with sick or disabled parents, as well as orphans and children whose fathers were at the front or whose mothers were working or pregnant, also became eligible. In reality, few questions were asked if a parent wanted to send a child to Sweden.

During World War II Sweden's neutrality saved it from the destruction suffered by other European countries. One of the richest nations of the world at that time, Sweden was a major source of relief and assistance to needy children in many countries. Strong traditional ties linked Finland and Sweden, and in many of the larger cities, for example, in the Karelia region, Swedish was also spoken among upper-class families. Agencies in Sweden also traditionally took Finnish children for the summer holidays, and under a similar arrangement invited the first groups of children for the summer of 1941. The evacuation of children to Sweden had a precedent: during World War I more than 17,000 children from various European countries were sent there for temporary care.[26] During World War II, thousands of children from other countries, such as Holland, Norway, Estonia, Latvia, and Lithuania, were also evacuated to Sweden for what was expected to be a brief period of recuperation lasting several months. However, many of these children who were also placed with foster families, remained in Sweden for many years.[27]

The Swedish donors, the Finnish government, and the Finnish parents had differing ideas about the need to evacuate the children. The Swedish evacuation organizers made their offer out of concern for the children's physical safety. The Finnish

Social Ministry, which initially sanctioned the evacuation, justified its decision by pointing out that Karelian mothers were unable to care for small children when they returned to rebuild and that food was scarce. Parents' decisions were influenced by the existence of daily hardships and the possibility of greater opportunities for the child elsewhere. The father was usually at the front, and many of the evacuated children were sick. Family size was also an important issue. Most families had five to ten children; thus the evacuation of one or several children did not destroy the family unit.

Pertti Kaven, in his study entitled "Evacuation of Finnish Children to Sweden During World War II" expresses doubts that the children's personal safety was at risk.[28] The fighting in Finland was localized and the children who were evacuated came from families that had been resettled rather than from a population under siege. Kaven also questioned the lack of food as justification for the evacuation. Food rations were provided. People did not starve in Finland during the war. In addition to providing for its own population, Finland also maintained 68,000 displaced persons from the Soviet Union. He suggests that the evacuation was probably accepted simply because it was offered under pressure from influential people.

It was assumed from the beginning of the evacuation that the children would return to their homes at the end of the war. Some children even returned home during the war and at least eleven percent were evacuated twice.[29] The largest number of children, 17,756, returned in 1945, and 13,249 more returned during the following two years.[30] Some 15,000 children (about twenty-two percent of the children who left), however, did not return to Finland. Some families lost interest in their children. Some Swedish families, not wanting to give up the child who had become a part of their family, adopted the child, sometimes without the knowledge of the biological parents. A survey of 600 Finnish families whose children had not returned showed that only one-third of the families wished the children to return. Many families believed the children should remain where there were better education and job possibilities.

Public outcries for the return of the children occurred several times. Demands for their return were based on the arguments that the evacuated children belonged to Finland (in a nationalistic sense). A special fund was established in Finland to assist poor families with legal fees in their attempts to have their children returned, and various cases went to court to determine the relative interests of the children, their foster parents, and their biological parents. In at least two leading cases the Swedish Supreme Court ruled in favor of the child remaining with the foster parents. In *Wahlstedt* v. *Vellonen,* the court decided (7 March 1951) that in spite of the pledge by foster parents to return the child to Finland whenever its mother so demanded, it was in the best interest of the child not to return against its wishes.[31] In another case (Nytt Juridiskt Arkiv 1946 No. 141) the court also granted the Swedish foster parents the right to adopt their Finnish foster child, emphasizing the best interests of the child.[32] In total, 48,628 children were evacuated through the official program, and an additional 15,000 children were evacuated privately. These more than 63,000 children represented no fewer than 7.2 percent of all children in Finland in 1941–1945.[33] The children evacuated were between the ages of one and fourteen and twenty to twenty-eight percent were less than five years of age when evacuated.[34] The length of separation from their families was, in some cases, just a few months; in other cases, permanently.

THE EVACUATION OF CHILDREN IN THE UNITED KINGDOM

During World War II, British children were evacuated to other locations within the United Kingdom as well as to other countries. On 3 September 1939, the day on which war was declared, the evacuation began. Within four days hundreds of thousands of families were divided. In total, 750,000 unaccompanied school children, 542,000 mothers with children under five, 12,000 expectant mothers, and 77,000 other adults in England and Scotland were dispersed, mostly within four days, from large urban centers to small villages and rural areas across England.[35]

The evacuation's planners assumed that children up to the age of five would be sent with their mothers, and that children between five and fourteen years of age would be moved by school units to rural areas where the children would be taken into private homes and looked after. The evacuation was to be voluntary, but those designated to receive children or mothers with infants were to be compelled to do so, with local officials determining the space available in every house. Another major group of children included those under the age of five whose parents were unable to remain at home. They were to be placed in residential nurseries. The evacuation scheme involved about 400 residential nurseries before the war's end.[36]

The evacuation was organized as a civil defense strategy, and was discussed and planned for more than a year before it was carried out. Planners assumed that large-scale bombing was a possibility once war broke out. The evacuation was seen as a way to save the lives of helpless civilians, prevent panic that would have hampered military operations, and lessen the problems of distributing food and essential services.[37] At the time of the evacuation, however, the general public was not endangered, and as time was to prove, bombing did not begin until some seven months later.

The initial evacuation was not successful even on its own terms. Fewer than half the number of people planned for were willing to evacuate, and no sooner was the migration accomplished than its reversal began.[38] By January 1940, little more than three months after the evacuation started, eighty-six to eighty-eight percent of the mothers with young children and forty-three percent of unaccompanied school children had returned home.[39] In retrospect, the evacuation plan was criticized for its focus on transportation without adequate consideration of the services needed or the psychological and social consequences.[40]

In addition to the massive evacuation of children to the countryside, children were also sent from England to Commonwealth countries and to the United States. In response to offers from Canada and the U.S., an organization called the Children's Overseas Reception Board (CORB) was established in England in 1940. The idea of sending children to other countries was popular. Within weeks, the organization received 211,000 applications from parents.[41] Once the parents had applied, an acceptance form was sent, constituting a contract between the board and the parents, with the parents agreeing to make weekly payments toward the upkeep of the children. The board agreed to send the children overseas, be responsible for their welfare and education, and bring them home as soon as practicable after the war.[42] Although plans were underway for the transfer of tens of thousands of children to other countries, only 2,664 children were sent before the program was terminated. The scheme ended after a ship, the *City of Benares,* carrying some of the children was torpedoed and sunk on 17 September 1940; seventy-three of the children died.[43]

In total, more than 12,000 children were sent out of England for the duration of the war to Canada, the United States, Australia, New Zealand, and South Africa. In addition to those sent by CORB (2,664), more than 10,000 children, mostly from middle-class professional families, were sent overseas privately.[44]

The evacuation of the children both within and from England raised many questions about the care and placement of unaccompanied children. Several critical studies of this experience, including the observations of Anna Freud and Dorothy Burlington on the effects of war on children, have contributed significantly to our understanding of the psychological needs of children, factors to be considered in the care and placement of children separated from their parents, and the consequences of separation for families.

3

Greek Civil War

The roots of the Greek civil war in 1947 lay in the long-standing ideological differences among Greek political parties. At the close of World War II (the German occupation forces had withdrawn at the end of 1944) a coalition government of opposing parties was briefly established. The coalition collapsed soon afterward, however, when civil war broke out to decide whether the existing government or the opposing socialist and communist parties should control the country. A near victory by the socialists and communists was stymied by British intervention in 1945, leaving a right-wing government in power. There followed several years of peace. In 1947, under the direction of General Markos, a leftist Greek guerrilla movement based in the mountainous areas of northern Greece attempted to take control of provincial areas and overthrow the elected national government. Civil war continued from February 1947 until 16 October 1949, when the insurgents conceded failure. They withdrew from Greece into neighboring countries that had supported them, mainly Albania, Bulgaria, and Yugoslavia.[1]

The people of Greece suffered greatly during World War II as a result of battles, massacres, destitution, and famine. At the end of the war there were approximately 50,000 "unprotected children,"[2] for whom the first foster home program in Greece was established as well as twenty special villages for over 18,000 children. The killings and torture during the civil war further aggravated the hardships. There was total destruction of more than 25,000 farm houses; the widespread killing of cattle, oxen, and horses; destruction of infrastructure and schools; and massive population displacement. As a result of the civil war more than 700,000 people fled south from the northern provinces toward Athens and the security of government-controlled areas.[3] There they lived in refugee camps, unfinished buildings, or whatever other shelter could be found; many families were destitute. The health standard of the general population in Greece at that time was considered the lowest in Europe and the general conditions were worse than they had been during World War II.[4] In addition to those who fled south, many people—particularly those who supported the guerrilla movement or who were treated harshly as a result of government suspicion—moved north across the border into the security of Yugoslavia, Bulgaria, and Albania. Even in 1945 more than 20,000 Greek refugees, most of Slavic extraction, were living in Yugoslavia as a result of alleged repression by the Greek Government.[5]

During the civil war, the unaccompanied Greek children became an international humanitarian and political issue. The Greek government charged that members of the guerrilla movement had carried out a census of children between the ages of three and fourteen in various areas in northern Greece and then forcibly moved thousands of them to Albania, Bulgaria, Yugoslavia, and other countries for "reeducation." The government asserted that children were being "abducted" in order to terrorize Greek families into supporting the guerrillas and to indoctrinate the children with communist ideology. The government claimed that the guerrillas were destroying the Greek people by alienating their children and disrupting agricultural production by forcing families to flee from the land to the towns in order to protect their children.[6] The Greek government considered this to be genocide and the Greek public made it an important national issue. A national day of mourning was proclaimed on 29 December 1949, and the public rallied to support the idea that children belong to their country of origin and that education in communist countries outside the home country "taught the children to become enemies of their country, religion, and parents".[7]

A United Nations Special Commission found evidence that a census of the children had been carried out by members of the guerrilla forces and that children had been moved by truck, bus, and train from Greece to Albania, Bulgaria, Yugoslavia, Czechoslovakia, Hungary, and Poland. Interviews with parents and area residents and radio reports in countries receiving the children confirmed that the movement of children had taken place from January through April 1948. The governments of the recipient countries, however, defended the movement on humanitarian grounds: children in guerrilla-controlled areas had been saved from the risks of war and famine and could receive education not available to them in Greece and find a haven from political persecution.[8] In Yugoslavia the children were living either with a relative, with other Greek refugee families, or were placed in children's homes often run by the Red Cross Societies.

Public opinion in Greece was different from that in the host countries on whether the children had been taken by force or with the approval of their parents. Through interviews with parents and area residents, the UN observer groups found that some parents had agreed to the movement of their children, either out of sympathy with the guerrillas, or because of poverty and lack of schooling for their children, or from fear of the dangers of war. The majority of parents, however, had opposed the movement of their children. The interviews suggested that most parents who agreed to send their children had done so out of fear of reprisals.

To prevent the guerrillas from abducting more children the Greek government organized its own evacuation of children to the interior of Greece. On 20 March 1948, for example, approximately 5,500 children from Macedonia and 5,000 from Thrace were moved. At least 14,500 children were moved in the first wave. Some parents took their children away themselves to prevent evacuation by the government. These were placed in children's villages, each called a *paedopolis,* modeled after the newly created Pestalozzi village in Switzerland. Within a few months some 18,000 children were living in fifty-two villages established in unused hospitals, schools, and abandoned homes. Children were divided into groups of fourteen to twenty-five, each under the responsibility of a female monitor and a young leader. At that time the goal was to give the children direct responsibility for the life and organization of their *paedopolis.* Also, to occupy the children, apprenticeships were offered in agriculture, carpentry, and shoemaking. At the end of the war, children were returned to those parents who could be located, but at least 3,000 continued to live in the children's villages until they reached the ages of sixteen to eighteen.[9]

By the end of April 1948, the Greek government presented to the United Nations Special Committee on the Balkans the first list of some 1,000 children reportedly abducted, requesting the United Nations to intercede. The repatriation then became an international issue. On the basis of the assessment carried out by the special observer group mentioned above, the United Nations General Assembly on 27 November 1948 recommended the return of the Greek children in a resolution that read as follows:

The General Assembly

recommends the return to Greece of Greek children at present away from their homes when the children, their father or mother or, in his or her absence, their closest relative, express a wish to that effect;

invites all the Members of the United Nations and other States on whose territory those children are to be found to take the necessary measures for implementation of the present recommendations;

instructs the Secretary General to request the International Committee of the Red Cross and the League of Red Cross and the Red Crescent Societies to organize and ensure liaison with the national Red Cross organizations of the States concerned with a view to empowering the national Red Cross organizations to adopt measures in the respective countries for implementing the present recommendation.[10]

The International Committee of the Red Cross (ICRC) and the League of Red Cross Societies (LRCS) responded by attempting to help with repatriation. The Geneva-based Red Cross organizations accepted from the Greek Red Cross lists of children who had reportedly been moved from Greece and whose repatriation was requested by parents or relatives. The national Red Cross organizations in countries harboring children were asked to provide the names of those who had arrived. This straightforward approach of matching the names of children claimed with the children available was unsuccessful, however. For more than four years concerted efforts were made to overcome various obstacles, including the refusal of the host countries to participate in repatriation. The Red Cross issued annual progress reports to the UN Secretary General, and the General Assembly continued to adopt resolutions recommending the continuation of repatriation efforts by all parties involved.[11]

The refusal of the socialist countries was complicated by the strained political climate between these countries and others, and by the lack of good will among all governments and agencies involved in the repatriation effort. The various national Red Cross organizations tended to reflect the different positions shown in the attitudes of their respective governments. Socialist countries resisted repatriation by claiming concern for the well-being of the children. Authorities questioned whether it was in the best interests of a child to return to Greece when its parents had been opponents of the regime. The government claimed that the children had left the country voluntarily at parental request. Their homes had been destroyed in the fighting, and some parents were in prison or still fighting. They accused the Greek Government of using the children for political and electoral purposes and raised questions about postwar conditions in Greece. They suggested that the problems of refugees and homeless children still in Greece should be solved before returning the children living in other countries.

Points raised by the socialist countries often reflected their uncertainty about conditions in Greece, especially whether children there were being imprisoned, put in concentration camps or labor colonies, or killed as a result of political discrimination. Yugoslavia also took the position that repatriation need not be pushed further since the children were happy where they were. On the basis of these various concerns (along with discomfort about possible problems with the Greek Government when relatives came forward to claim children and the complications that would arise from multiple claims made by various relatives living in different countries), the socialist countries declined even to send lists of the Greek children in their care.

The Greek claim to the children and the refusal of the socialist countries to return them were ostensibly based on different considerations. Greece based its request for the children's return both on the desirability of family reunification and on a national claim to the children. Those resisting repatriation based their position on humanitarian grounds and on what were said to be the best interests and wishes of the children, but political considerations were paramount.

In a progress report to the UN Secretary General, the ICRC and LRCS summarized the major obstacles to their efforts to repatriate the Greek children as follows: The first obstacle was the total and regrettable absence of constructive cooperation by the majority of participating governments and agencies. The second problem arose from lack of agreement on the very definition of a child. The Greek Red Cross Society, following Greek law, assumed that a child was a person under twenty-one years of age and, in accordance with local customs, included young unmarried women living with their parents even if over the age of twenty-one. The Greek Red Cross Society also held that the relevant age was that of the child upon leaving Greece. This definition, particularly as it related to young unmarried women, caused conflict. The third problem was the credibility of requests for repatriation. While the essential condition for the return of a child was to be an "expression of a wish to that effect," there was disagreement on the form such an expression should take. The socialist countries maintained that claimants, even relatives, were being compelled to make the requests.

The fourth problem, described as the most serious, arose over disputes about who was entitled to claim the return of a child and where the return should take place. It was assumed that in the absence of the father or mother the nearest relative could request repatriation. This implied that the person making the request would, in fact, be the sole survivor of the child's family, and therefore the only person with whom the child would find a "home" again. In reality, multiple claims were made for some children, at times by distant relatives who made claims for children whose parents had themselves left Greece. Since it was recognized that families had dispersed to several countries, reunification should include sending the child to the country of the family's residence, even if socialist.

The governments of host countries demanded safeguards as a prerequisite to repatriation. Czechoslovakia, for example, requested an authentic document fully establishing the identity of the child. Those applying for the repatriation of a child had to produce an authentic document establishing their own identity, their relationship to the child, and, if they were not the parents, the reasons for their action. They also had to declare that they were making their request freely and not under the pressure of any threat. In addition, a guarantee was required that the repatriated Greek children and their relatives would not be subject to any reprisals whatsoever.

The Greek government gave assurances that no discriminatory measures of any kind would be taken against repatriated children and that the few who could not be sent directly to their families on returning to Greece would be housed and maintained either by the Greek Red Cross or by other Greek charitable organizations. In Greece, the Red Cross delegates individually interviewed claimants to verify the authenticity of their requests and visited the homes of several claimants to make sure that conditions were satisfactory. They also drew attention to the "moral and emotional effects which prolonged absence was having on these children and on their relatives." Nonetheless, these measures were not considered adequate by socialist governments holding the children.

The repatriation of the children was also obstructed by such practical issues as variations in the spelling of names, in some cases the result of translating names from Greek to Roman letters or the anglicizing of names for those parents who had migrated to Australia. There was confusion caused by the existence of different dialects, incomplete addresses, a lack of exact knowledge of birth dates and birthplaces, and uncertainty about the whereabouts of some children who had been moved between receiving countries such as from Albania to Hungary.

In 1950 the Swedish Red Cross was invited by the Yugoslav Red Cross to document all children. This process took over a year but it identified 722 children who wished to return to Greece. Some were children whose return had been requested by parents and some were not on any list but wished to return. This assessment provided the basis for the repatriation of the children from Yugoslavia.

The exact number of children removed from Greece is unknown. The Greek government claimed that 28,000 children had been abducted. In 1948 the national Red Cross Societies indicated that over 23,000 Greek refugee and displaced children were living in neighboring countries, including Albania (2,001), Hungary (3,000), Rumania (3,801), Czechoslovakia (2,235), Yugoslavia (10,000), with some in Bulgaria and the German Democratic Republic.[12] In total, the Red Cross organizations in Geneva received applications for the repatriation of 12,172 children by October 1951. The first group of twenty-one children were repatriated from Yugoslavia to Greece on 25 November 1950, and four days later, all twenty-one children were handed over to parents who had made written statements, authenticated by a competent official, declaring that a child had been returned. By May 1952, four years after the initial movement of the children, Yugoslavia was the only country from which children had been repatriated and the number totaled only 469.[13]

Some of the children who had been moved to the Eastern European countries in 1948 were the sons and daughters and next-of-kin of Greeks who had migrated to Australia before the war.[14] In Greece at that time it was usually the father who migrated first. After securing a job, finding housing, and saving enough money, he sent for his family. Many families were separated by the outbreak of World War II which cut all communications and prevented all such reunions. More than 700 family reunion requests were received by the Australian International Social Service, which by 1951 had managed to reunite 130 children with their parents and relatives in Australia. For most families, the reunion took place after more than ten years of separation.

As political conditions changed, repatriation from Eastern Bloc countries to Greece became easier. In 1953, 4,611 adults and children returned to Greece from

Yugoslavia and Romania, and in 1954 an additional 3,927 adults and children returned from Bulgaria, Hungary, Poland, Romania, Czechoslavakia, and Yugoslavia.[15] After this, those who had been children in 1948 and who attempted to return to Greece were refused entry on the grounds that they had been trained in communist countries and might be a security threat. For years, this policy provoked heated public debate in Greece. It was not until the 1980s that people who had left in 1945 were again permitted to return to Greece.

4

Korean War

Over a ten-year period, from 1945 to 1955, the number of unaccompanied children in Korea rose from a few thousand to more than 100,000, reflecting the extreme difficulties of the times. The first five years, from 1945 to 1950, saw massive social upheaval and the migration of large numbers of people. From 1950 to 1953, Korea was devastated by war which was followed by several particularly difficult years after reconstruction began. The existence of unaccompanied children and the effects of the war on family separation are best understood in the context of the overall emergency.

At the end of World War II the U.S.S.R. and the United States jointly liberated Korea, thus ending thirty-five years of Japanese domination. There was, however, no agreement on the type of government to be established. As a result, the Soviet command organized the Executive Committee of the Korean People in the north, and the U.S. military government in Korea established the Representative Democratic Council in the south. Both North Korea and South Korea asserted that there should be only one Korea, with each believing (to this day) that its own system of government was best and that the other was unacceptable. Five years of economic, political, and social upheaval led to mass movements of people, as Koreans returned from Japan and Manchuria and tens of thousands more fled from North to South Korea.

The Korean War began at dawn on 25 June 1950 with the surprise invasion of South Korea by North Korean forces. The invasion was condemned internationally and United Nations forces were called in to support the south. During the war, the north was supplied by the U.S.S.R. and supported by over 1.2 million Chinese combat troops. The south received military aid from sixteen western countries, the United States primary among them. At the end of the Korean War, the territory controlled by the north and the south was almost exactly the same as at the beginning—marked approximately by the 38th parallel. The war ended with the signing of an armistice on 27 July 1953.

During this three-year conflict approximately 6,000,000 people were killed or died from other war-related causes. The staggering loss of life was due in part to a military strategy adopted by both sides which tried to inflict the heaviest casualties possible. People in the south suffered extensively. More than 425,000 men in the South Korean and UN forces were killed, wounded, or captured, and close to 500,000 South Korean civilians were killed or died from wartime hardships. An estimated 84,000 civilians were kidnapped and forced to go to the north where many were pressed into military service. At the end of the war, 300,000 people were listed as missing. The people of the north suffered even more. As many as 3,000,000 civilians died in the north in addition to some 2,000,000 military casualties, sixty percent of whom were Chinese.[1]

The war had devastating effects on most families. Many families separated during migration or because of military duty or death. Hunger and undernourishment became commonplace for much of the population because of family displacement, loss of income, and disruption of food production. The fighting lines swept back and forth across the country from the extreme north to the extreme south, dislocating many people. Overcrowding, unsanitary conditions, and the lack of medicines and health facilities caused epidemics of smallpox, typhoid, and typhus. Physical damage, including the destruction of thirty-three percent of the housing, forty-three percent of industry,[2] and eighty percent of the hospitals, increased hardships dramatically. At the end of the war the number of refugees in South Korea alone was approximately 5,000,000, including an estimated 830,000 people from the north.[3]

Unaccompanied children were found along the roads, in refugee camps, and in shelters. Thousands of them lived on the city streets and survived by begging or stealing. Public social services failed. In the chaos of the war years, the Ministry of Social Welfare was concerned almost exclusively with providing emergency relief supplies. Existing orphanages were gravely overextended. The problem of organized care was exacerbated by the repeated forced movements of the population, which meant either moving the children as well and relocating facilities, or releasing the children. As early as September 1951, the United Nations Korean Reconstruction Agency (UNKRA) estimated that 100,000 children had been orphaned or separated from their families.[4]

Special assistance for unaccompanied children came from state orphanages, military units, churches, and locally organized orphanages. One can only speculate on the amount of assistance unaccompanied children received from unrelated adults, since few observations were recorded. Korean families did not traditionally take in unrelated children. Although only a small percentage of the population was Christian, local churches disbursed millions of dollars from international church organizations. Churches often became distribution centers and refuges for displaced persons, including the inevitable unaccompanied children, and churches sponsored most orphanages established during the war. The military also identified and assisted unaccompanied children. Soldiers often found unaccompanied children and took a personal interest in helping them. Many children attached themselves to military units, as described in this American soldier's account.

> During all this period we were gathering stray children. For the sick and obviously homeless kids we had set up a big tent with cots in it—entirely against regulations, of course. They infested the camp like ants, becoming so numerous that at last we had to pack them to Seoul to deposit them in a central orphanage. This was to become the Fifth Air Force Orphanage, and the whole organization contributed toward maintaining it.[5]

In the early phases of the war, members of the armed forces organized orphanages and spent much of their noncombat duty providing for the children. In Seoul, for example, in 1953, seventeen out of twenty-nine orphanages received supplementary assistance from military units.[6] Examples of such institutions include the Pusan Free Pediatric Clinic, which housed about 200 children in a makeshift building, one third of whom were sick or wounded. The Munske Children's House in Seoul was established for 450 children gathered by the UN troops. The Cheju and Wenju orphanages housed 400 and 778 children respectively. These institutions were usually set up in great haste and in the face of enormous need. Military personnel assisted in evacuating 1,000 children from the Seoul Central Orphanage to an off-coast island when the city fell to the communist forces. Reestablished on a virtually barren island, the orphanage depended totally upon supplies donated by the government, the United Nations, and the military. Within three months of the move, 200 children died for lack of proper food and medical care, many from whooping cough.[7] For most orphanages, essential supplies and staff were in critically short supply.

When the war began, bringing the children together for shelter and to distribute at least minimal amounts of food seemed the only alternative to letting them fend totally for themselves under impossible circumstances. The Korean Relief Law stated that homeless children up to the age of fourteen must be cared for in orphanages. A Ministry of Social Affairs regulation extended the age limit to eighteen.[8] Under Korean law at that time children became adults at the age of thirteen. Orphanages became the standard conduits for emergency services to needy children. Numerous temporary orphanages were established near relief centers and military compounds, not necessarily because more unaccompanied children were to be found there, but because the orphanages depended on nearby support. It was hoped that such orphanages would shelter and keep the children alive during the emergency and that coordinated long-range programs would replace the orphanages after the war.[9] The assumption, sadly, proved incorrect.

The largest number of orphanages for unaccompanied children were founded by churches and individual Koreans. Programs that were first established as temporary emergency measures were formalized with time, usually by a donation of property from the person who had organized the assistance. The orphanage director was therefore also the property owner. Orphanages were granted a government license on proof of available property (regardless of size or suitability) and of sponsorship. After licensing, an orphanage could receive relief goods on a preferential basis and privately solicit international assistance. The availability of money and goods was, of course, critical to the care of many children; but by multiplying the number of orphanages, it caused long-term problems.

Up until 1955, the children in institutions were primarily victims of the partition of Korea and of the subsequent war. They were therefore called "war orphans." Some children were indeed orphaned, without known parents or other adult relatives. Others had been abandoned or accidentally separated from their families. Many children came from poor families. Children were often left in the institutions by indigent widows who found it difficult to support themselves and, under the prevailing customs, had very little hope of remarrying. By September 1952 more than 293,000 widows were supporting over 516,000 children under thirteen years of age.[10] In addition to children in institutions, there were estimated to be 10,000 street children in South Korea after the war; approximately half seemed to have no families.[11] The street children posed particularly difficult problems. Having learned to survive on the streets, they had difficulty in adjusting to institutional life. In cities where there were periodic collections of street children by authorities, twenty to forty percent ran away within the first two weeks only to be picked up again and again.[12]

In 1945 there were only thirty-eight child welfare institutions caring for 3,000 children in an undivided Korea. Because of the social upheaval in the years between

1945 and 1950, the number of orphanages for "lost and abandoned" children in South Korea alone rose to 215, with a population of 24,945.[13] During the three years of war the number of institutions nearly tripled to 440, and the number of residents quadrupled to 53,964.[14] More orphanages were established in the two years after the war, so that by 1955 there were 534 institutions with a population of 54,927 children.

The quality of care provided to the children varied with the institution but most observers described orphanages with such phrases as "completely inadequate" or "criminally ill-run." A United Nations Korean Reconstruction Agency memo described the institutions as "totally inadequate in capacity, type, and standards of care to provide even minimum requirements for needy children." After the war international donors expanded the institutions in an attempt to improve conditions. "Reception centers" for the initial examination and screening of the children were established. Training for staff and vocational programs for the children were started, but problems of inadequate funding, a lack of trained personnel, poor facilities, and most importantly, the problems of providing for the developmental needs of children in any large institution, were never overcome. Deplorable institutional conditions continued through the 1950s and 1960s. The difficulties often grew worse because foreign funds, easily obtainable immediately after the war, became increasingly scarce in the following years.

While many who ran the orphanages were sincere in their efforts, others set up such institutions as an easy way to raise funds for their personal use. It was usually difficult to correct poor management because the institutional directors were also the founders.

In the first years after the war there were no organized programs to trace the families of the unaccompanied children,[15] although there were several independent efforts to reunite families. Children occasionally left orphanages to search for their families, but orphanage administrators made no concerted attempts to search on the behalfs of the children.

By 1955 the number of orphanages had temporarily peaked, and in the next two years the number dropped slightly to 496 institutions with 53,592 residents.[16] The trend then was reversed and the number of both institutions and residents climbed steadily until 1967 when there were 602 child welfare institutions with a population of 71,816.[17] Table 4-1 shows the rise in the number of institutions.

Prior to 1955, the major reasons for admitting children to institutions were related to the effects of the war. After 1955, the children tended to separate from families for other reasons. As parent-child separations and the care and placement of unaccompanied children in the postwar period were directly related to the services established during the emergency period, these services are worth reviewing.

The abandonment of very young children increased both during and after the war. In 1955, 715 children were left on orphanage doorsteps, in railroad stations, on the streets near police boxes, or anywhere else where quick discovery was ensured. This number peaked in 1964, when 11,319 children were abandoned. Between 1955 and 1970, a total of 80,520 children were abandoned.[18] Most were babies less than one year old, with about twice as many girls abandoned as boys.[19] Table 4-2 shows the recorded increases in child abandonment in Korea between 1955 and 1970.

The abandonment of children had many causes. During and after the war the number of illegitimate births increased and the children of these births were abandoned. To this day in Korea, mothers of illegitimate children are likely to be dismissed

Table 4-1. Child Welfare Institutions*

Year	Total Number	Total Population	Year	Total Number	Total Population
1945	38	3,000	1968	576	66,211
1950	215	24,945	1969	562	61,380
1957	482	48,594	1970	458	58,281
1958	526	51,630	1971	466	44,550
1959	503	53,016	1972	430	40,014
1960	511	56,042	1973	385	36,176
1961	542	55,385	1974	365	34,804
1962	536	52,815	1975	350	32,996
1963	541	56,494	1976	343	31,147
1964	537	61,963	1977	333	29,050
1965	565	69,487	1978	307	26,310
1966	587	71,709	1979	299	24,616
1967	602	71,816	1980	277	23,385

*Miller, op. cit., p. 19, and Chi, op. cit., p. 7.

from their jobs and lose their social standing, and illegitimate children are still treated as social outcasts. After the war, the breakdown of traditional values, rapid migration to the cities, the changing role of the family, and increased mobility also contributed to widespread child abandonment.[20]

Poverty-stricken families would often abandon their children, knowing that orphanages, however deficient, offered better opportunities than did average lower-class life. In an institution, children at least received the schooling and clothing that the parents could not themselves provide. Social workers confirmed that many parents abandoned children only because they believed the children would be better off in institutions.[21] This was later verified when parents reclaimed children, particularly if an institution was closing. A further reason for abandonment was the general lack

Table 4-2. Child Abandonment in Korea, 1955–1970*

Year	Total Number	Percentage Under One	Percentage One-to-Six	Percentage Over Six
1955	715	61.3	24.5	14.2
1956	1,425	66.5	26.4	7.1
1957	2,506	56.5	25.6	17.9
1958	2,113	55.2	26.9	18.0
1959	2,415	51.4	33.6	15.0
1960	2,537	51.5	31.2	17.3
1961	4,453	45.8	29.2	24.9
1962	4,646	40.1	30.6	29.3
1963	8,207	31.8	34.5	33.7
1964	11,319	24.6	43.8	31.6
1965	7,866	20.9	48.6	30.5
1966	7,284	21.2	47.7	31.1
1967	6,526	24.5	48.4	27.1
1968	5.976	21.6	54.9	23.5
1969	5,743	26.2	52.1	21.7
1970	5,788	29.8	52.7	17.5

*Miller, op. cit., p. 20.

of social services. Only by abandoning their children could some families secure help for them.[22]

For more than seven years after the war little effort was made to find alternatives to institutional care. No foster home programs existed in Korea, and Korean families did not formally adopt children. Intercountry adoptions of children did, however, increase every year.

Such adoptions during emergency situations had occurred earlier in other countries, but the first large-scale intercountry adoption programs began at this time in Korea. They grew out of concern for the illegitimate mixed-race children, most of whose fathers were American soldiers. Estimated at 8,000 or less by 1970,[23] these children were often abandoned as infants and were socially stigmatized. They were usually denied schooling and the various opportunities available to other children. As early as 1953, American agencies began moving these children from Korea to the United States for adoption by American families. In response to significant international interest in the adoption of Korean children, and recognizing that thousands of children still lived in Korean institutions, at least six American and Korean agencies founded international adoption programs. With government approval, these programs expanded to include Korean children who were not of mixed race. The number of placements grew annually from four children in 1953 to over 1,800 in 1970.[24] The total number adopted during this period was approximately 11,000,[25] with children placed in the United States, Sweden, Norway, Switzerland, Denmark, Belgium, the Netherlands, France, and at least thirteen other countries. Waiting lists of willing families existed in most of these countries.

Intercountry adoption had many critics, however, some of whom questioned the ethics of moving children between nations and who were concerned about the psychosocial effects of cross-cultural placement. Although well intentioned, the adoption agencies had insufficient social work staff and operated outside established social welfare services. Outside social workers considered the existing safeguards designed to protect the interests of the children to be inadequate. Questions were also raised about the removal of children without sufficient consideration of the circumstances under which the children had separated from the parent, as well as about the lack of counseling and support for the unmarried mothers who were giving up their children. Intercountry adoption was thought to encourage abandonment because some mothers considered foreign adoptions a golden opportunity for their children. There was criticism about the screening and selection of adoptive families and the lack of follow-up services. The adoption agencies were also criticized for not training more Korean social work staff, for not directing their attention toward preventive measures, and for not helping Korea to deal with the children's problems within the country itself. There were objections to the continuing use of fund-raising advertisements, even many years after the war, depicting children in war conditions and purposely mislabeling them as "orphans."

Criticism of the orphanages, the continuing abandonment of children, and the existence of intercountry adoption programs led to revisions in adoption laws and improvements in services. A law permitting the adoption of children by Korean families was promulgated in 1960, and in 1961 a modest program was begun to remove children from institutions and place them with families of relatives or with foster families. In that year, twenty children were returned to their families.[26] Organizations such as the Foster Parents Plan and Save the Children began to shift their emphasis toward preventive measures by supporting needy families. Adoption agencies broadened their services, developed more acceptable procedures, and incorporated more professional staff. Korean foster placement and adoption programs were started, as well as counseling and family planning services. After 1962 (the first year that data were available), the number of adoptions by Korean families increased annually from 833 adoptions to 1,724 in 1970.[27]

In summary, although the orphanages and adoption programs established for unaccompanied children during the emergency period of the Korean War provided essential services to needy children, they also had serious shortcomings. Criticism of both the orphanages and the international adoption programs were based on similar grounds; services provided were often of poor quality and were so narrowly defined that they neither addressed the causes of parent-child separations nor allowed alternative care; both program approaches had unforeseen social consequences that may themselves have encouraged further separations.

5

Hungarian Revolt

Between 23 October 1956, when the Hungarian Revolution began, and the end of February 1957, when the borders to the West were effectively closed, more than 190,000 people fled from Hungary to Austria and at least 17,000 fled to Yugoslavia. Among these refugees were at least 6,000 unaccompanied minors (approximately three percent of the total), most between the ages of sixteen and eighteen, with very few under thirteen. What is known concerning their care and protection clarifies the social and legal issues that must be dealt with in regard to unaccompanied children— issues that have been raised in many subsequent emergencies.

The flight of the unaccompanied children from family, community, and across a national boundary was largely impulsive, as it was for most Hungarian refugees. Decisions were made with little planning or reflection. The revolution itself had been unanticipated. It was rooted in long-standing grievances and resentment against the communist government in power. The revolution ignited on 23 October when a peaceful demonstration in Budapest turned violent after police fired on unarmed people. Another action a few hours later by Hungarian police, which was supported by Russian tanks and soldiers, further united the Hungarian people against the regime. Workers and students took up arms in a spontaneous national uprising with bitter fighting occurring in the streets. The communist government collapsed and within a few days a new coalition government was established. Hungary declared itself neutral and withdrew from the Warsaw Pact. The government demanded the withdrawal of Soviet troops, free general elections, freedom of speech, and the abolition of the secret police. However, a counter-government was also formed, which asked the U.S.S.R. for military support. On 4 November 1956, twelve days after the uprising had begun, the U.S.S.R. mobilized large numbers of troops and tanks throughout Hungary and brutally suppressed the revolution. Members of the new government were executed, and trainloads of Hungarians were deported to the U.S.S.R. Six days later, the fighting ceased and the revolution collapsed.

Most of the unaccompanied minors who fled Hungary were either students or young workers; the majority of all refugees were single males between the ages of eighteen and forty.[1] Interviews done by social welfare workers confirmed a wide variety of reasons for the impetuous flight of the unaccompanied children. Many fled in fear of reprisals for their participation in the revolution. Most "street fighters" were said to have been students and workers under twenty-two years of age. A thirteen-year-old boy who fled to Austria, for example, had participated in the fighting and had spent three weeks in jail. Such stories tended to confirm the stereotype of young Hungarian escapees as "freedom fighters," playing heavily on the emotions in the short-lived but intense international response to the Hungarians by much of the noncommunist world.

But the unaccompanied minors fled Hungary for other reasons as well. At a conference in Vienna in 1957, the Chief Medical Officer of a school noted that many of the Hungarian children were "fighters by nature" to whom the revolution had given the opportunity for adventure, the chance to use real weapons, and an urge to conquer and annihilate. Some children were escaping from bad home conditions; others were drawn by the thought of distant countries and the call of the "Golden West." Some were weak, passive, and easily influenced and left home to follow the rest; these were often homesick. "Only a few could give coherent motives for their flight. It was but gradually that the real reasons emerged, in private conversation, not in front of a committee."[2] Because the motivations of the unaccompanied children varied greatly, there was a need for child welfare workers to make an assessment of each case, with regard for both immediate care and protection and the child's future.

The Hungarians who reached Austria in the winter of 1956 arrived on foot, more than 10,000 people per day, bringing very little with them. The exodus began on the day after the fighting started in Hungary; unaccompanied children were among the first refugees. One early group, for example, included 300 students and professors. Providing food, shelter, sanitation, and clothing for such large numbers of people was in itself a major undertaking, and workers often described the situation as chaotic. The arrivals were housed at more than forty different locations, the majority of them in what were called "reception camps" in the Vienna area, or in the more permanent refugee camps that had been used since the end of World War II. Some of the facilities were abandoned Soviet and American military installations that were quickly refurbished. The conditions in many camps were horrible, with as many as sixty or seventy men, women, and children jammed into small rooms. Smaller camps were usually better. Refugees were also sheltered in holiday camps, hotels, inns, and private homes. Initially, no special services were provided for unaccompanied children except for many who had been students. Popular efforts to continue the education of students resulted in the opening of special schools and the acceptance of some Hungarian minors into Austrian schools.

The Austrian government requested international assistance. The Intergovernmental Committee for European Migration (ICEM), responsible for organizing the international movement of refugees; the United Nations High Commissioner for Refugees (UNHCR), responsible for protection of refugees; and the United States Escapee Program (USEP) provided the largest programs. Direct services were provided by about one dozen national and international voluntary agencies, most of which were already in Austria providing services begun after World War II. Voluntary agencies were also handling the emigration and resettlement of refugees on a day-to-day, case-by-case basis for the international organizations and individual governments.

Many nations responded immediately to the plight of the Hungarian refugees, providing millions of dollars in cash and goods and, most importantly, agreeing to accept them as emigrants either for temporary asylum or for eventual permanent resettlement. Within three weeks of the first exodus into Austria, thousands of Hungarian refugees were moved around the globe, often without being asked anything more than whether they wished to go. By June 1957 more than 140,000 Hungarians had been moved to thirty-five different countries.[3] Unaccompanied children were generally not moved because it was felt that emigration might threaten family reunion. Nonetheless, more than 2,500 unaccompanied minors did leave during the first three months, often temporarily attached to other families.

In January 1957, when immigration into Austria ceased, various agencies redirected their efforts from providing emergency goods to providing services for children. They concentrated on three immediate problems: the emigration of the children before proper procedures had been followed or plans made; the undesirability of the children's remaining in camps; and the need for individual planning for each child.

LAW AND PROCEDURES

Some national and international refugee service agencies feared that prompt emigration of unaccompanied youths might hinder sound resettlement planning or eventual reunion with their families. They also suggested that emigration be contingent upon confirmation of suitable placement. On 28 February 1957, the Austrian government, responding to these concerns, blocked further international movement of unaccompanied children until emigration procedures were developed that considered "the best interests of the child."

To establish procedures the Austrian government and the UNHCR had to clarify various legal issues. First was the status of each party. Although UNHCR had a protective role, it was not the legal guardian of the unaccompanied children. Since it could not be appointed guardian it therefore could not decide on the children's residence or removal. The basic responsibility for the children rested with the Austrian government as the government of the country of asylum. It was also determined that the children could be granted the status of refugees and that each case had to be considered on its own merits. Each minor was to have a legally appointed guardian in the absence of parents. Also at issue was whether the minor's wishes had legal standing. After consideration of various national laws and practices, UNHCR confirmed that children's wishes should be considered in all decisions regarding their fate and that children should be provided the opportunity to express their own opinions. Unaccompanied children were permitted to emigrate to other countries if an acceptable social welfare investigation had been completed, if the child himself expressed the wish to emigrate, if the child's parent or guardian consented in writing (usually through letters), and if there were no reasons to doubt the soundness of the emigration project. If a conflict existed between the wishes of the parents and that of the child, both were given the opportunity to present their case and have it decided in the child's best interest. Before repatriation could occur the merits of each case were to be scrutinized and the child's wishes considered; both UNHCR and the Austrian government agreed that no child would be repatriated involuntarily if considered to be a refugee.

Most unaccompanied children corresponded with their families in Hungary who advised them on decisions regarding their return or resettlement to another country. If a child could obtain a letter confirming parental authorization to emigrate, neither the Austrian government nor UNHCR introduced obstacles. Children who could not get parental authorization because of death or disappearance were provided a court-appointed guardian to advise them.

In keeping with these principles and considerations, the emigration procedure for the unaccompanied children was as follows. Voluntary agencies prepared case histories for the children under their care. Otherwise, social workers from the International Social Service (ISS), under contract by the UNHCR, compiled a file which included a personal history, parent's or guardian's authorization, an evaluation of the prospective sponsor (done in the host country), and a recommendation by a social worker as to the desirability of migration. The final recommendation on emigration was prepared by ISS social workers on behalf of UNHCR. These procedures were modified after several months to permit emigration from Austria without such evaluations of the sponsor because they could not be vouched for. Although such evaluations were still considered essential, the responsibility for these was transferred to authorities within the country of resettlement.

MOVEMENT FROM CAMPS

As noted earlier, from the beginning of the influx into Austria, some unaccompanied minors had been placed in schools but most lived among the larger refugee population in whatever facilities were available. UNHCR and the social welfare agencies involved agreed that the camps were unsuitable for unaccompanied youths. Authorities repeatedly mentioned the destructive environment of the camps with their lack of vocational opportunities and the dangers of idleness and promiscuity. National and international voluntary agencies had sponsored or set up throughout Austria many residential centers, each with a capacity for thirty to seventy children. Children were also placed in schools and homes, and in Yugoslavia, a children's village was established. In spite of these concerted efforts hundreds of unaccompanied children remained in camps and many openly opposed having to leave. In April 1957, the government, in agreement with UNHCR's wishes, began removing the children from the camps to homes or institutions where they could receive education or training until plans could be made for each of them individually.

INDIVIDUAL ASSESSMENT AND PLANNING

To provide individual assessments of the children, prepare files and recommendations, and encourage movement out of the camps, UNHCR contracted the ISS to provide case work services for the "unattached" Hungarian minors. From March to October 1957, ISS social workers interviewed 1,004 unaccompanied youth in twenty-three camps and other locations. The interviews revealed that approximately half of the children wished to resettle in another country. Of the remaining youths, twenty-two percent were recommended for placement in homes and twenty-five percent for residential schools in Austria. Almost one-third of the children were classified by the social workers as having no definite plans and needing further individual counseling.

Each child required individual assessment as each had unique problems requiring special consideration. Only after four or five interviews were some children willing to share information about themselves or their families. Social workers did not find the "indomitable fighters of freedom" depicted by the press, but rather "youths in distress" who were frequently depressed and confused. Many children in camps had already been moved four or five times, often after unsuccessful placements in schools and institutions. Many arrived with totally unrealistic expectations about what "freedom" would mean and what options would be available. When they saw that their dreams were unattainable they became discouraged and disillusioned. These difficulties were compounded by the lack of planning in the early phases of the program, when each agency presented to the youth whatever scheme it had devised, many of which were unrealistic. In addition, the children had been given little opportunity to express their own wishes, aspirations, and problems. The interviews by social workers, while essentially for documentation purposes, also had therapeutic value.

At the beginning of May 1957 there were 3,665 unaccompanied adolescents between fourteen and eighteen years of age among the 35,000 Hungarian refugees still in Austria. Of the unaccompanied youth, 2,494 were boys and 1,171 were girls.[4] By October, many unaccompanied children had been assisted to emigrate, at least forty-nine had been repatriated to Hungary, and many had been placed in homes and institutions in Austria. One year after the influx began, at least 625 unaccompanied youths remained in camps. Social workers considered "a Children's Village with trained personnel oriented to the problems of these children" to be the ideal solution.[5] They felt that a positive and objective setting would allow individual work with the youths over a period of time. Yet there existed no program that was willing to provide such a service.

REPATRIATION AND FAMILY REUNION

The Hungarian government asked for repatriation of the unaccompanied minors and promised amnesty from criminal proceedings for their having left the country illegally. (Criminal charges on other grounds were still possible, however.) Some of the adults and unaccompanied children who fled initially to Austria later requested repatriation. By June of 1957, official repatriations totaled 4,767, and many persons undoubtedly returned on their own. Repatriation continued for some years. The number of unaccompanied children who returned to Hungary is uncertain, although at least 162 were repatriated from programs in which ISS was involved.[6] While unaccompanied children alleged that faked telephone calls and messages were being used to entice them back, the policy of the Hungarian government was that unaccompanied children should be repatriated on the basis of the freely expressed desire of the parents and the interests of the individual child as determined by the country of residence.

The reunion of families outside Hungary was more difficult. Refugees were obliged to leave behind wives and, in some cases, young children, rather than expose them to the risks of crossing the border in midwinter. Members of the same family often crossed the border at different times and were resettled in different countries before they had the opportunity to reunite. Until postal communications with other countries were restored, ICRC broadcast 16,155 requests for news of families over Radio-Inter Croix Rouge and transmitted 41,722 written messages.[7] Immediately after the uprising ICRC had also unsuccessfully attempted to reunite families through repatriation both to and from Hungary. There were claims from parents in Hungary who requested the return of their children and claims from parents who had crossed the border requesting that the children join them. The Hungarian government at first demanded the return of the children claimed by their parents and at least initially did not allow children or wives to leave Hungary to join other close relatives.

RESETTLEMENT

Unaccompanied Hungarian minors were resettled in two ways. First, the children moved as part of larger refugee movements, often by attaching themselves to families in camps. More than 800 unaccompanied minors arrived in France, at least 466 in Great Britain, 350 in Italy, 188 in the Netherlands, and approximately 1,000 in the United States. Second, smaller numbers were resettled through official resettlement programs for unaccompanied children established later.

The reception and care of the unaccompanied Hungarian children varied from country to country, yet there were some similarities. The most obvious similarity was the lack of preparation, thus showing the need for special procedures and services.

When Hungarian refugees were accepted into Great Britain in the evacuations of November and December 1956, unaccompanied children were not expected. A Home Office memorandum issued in January 1957, after more than 12,500 refugees had already been admitted, estimated that only thirty-five children had arrived without parents or guardians. Because of the language barrier and the fact that the children had attached themselves to families while in Austria they were not immediately discovered. That number grew to over 450. Responsibility for the unaccompanied minors was guided by a Home Office suggestion that those up to the age of seventeen should be: (1) in the care of a recognized voluntary organization, (2) in the care of the local authority, or (3) in an approved private home—which meant notifying the children's officer of the local authority. Officials tried to find foster homes for the youth or to place them in hotels or other established youth programs. Children above the age of sixteen were found private accommodations near their jobs and were given a supplement to their salaries if they earned too little to support themselves. A voluntary agency made arrangements for the children, visiting them regularly and ensuring that they were attending school.

Of the 32,000 Hungarians admitted to the United States approximately 1,000 of them were unaccompanied minors who arrived in the evacuations of the first months. Because every effort was made to get refugees out of reception camps as soon as possible, decisions concerning unaccompanied children were hastily made. Most refugees stayed in reception centers no longer than ten days. No federal guidelines governed the care and placement of these children, and each voluntary agency responsible for the settlement of refugees (and consequently for unaccompanied children) established its own procedures. Social workers interviewed most minors and placed them wherever they found people willing to take them and "presumably supervise" a foster family placement. Only one agency made it a policy to inquire through a local social welfare agency into the quality of care relatives could offer before the unaccompanied children were sent to them. It was not clear who had legal and financial responsibility for the minors. After minors were moved, public authorities were often not aware that the children were in their area and therefore did not assume the normal protective and supervisory roles necessary.

In response to concern for the children expressed in Austria and to the many offers received by the U.S. Department of State from relatives of unaccompanied Hungarian children to sponsor and provide homes for them, a special program began bringing unaccompanied Hungarian children to the United States between June and December 1957. This program (under the direction of the Immigration and Naturalization Service) was as rigid as the former programs had been lax. Both the INS and participating voluntary agencies carried out the legally required home studies on each potential sponsor. Most important was the ruling that only those unaccompanied minors would be admitted whose sponsors were both relatives and U.S. citizens. This eliminated the possibilities of placing the minors with recently arrived Hungarians, placing them in the homes of nonrelatives, or making other arrangements for them. As a result, forty-four percent of the children in Austria who wished to emigrate to the United States, had relatives here, and were therefore recommended for emigration were rejected either because their relatives would not accept them or because the relative-sponsors were disapproved of by either the social welfare agencies or the INS. By the time this program ended, only 136 children—119 from Austria and 17 from Yugoslavia—had been admitted.[8]

6

Cuban Revolution

On 1 January 1959 the Cuban Revolution succeeded the previous government and brought Fidel Castro to power. By December 1980 about 691,000 people had fled the revolutionary regime for the United States.[1] Among them were approximately 18,000 unaccompanied children.

Cubans entered the United States during three periods, each one the outgrowth of a political crisis in Cuba. Most of the unaccompanied children were among the 280,000 Cuban refugees who fled to the U.S. during the first exodus, which lasted from January 1959 until the Cuban missile crisis in October 1962. During the second exodus, between 1965 and 1973, an additional 273,000 Cubans arrived in the U.S., many of them parents seeking reunion with children who had gone there during the first exodus. The third exodus, which was comprised of about 125,000 emigrants, took place between April and December 1980 and included 2,000 to 3,000 unaccompanied minors. Cubans also left for Spain and for Central and South America, however this discussion is limited to the larger number of unaccompanied children who went to the United States.

Unaccompanied children were surely among the 26,527 Cubans who entered the U.S. during the first six months of 1959, although there is no record of their entry. Neither the authorities nor the voluntary agencies seem to have investigated or even considered the possibility of their presence. Unaccompanied Cuban children were first identified in November 1959, almost one year after the revolution, when several came to the attention of the courts and of child welfare agencies in Miami. Some children were reportedly placed with emigrating strangers at the Havana airport when parents were denied permission to leave. Further investigation by social welfare agencies confirmed the need for special services for unaccompanied children, both for those who had already arrived and for those who were expected to come. Programs were hastily established and the number of unaccompanied Cuban children to receive services quickly increased from just three in November 1959, to 174 by the end of January 1960. By October 1962, when Cuba temporarily stopped all emigration, at least 14,000 unaccompanied children had entered the United States.

Many Cuban middle-class families feared the possible consequences of the revolution. In 1960 parents were even uncertain as to whether they would be allowed to raise their own children. These fears were often fueled by rumors, such as one that children above the age of three would be removed from families to live in government-run dormitories. Parents opposed state indoctrination of their children and worried that they would be sent to the Soviet Union for training. Many parents wanted their children educated in religious schools. Sometimes children were sent to the United States because their physical safety was at risk, particularly when they or their parents had been involved in the resistance activities. A number of parents remained in Cuba hoping to maintain their homes and businesses, but sent their children ahead of them to establish a foothold in America. By guaranteeing care, financial support, and education to refugee children, newly established reception programs surely influenced parents (at least until 1962) to send their children to America alone.

A project called Operation Pedro Pan assisted thousands of unaccompanied Cuban children.[2] Devised by an American headmaster in Havana to assist Cuban parents who wished to send their children to the United States, but who had neither relatives nor friends there, the program became something of an underground railroad for children. Program organizers in Cuba assisted these families. American corporations donated the airfare, sending the money to Cuba through private channels. The children were often flown through neighboring countries to confuse the Cuban government about their true destination. American immigration authorities made special visa waivers available, sometimes with the assistance of other embassies. At the Miami airport the children were met by social workers from the Catholic Welfare Bureau, Children's Division, a licensed child welfare agency that assured assistance to all unaccompanied Cuban minors. On arrival, the children were either reunited with their relatives or friends, or, when this was not possible, placed by an affiliated agency of the Catholic Welfare Bureau. At the inception of this project, its organizers assumed that only several hundred children at most might leave Cuba in this way. It was believed that the movement of children out of Cuba would very likely be stopped by Cuban authorities. But this did not happen until the Cuban missile crisis occurred more than two years later. Within that two-year period, social workers met about 14,000 unaccompanied children.

Families from all economic levels were represented, but most of the unaccompanied children came from middle-class families and had attended private Catholic schools. Approximately ninety-five percent of the immigrants were Catholic. Some families undoubtedly separated hurriedly when the opportunity to leave suddenly arose, but the exodus pattern suggests that most separations were not impulsive. Some unaccompanied children were among the refugees to make their passage to the United States by small boats across the ninety miles of ocean that separates the two countries; most, however, arrived on commercial airline flights directly from Cuba to Miami or circuitously through neighboring countries.

When the presence of unaccompanied Cuban children first became known in Miami in November and December 1959, local social welfare agencies invited the federal government to participate in establishing a special program. The guiding principles were to be as follows: the organizations responsible for the care of unaccompanied children should be licensed child welfare agencies, the religious heritage of the children should be safeguarded, and foster care should be adequately funded.[3] Through the office of the Children's Bureau, the federal government negotiated and supervised a contract with the Florida Department of Welfare of the State of Florida to develop an emergency program. The Department of Welfare awarded contracts to one international and three local agencies to provide foster care. Of the agencies contracted, two were Catholic, one was Protestant, and one was Jewish. Agencies cared for children according to their religious affiliations, although most of the children were Catholic.

An unaccompanied Cuban refugee child was defined by the U.S. government as someone who, while meeting the definition of a refugee, is "a child in the Miami impact area at the time service is initiated, whose parent or relative cannot provide care and supervision for him, and who is in need of foster care."[4] Procedures were developed to record each child's arrival, location, and type of care provided until final discharge from the program. Most important for the agencies was that the federal government agreed to pay for foster care. For the first time, U.S. government money supported arriving refugee children, enabling voluntary agencies to assume responsibility for many of them.[5]

When the programs began in November 1959, unaccompanied minors arriving without friends or relatives were placed in local foster homes. Cuban foster parents were chosen to provide both emergency care and permanent placement and were given a monthly stipend. Florida child welfare services made the home studies. The number of arriving children soon far surpassed the number of local foster homes and the residential space available in local facilities. Thereafter, children were sent to child welfare agencies in other states for placement in foster homes, group homes, and institutions.

The agencies caring for Cuban children had to be approved or licensed by welfare departments in their respective states and had to make social casework services available to the child. About eighty-five percent of the children were sent to Catholic child welfare organizations or institutions. As mentioned before, the others were assisted through Protestant, Jewish, or nonsectarian organizations. By 31 October 1962, the Catholic Welfare Bureau had provided some assistance to 7,464 of the 14,000 unaccompanied children it had received at the Miami Airport.[6] As of that date, in all programs, some 4,010 Cuban children were still under agency supervision; 1,400 were in foster homes; 2,500 were in local institutions for dependent and neglected children, small group homes, or private boarding schools. It has been reported that every effort was made to keep brothers and sisters together.

The children were between the ages of six and nineteen; most were over fourteen. Approximately two-thirds were boys. The majority were from "comfortably off" middle-class families and were, on the whole, well-behaved, good-natured, law-abiding, and often very bright. Because they expected to be in the U.S. for only a very short time before returning to Cuba, many, at first, were not inclined to learn English. They clung to this attitude even after placement in foster care. Their attitude changed when they realized that by knowing English they could assist their parents when they arrived. Children who remained in the Miami area, however, continued to speak Spanish and were often cared for by Spanish-speaking agency personnel.

Almost all of the children had lived with their families prior to separation, and most had been prepared to "be sent to school" in the U.S. From the moment of their arrival they were reported to be in frequent contact with their parents by phone and letter, and parents continued to participate in their children's decisions. On arrival nearly all of the children began immigration procedures for their parents. At that time, any Cuban with a first-degree relative in the U.S. could receive authorization to emigrate with a waiting period of three months to two years.

Questions concerning the legal custody and guardianship of Cuban children arose early. Parents, agencies, and governments chose to view the exodus as the movement of Cuban children to school in the U.S.—not an uncommon practice for middle-class Cuban families. Because it was assumed at first that things in Cuba would return to normal quickly and enable the children to return home shortly, custody and guardianship remained with the parents. Efforts were made initially in Miami to establish a Cuban school, with Cuban textbooks and teachers, so that the children would not be at a disadvantage when they returned. When the Cuban missile crisis erupted, hopes simultaneously faded for repatriation of minors and for reuniting parents with children in the U.S. Therefore, attempts were made to find foster homes for the younger children, but group homes remained the most common solution. In 1964, a system was devised to bring the parents of unaccompanied Cuban children to the U.S. through Spain and Mexico but only a small number of families took advantage of it.

On 28 September 1965, Premier Castro unexpectedly announced that any Cubans with relatives in the U.S. could leave after 10 October. A Cuban port was designated for embarkation and hundreds of small boats, many considered unseaworthy, left Cuba for the United States. The ocean crossing was very risky and to avert great loss of life the American government proposed an orderly departure program, which was subsequently accepted by the Cuban authorities. After a temporary sealift the Cuban port was closed and refugees left on regular airline flights. A Memorandum of Understanding established priority for departures of immediate relatives, defined as parents of unmarried children under the age of twenty-one, spouses, unmarried children under the age of twenty-one, and brothers and sisters under the age of twenty-one. During November 1965, 4,598 people left on the sealift; then between 1 December 1965 and 30 April 1967, an additional 62,861 refugees arrived by airlift. Among these were one or two parents for some 2,436 unaccompanied children under twenty-one years of age.[7]

By 30 April 1967, at least 142 children's agencies located in 110 cities across the U.S. had participated in the program for unaccompanied Cuban children. As of that date, 8,331 Cuban children had received federal assistance, 375 remained in foster care, and 419 were still receiving public assistance in Miami "guardian" homes. Federal expenditures for the children to 30 April 1967 totaled $28,531,489.08.[8] Besides those children who had received assistance from agencies, approximately 5,000 to 7,000 more arrivals had been cared for by relatives and friends.

The total number of children reunited with their families in the U.S. is not known, nor is the number of children who returned to Cuba to rejoin their families. Of those reuniting in the U.S., some had been separated for only a few months, but most had been separated three to four years. Some children could not be reunited with parents because the parents either chose not to, or were unable to come to the U.S. Some parents had divorced and established new living arrangements not conducive to reunion, and some children did not want to join parents after such a long separation.

The third exodus from Cuba known as the Mariel boatlift began on 6 April 1980, when more than 10,000 people entered the Peruvian Embassy in Havana demanding asylum. The situation lasted until September of that year. After diplomatic difficulties between Cuba and Peru, Cuban authorities finally allowed those seeking asylum to leave for Costa Rica. The U.S. government announced it would accept 3,500 as refugees. Then, just as at the beginning of the second exodus in 1965, Cuban authorities unexpectedly announced that anyone could leave Cuba from a designated port, this time from Mariel in the north. A large flotilla of small craft privately hired by residents of the Cuban community in the U.S. set sail immediately. The boats from Florida picked up separated family members and friends. Cuban authorities also directed each departing boat captain to take additional persons who were considered undesirable: inmates of jails, prisons, mental institutions; homosexuals, prostitutes, and street people.

In the United States the large number of people involved in this exodus shocked officials. In the first week, over 5,000 Cubans arrived in Key West, Florida. As numbers mounted and U.S. officials became aware of the refugees' backgrounds they declared the massive boatlift illegal and threatened boat owners with fines but these measures did little to stop the influx. In one five-day period alone approximately 27,000 people arrived. By 1 July, 115,000 Cubans had entered the U.S. and by the end of that year the number had increased to 124,500. Among those entering were 2,000 to 3,000 unaccompanied minors.

New arrivals were screened by immigration officials and initially those with relatives or other sponsors were allowed to leave the reception centers on parole. The remainder were housed at Eglin Air Force Base in Florida and Fort Chaffee, Arkansas. The number of new arrivals strained facilities and facilitators, and additional military bases were opened as reception centers. In all essential aspects these were detention centers. Unaccompanied minors were initially transferred to Miami for final processing and placement, but this arrangement was short-lived. Thereafter they were held with other Cubans in the main reception centers—Fort Chaffee; Fort McCoy, Wisconsin; and Indiantown Gap, Pennsylvania—where they remained for approximately four months.

This time there was no coordinated response by the federal government, state social welfare agencies, or the voluntary agencies that had followed earlier arrivals of unaccompanied minors. The federal government ruled that the arriving Cubans would not be granted refugee status en masse, but would be regarded as individuals seeking asylum, and an assessment of each claim would be required. The Cubans were given a new classification, that of "entrants," while the process was pending. Under this label they had no legal rights to public assistance and no positive assurance of permanent resettlement. This was very important in the care and placement, or lack of such, of the unaccompanied children: without legal clarification and assurance of federal assistance, states and agencies were not willing to assume responsibility for these children, particularly if institutional care or other long-term assistance was required. From 20 April 1980 to 10 October 1980, when legislation was enacted, there was no federal agency legally authorized to deal with these Cuban children and their care was not entrusted to a child welfare agency. No specific registration of the children was implemented and only a random survey of the children was carried out in the first two months. Between July and October, a moratorium on the resettlement of minors delayed the placement of about 300 of them, even of those with relatives willing to take them.

Complicating the care and protection of the unaccompanied minors in this exodus were their unusual characteristics. The minors were predominantly male and close to eighteen years of age. A random sample showed that thirty-seven percent had been institutionalized; forty-one percent came to the United States because they were given a choice of leaving Cuba or serving more time in jail; seventy-four percent were labeled by those making assessments as having some form of psychopathology, for nineteen percent it was considered to be "more than a little bit."[9] Child welfare agencies and state services were less willing to assume responsibility for minors with such special problems particularly when it was uncertain that anyone would eventually assume the legal and financial responsibility.

Within the camps the physical protection of the minors was a major problem because they lived among seasoned criminals who often armed themselves with homemade weapons. Sexual abuse was a serious concern. At the Indiantown Gap camp officials tried to segregate unaccompanied children but there were still cases of abuse by homosexual adult males. At Camp McCoy there were reports of gang rapes of the minors. Eventually, in all three camps unaccompanied children were segregated from the general camp population. It was also necessary to segregate violent minors.

The procedures followed for release of the minors also failed to protect them adequately. Initially, children were released to anyone expressing a willingness to care for them, even where such accepted child welfare procedures as home studies were lacking. A simple claim allowed anyone to gain custody without an official verification of relationships first. One "uncle" reportedly claimed twelve or thirteen children.

Legislation enacted on 10 October 1980 resolved many of the legal, financial, and bureaucratic problems that had complicated the care and placement of the unaccompanied minors. Cuban entrants were henceforth to be treated as refugees, thus allowing them to become eligible for benefits. The Immigration and Naturalization Service agreed to retain legal responsibility in reception centers, then parole the minors (temporarily transferring responsibility) to the Office of Refugee Resettlement, then legally assume responsibility again until a state or local agency could take over. Federal funding made placements possible through voluntary agencies and state services.

Finally in December 1980, eight months after the influx, the federal government issued guidelines specifying responsibilities and criteria for state child welfare agencies assisting unaccompanied Cuban children within their jurisdictions. Now responsible for the care and placement of the minors, state agencies were directed to consider special needs and cultural factors in determining placement. While either individual or group care could be selected in accordance with the circumstances, large remote facilities were to be avoided. The minors were placed in foster homes, group homes, independent living arrangements, and, in some cases, institutions.

In summary, the problems encountered by unaccompanied Cuban children who entered the United States are very similar to those faced by unaccompanied children who crossed national borders in other emergency situations. The assumption that the children could be temporarily evacuated and then returned within a short period proved to be unrealistic because of unforeseen political complications. For many children, expected family reunions never materialized and short-term placements turned into long-term arrangements. The problems unaccompanied children among the Mariel entrants faced demonstrate the need for established legal procedures ensuring the care and protection of all children, regardless of their legal status. The experience of the 1960–62 arrivals demonstrates the usefulness of existing child welfare agencies in providing services to displaced children. By contrast, the problems faced by the unaccompanied children who arrived in 1980 illustrated the difficulties that occur when existing children welfare agencies are ignored.

The experience of the Cuban unaccompanied children also raises questions about the social and psychological consequences of intercountry displacement. Literature on these children contains little information about the emotional effects of separation on them. Most authors describing these programs only allude to the problem: "A remarkable number adjust to their new circumstances with relative ease, but others develop situational anxieties";[10] "While some of them have been found crying at night by their foster parents, relatively few have been reported as showing symptoms of severe emotional disturbance . . . ";[11] "Many of the younger children did not understand why they had to be separated from their parents . . . As the years passed and the children had not been reunited with their parents, some of the older children became confused, resentful, and often hostile because of the prolonged separation."[12] But these references were written as casual observations. Study has not documented the emotional experience of these children or assessed the effects of the family separation on the rest of their lives.

7

Nigerian Civil War

The total number of children separated from their families during the Nigerian Civil War is not known, as large numbers died during the conflict and many others spontaneously joined other families. At one point during the war the number of children separated was estimated to be 100,000. At the end of the civil war more than 40,000 were in hospitals, sickbays, and feeding centers, and an additional 5,000 had been evacuated to neighboring countries.

The civil war broke out in July 1967 when the eastern region of the country, calling itself the state of Biafra, seceded from Nigeria. For the duration of the war the Nigerian government blockaded the area by sea, air, and land. People within the war zone thus lived within an embattled enclave which was slowly reduced in size by military advances until near the end of the war, when the area defended was less than one-fourth its original size. The blockade and the military advances had at least two major consequences for the civilian victims: dislocation and food shortages. Almost one-half of the eight million Biafrans became refugees within their own region. In January 1970, with the collapse of the Biafran defense, the signing of a peace treaty ended the war.

The predominant issues in the care and protection of children during this conflict were inextricably linked to the food shortages and the famines faced by the general population. During the war starvation was used and defended as a military tactic. At least one million civilians died from starvation and related diseases during the thirty-month conflict.[1] Of eight million Biafrans, forty to fifty percent were children under the age of fifteen.

Famine conditions were not constant, however, but developed gradually and then abated. Nor did everyone suffer; the famine most severely affected women, children, and the elderly, as well as refugees, who obviously had limited access to local resources. Unaccompanied children were certainly more vulnerable than other children. The International Union of Child Welfare (IUCW) observed that after the general nutritional status of most children had improved, it was among the unaccompanied children that the scattered cases of *kwashiorkor,* a severe form of malnutrition, persisted.

As might be expected, the problem of local care for unaccompanied children became more pressing during food crises, and then lessened during periods when the local community was better able to cope. The first famine developed within a year after the fighting began and peaked in August–October 1968. During the worst periods, food shortages were so severe that relief agencies might have no food at all to distribute, even to small children brought to them in the final stages of starvation. An estimated 500,000 children under the age of ten died of starvation and related diseases in 1968 alone.[2] The second famine followed in the fall of 1969.

Within Nigerian society, children who were orphaned or separated from their parents were traditionally cared for by members of the extended family. This cultural pattern explains the comparatively small proportion of unaccompanied children in relation to the affected population during the war. Families also cared for unaccompanied children in need of help who were unrelated to them. Children were also placed in foster homes through the existing social welfare agencies and in orphanages through private arrangements. Unaccompanied children were sometimes picked up along roads and in camps and villages. Most were found in hospitals, clinics, and feeding centers into which they had wandered or been taken in search of food. They were referred to by the Nigerians as "unclaimed" children and by international relief workers as "orphans" or "abandoned" or "detached" children. As noted in field reports, the causes of family separation included death of family members, accidental separation during sudden evacuations, and placement of children by family members in sickbays and feeding centers where the children might receive food that parents, themselves starving, were unable to provide.

Relief workers found it particularly difficult to provide adequate care for unaccompanied children during much of the war. As the children's food needs were often paramount, many of them were fed and accommodated in the clinics and sickbays where they were found. Yet the already overextended staff and facilities could provide only limited assistance. Sometimes emergencies caused sudden relocations of even these exhausted facilities, during which resident children were sometimes separated from family members.

Before the civil war orphanages did not exist, as they were considered incompatible with the Nigerian extended family tradition. During the war, however, individuals, agencies, and missions set up 120 orphanages to care for unaccompanied children. Many orphanages took children of all ages, but some, often called "motherless baby homes," cared only for infants and very young children. Some were authentic orphanages but a large percentage were probably exploited by the local adult population, which obtained relief food for itself or its children there. A majority of children in institutions were believed to have been placed there by parents or other relatives for the specific purpose of obtaining food. Investigations showed that many times it was the management and staff who benefited most from available food. There were other problems. The care provided was reportedly very poor and the mortality rate was high. In the worst periods, a single room built to accommodate 30 children housed as many as 120. Many orphanages were described as filthy, with children uncared for, and staff showing little initiative or concern. The problems seemed so serious at the end of 1968 that the field staff discouraged the proposed establishment of new internationally-supported orphanages, citing not only staffing problems and food shortages, but the impossibility of protecting the institutions in the event of fighting. It would have been next to impossible to suddenly move several hundred children under emergency conditions it was reasoned.

Over the course of the war approximately 5,000 children were evacuated to neighboring countries, mostly to Gabon and the Ivory Coast. The evacuation occurred only because of an airlift established to bring relief goods into the war-affected area.

In January 1968 the first relief planes, called mercy flights, flew food and medicines to the civilian population within the war zone against the Nigerian air blockade. Only at night could the flights hope to avoid detection by the Nigerian Air Force; eight relief planes were shot down, however, and the pilots and crew killed. In August 1968, in response to the famine, churches, agencies, and international relief organizations expanded the airlift to ten flights a night and delivered a total of as much as 100 tons of relief food and medicines. The airlifts influenced the possible evacuation of children in two contradictory ways: empty planes could carry children out of the war-torn area, but successful delivery of food and medicines made evacuation of the children less urgent.

The evacuation of children first came to light when pilots and crews told of people thrusting babies and small children into the planes. Some children were flown out. The idea of evacuating starving children on empty planes quickly gathered momentum. The first group of children was sent to the island of Sao Tome by Caritas International that same month. The Biafran Red Cross Society, the Order of Malta, the French Red Cross, and Terre-Des-Hommes soon began evacuating children. By January 1969, about 2,250 children had been flown out of Nigeria, most to Libreville, Gabon, where a large residential center was established for them. The evacuated children were often very young; a majority came from large and impoverished families; and most were classified as moderately to severely malnourished, with 1,000 needing treatment for tuberculosis. A minority were reported to be children of authorities. These children were not usually malnourished.

The publicity given the initial evacuation stirred international enthusiasm for the removal of even larger numbers of children. In the Cameroon, which borders the Biafran area on the east, church leaders, local medical officers, and voluntary agencies proposed to establish reception centers for "suffering and starving children," to provide medical screening at local hospitals, and to organize foster and group care in local facilities while the children were recuperating. Many offers came from other countries as well. In separate actions private individuals in France and the prime minister of Canada publicly announced plans for large-scale airlifts to pick up children who were to be flown to the two countries for adoption or at least temporary care. While a few children probably reached Europe, large airlifts from the outside never happened. While the offer from the Cameroon seemed to many to be the best option, the Biafran authorities rejected it as well as the others, refusing to allow children to be evacuated to countries that did not politically recognize Biafra.

Many child welfare agencies and concerned individuals both in Biafra and elsewhere opposed evacuation. Soon after it began, a senior lecturer in pediatrics at the University of Biafra suggested that the project had been undertaken as a "panic measure" because of the prevailing gloomy military situation.[3] He believed the children could be cared for within Biafra instead and drew attention to the inadequate arrangements made for the children that were moved, noting specifically that children were hastily selected on the very day of the evacuation and removed from hospitals, sickbays, clinics, and villages without adequate consultation with parents, guardians, or medical staff. Personnel in charge of the evacuation kept very poor records of the children (and in some cases, no records at all). Labeling of children for identification purposes was so inadequate that some children could not be identified afterward. Many rumors arose among the public, which had little information about the evacuation program. Nigerians in the war-affected area called for an immediate assessment of the evacuation of children.

The advantages and disadvantages of evacuation were reviewed. The primary advantage was obviously the chance for children to get adequate food and medical care in a war-free atmosphere. Evacuation would also offer favorable publicity for the Biafran resistance movement. However, the assessment team thought the disadvantages outweighted the advantages. It noted the physical hazards, especially the risks involved in moving very sick children. During the war at least 500 children, many severely malnourished, died while being evacuated or within ten days thereafter. Frequent air raids on the airport also threatened their safety. Other disadvantages were the possible psychological "estrangement" of the children from their environment and the difficulties in adjusting when they returned. The team considered the psychological disadvantages for parents (and other relatives) to be important, noting that parents supported evacuation when the local situation was at its worst and the lives of their children seemed to be threatened; but several months later, when general child health had improved within the area, parents mourned the absence of their children and some were convinced they would never see them again. Another disadvantage was the difficulty in providing the continuing care and supervision necessary for the resettlement and rehabilitation of the children on their return. The assessment team also deplored the use of very limited material resources and personnel to provide special care for a few thousand children, when even a small percentage of the same resources could have helped many more of the four million children affected. The novelty of the evacuation of a small number of children was obscuring the greater needs of all affected children.[4]

For all these reasons Biafran authorities discouraged the evacuation of children, recommending that only children who required medical services locally unattainable, such as heart surgery, should be evacuated. But exceptions were recognized. If the food situation deteriorated or if the military situation warranted it, it was said that all affected children should be evacuated. The number of affected children would be in the hundreds of thousands, however, and a total evacuation was recognized as impossible. Yet the evacuation of children was not entirely prevented. A Biafran government board supervised the evacuation of the children by voluntary agencies in a program called the Biafran Children Relief Scheme Abroad. The board developed selection guidelines, a screening process, and consent and record forms.

Rather than evacuation, the authorities advocated a coordinated on-the-spot child welfare program. They recommended the establishment of orphanages, nutrition centers, sickbays, clinics, and mobile medical and nutritional units to help the children, and proposed that all children be registered and that local child welfare officers be appointed to visit children in homes. Many relief and child welfare agencies agreed and supported these recommendations. The International Union of Child Welfare (IUCW) was then the coordinating agency for many of the international child welfare agencies and had child welfare staff members working in Biafra. Within a month after the first evacuation, IUCW Headquarters in Geneva issued a position paper to all its member agencies concerning the removal of children from the affected area: most important was the delivery of massive relief, specifically foodstuffs and medicines, to the population, and to the children first of all; material and technical assistance were to be provided within the region to ensure that abandoned and orphaned children were cared for; and eventually, the children were to be reintegrated into families.

Supported by such agencies as International Social Service, UNICEF, and the Canadian Save the Children, IUCW offered guidelines for assisting children in war-affected areas. It urged that children be provided for locally, unless necessary relief aid was unavailable, and that evacuation of children to countries outside Africa be ruled out altogether. Children with medical problems too difficult to treat locally or in a neighboring province should be evacuated to neighboring African countries with their mothers or at least adults from their own ethnic group. The most appropriate program for the care of children, including "orphans," was said to be one that helped local people to provide assistance and assume responsibility for the children.

Within the war-affected areas during the crisis, the pressure from relief workers to evacuate the unaccompanied and the sickest children depended upon the current food supply as well as predictions of the future. At the end of 1968, for example, the general health of children was reported to be greatly improved, with edemas gone, and hair and skin color returned to normal. Children were laughing and playing again. Still, the situation was considered gloomy because a carbohydrate shortage was predicted within months. Therefore field workers, including IUCW staff members, continued to lobby for the evacuation of some children to other African countries, particularly "orphans," who were considered the most vulnerable.

Yet the situation did not deteriorate in the next several months as expected and renewed efforts to provide local care replaced efforts to evacuate children. An emergency food production program was initiated within the war zone and night flights were doubled in December 1968, with 200 tons of food and medicines delivered each night. In January 1969 the general nutritional state of the population had improved to the extent that officials considered closing the milk stations. Personnel at existing sickbays were able to care for all *kwashiorkor* cases. Schools, which had been closed for a year and a half, were reopened and a foster care program was established within Biafra.

By the summer of 1969, however, the food situation had again worsened. In June 1969 a Red Cross relief plane was shot down and the airlift of relief food was disrupted and never again matched earlier levels. By July of 1969 most children in the country were reportedly suffering from malnutrition, and the extent of *kwashiorkor* spread alarmingly even among adults. Just as had happened a year before, the death toll from malnutrition began to climb; by the end of July it was estimated at 1,000 a day. The number of displaced persons had also increased. By November 1969, 1,586 relief camps in the war-affected zone contained 1,049,256 displaced persons.[5] During this period, the evacuation of children to Gabon and the Ivory Coast was revived, and it continued at a low level until the end of the war in January 1970.

A majority of the unaccompanied children living in the emergency facilities at the end of the war probably reunited with their families through independent actions that either the families or the children themselves undertook. At the beginning of 1970 a survey indicated at least 40,000 unaccompanied children in orphanages, sickbays, and feeding and convalescent centers throughout the war zone.[6] As emergency services ended and hundreds of thousands of displaced people returned to their homes, many of the resident unaccompanied children were claimed by relatives or left home on their own. By June 1970 the estimated number of remaining unaccompanied children had dropped to 14,500, and by December the number was thought to be about 7,000.

When the war ended, Nigerian authorities made several major decisions about unaccompanied children. Orphanages established throughout the war-affected area were to be closed; children previously evacuated to other countries were to be returned to Nigeria; and all children were to be reunited with their families as soon as possible. The government invited the IUCW, in association with several other voluntary agencies, to assist in implementing these policies.

A tracing and family reunification program for both unaccompanied children within the war-affected area and those who had been evacuated was carried out between 1970 and 1973. When the program began very little information was available about unaccompanied children scattered throughout the war-affected area. Social welfare personnel in facilities where the children lived had kept few records. Although more documentation existed for the evacuated children, no one knew how many would return, where they had come from, how many were without parents or families, or how many were unidentified. Local social workers who were knowledgeable about the area, language, and people, and the population movements during the crisis, ran an intensive program. At its peak the project involved more than 500 Nigerian staff assisted by international staff. In many cases the families of the children knew where they were and reunions were easy. Some children gave precise information about themselves, their families, and homes; their relatives were easy to trace. There were, however, many children about whom little was known. Some had been separated from families at the beginning of the war. Many records had been lost or destroyed. Institutional staff had changed and many of the youngest children had been nicknamed because their real names were not known. A large proportion of the last children to be traced were under three years of age.

Officials and social workers tried in many ways to collect and disseminate information about separated children. They searched for villages on the basis of a child's faint description of a landmark. Children assumed to have come from a certain area were taken there in the hope of reviving a memory. Posters showing pictures of unaccompanied children were widely circulated, and information was disseminated through the press, radio, churches, local administrative bodies, and community leaders. A substantial number of children were identified when groups of children were taken to villages to be met by parents known to have lost children. Clues such as accents, physical features, and tribal markings were used to determine the identity of some children. As a result of these efforts approximately 6,000 children from orphanages and other emergency facilities in the war zone were reunited with their families. Most had been traced and reunited within eighteen months of the war's ending. Tracing for the most difficult cases continued until 1973.

The closing of the institutions met with resistance. In many cases, orphanage directors objected and instructed both staff and children to tell convincing hard luck stories to prevent departures of children. Other diversionary tactics were used such as giving names of fictitious relatives who, of course, could not be found. The basis for the resistance was believed to be the orphanage managers' reluctance to lose relief foods and other benefits. The decision to close the institutions stood, however, and children who were not reunited with their families directly from the orphanages were placed in newly established, temporary "reception centers" until tracing was successful or other permanent arrangements were made.

An intense program of tracing and social welfare assessment prepared for the return of the evacuated children to Nigeria. In the war-affected areas, the repatriation program began with the collection of all records of the children's movements. Public announcements seeking additional information about children believed to have been sent overseas elicited about 5,000 names. Social workers interviewed some 3,500 adults to gather essential information for the reunion of children with their rightful parents. In Gabon and the Ivory Coast, workers interviewed each child and recorded, whenever possible, the child's name, age, sex, place of origin, names of parents and other relatives, and other available details. Comparing lists of missing children with the files of those evacuated resulted in immediate matches of about 2,500 children.

Approximately 4,454 children returned from Gabon and the Ivory Coast over a ten-week period between November 1970 and February 1971. The children were between the ages of two and fifteen, with the largest numbers between the ages of three and five, and eight and ten years. Approximately half of the children were six years old or younger and had been away from Nigeria for one to three years. Approximately eight percent of the children returned with residual malnutrition.[7] On arrival, the children were dispersed immediately to one of five reception centers located throughout the war-affected area. There, personnel completed medical and nutritional evaluations and interviewed each child prior to family reunion. The procedure for the reunions varied from one reception center to another. Generally, on completion of the examinations, a social worker and a nutritional surveyor drove the child to its home, where they verified identities, assessed the home, and handed over the child to the parent(s) who signed a receipt. When personnel could not find parents, they reunited children with uncles, aunts, brothers, sisters, grandparents, or cousins. The pressure to reunite the children with their families quickly was great, both as a matter of policy and in order to limit the number of children in the reception centers at any one time. When families were known, children were usually reunited within two to five days after arrival. The existence of many evacuees' families was not known, however, and the extensive tracing required kept those children in reception centers for a number of weeks.

Some observers in reception centers described the returnees as carefree and happy and apparently glad to be back. Almost all older children, when asked if they would prefer to stay at the center, go home, or return overseas, said they wanted to go home. Among older children few stress-related behavior problems were reported, although the young children's first nights were often disturbed with bad dreams and crying. In one reception center sixty percent of the children became physically sick in the first three to four days.[8]

Social welfare workers adopted various verification methods to ensure that a child was returned to the correct family. Commonly, persons claiming to be a child's relative had to be vouched for by adults not related to them. Sometimes further proof was required, in which case personnel might examine birth certificates or other documents, compare early photos of the child with its present appearance, or compare the family's account of the lost child's past with other facts known about it. In more urban settings, such as Port Harcourt, parents were asked to come to the reception center to pick up their children and, after the proper documents had been completed, were required to swear in court that the information given and the claims made were genuine.

By the end of the tracing program, virtually all (98.3 percent) evacuated children had been reunited with their families. Of 3,922 children, only 327 were found to be full orphans, and they were reunited with their extended families. In total, including both local and evacuated unaccompanied children, the tracing and reunification program assisted between 10,000 and 11,000 children. The families of all but seventy-nine children had been found and the children reunited with them.[9]

Those who had dealt with the children thought they would have difficulty adjusting to their families; the evacuated children were considered to be more at risk than children who had remained in Biafra. Many of the children had left their families at a very young age and had lived for one to three years in institutions. The evacuated children, however, had received good food and care. They were accustomed to drawing water from a tap, sleeping on beds, going to school, and eating three meals a day. They returned to war-damaged areas where the standard of living was often very low. For many families food was still scarce. Many children received two meals a day of food that was foreign to them. Some spoke a different language because Nigerian staff from a different ethnic group had cared for them overseas. On return, many were expected to work with their families, rather than attend school. Children from institutions within the war-affected areas had not known any family life for several years, and the youngest children had known none at all. Most had passed through several institutions after family separation and had experienced considerable physical and emotional deprivation, but they had survived, often with minimal assistance.

Social workers thought many of the children who were reunited with families after leaving institutions in the war-affected area would need "frequent" follow-ups. For the poorest families, the social welfare office provided a food supplement when available. The effectiveness of the follow-up depended largely on whether officials had kept adequate records of the reunion and on the competence of the social welfare service involved. In the Umuahia division, for example, the social worker reportedly visited all children repatriated in her area and provided regular reports on their children's progress. An effective follow-up program proved difficult for many. Although statistics are not available, weekly field reports noted the return of some children to the institutions with the comment that "most came from bad homes."

From the inception of the evacuation program, a home visitation program was implemented to provide support to the returned children and their families and to assess their needs. Mobile follow-up teams, composed of a social welfare investigator, nurse, and nutritionist, visited the children's homes. The two most vulnerable groups of children, needing medical and nutritional supervision, were those returning to relatives (not parents) who had other children showing signs of malnutrition, and children who at the time of reunion required some medical treatment or follow-up for other specific reasons. By April 1971, a follow-up study of 3,000 returned children revealed that eight to ten percent were at severe nutritional/medical risk; most had been at risk when reunited. Readmission to reception centers, though not encouraged, was sometimes necessary. An analysis indicated eight short-term social and economic problems and twelve long-term conditions influencing the families' care of the children. These ranged from the parent's lack of employment or disability to ignorance of the foods the child needed or lack of interest in the child. The major source of nutritional problems was clearly the inability of families to provide adequate food in the face of poverty and unemployment.

In the follow-up surveys, social workers noted the initial difficulties of reunited children. The survey staff concluded that despite many problems, most of the children reintegrated well into their families. Older children sometimes had difficulty accepting rural village life, and children often wished to continue the schooling started during evacuation. Some older children refused to work. Sometimes younger children would demand the more expensive foods they had become used to and often would refuse or vomit up local foods. Surveyors noted that the general response of families was often to overprotect the children in an attempt to provide what was demanded, even at the expense of other siblings. Children without parents were usually well cared for within the extended family, although approximately half of the children found to be at risk after reunion were full orphans living with relatives; at least ninety-eight such children had initially been rejected by their relatives. Follow-up surveys showed that the thirty mentally handicapped and epileptic children who had been reunited were coping quite well.

After officials found that ten percent of the children were not receiving adequate food, they began a pilot program to assist 150 families for three months. Each family received a small cash grant and usually some food in the early stages of the program. The pilot project grew into a program for more families of children at risk. The aim in all cases was to make the family capable of earning enough money regularly to buy adequate food. It was also found to be absolutely necessary to distribute food to complement cash grants so that grants were not used for food, rather than income-generating activities. Often, simple cash assistance had not improved the family conditions or rectified problems causing children's poor health. Therefore, the program was expanded to provide money for craftsmen to replace tools lost in the war; for farmers to import laying chickens or for bulk purchases of seeds; and for training programs for unskilled workers. Special grants were provided to orphans cared for by the extended family.

By the conclusion of the program social workers had made over 5,000 home visits to some 2,700 families, of whom 1,200 were receiving financial assistance. A review of seventy-three percent of these families showed an economic recovery rate of over forty-five percent. At least ninety-nine children were readmitted to reception centers for treatment of severe malnutrition, despite the family assistance program.[10]

By September 1971 the parents or families of approximately 430 children had not been located. These children were referred to as the "hard core." The majority were between three and seven years of age because the youngest children were the most difficult to trace. Because of persistent efforts, at the end of the program in 1973, less than 100 of those children had not found their families. Most were placed in long-term foster care; a small number in institutional care. Policy makers throughout the war-affected area ruled out the adoption of these children as incompatible with traditional local child welfare practices. Long-term foster home programs for children were also new to this part of Nigeria, but social workers readily found homes willing to care for children permanently. (In practice, there was no effective difference between adoption and foster care as understood by parents taking in a child.) One of the most obvious indicators of the integration of the children into foster homes was that at follow-up, most foster children had taken their foster parents' names.

8

Vietnam War

A great many children have been separated from their families as a result of the protracted war and the political and social upheaval in Vietnam between 1954 and the present. Over the years, as war escalated, more and more families were subjected to its direct effects, including bombings, loss of life, poverty, repeated forced movement, life in refugee camps, and disruption of normal patterns of family life, all of which contributed to the separation of children from families. The Vietnam War is estimated to have caused 1,500,000 casualties of which 415,000 were civilians. The number of civilians wounded was estimated to be 900,000.[1] Millions were displaced. In 1968 alone, 1.5 million refugees were living in 850 refugee camps throughout South Vietnam alone.[2]

There is little data about the number of children who were unaccompanied in the earlier part of the war period, but we do have this report, dated 1966, from the Director of Public Health in South Vietnam:

> There are at present 77 orphanages, of which two are governmental, the others being privately run, often by religious orders, subsidized (as we have said) by the Ministry of Social Action. The number of orphans in care, according to a recent census, totalled 9,677: 1,351 from birth to two years old; 2,277 under six years old; 4,050 from six to twelve, and 1,999 from 12 to 17 years. Most of these orphanages are in the most precarious health conditions: overcrowding, shortage of water, little or no sanitary provision. The staff who have a role of capital importance are not often qualified; it is rare to find a nurse or a children's nurse, and there are frequent changes in personnel. Only the devotion of those in charge—often in religious orders—remedies this situation, with difficulty. Medical supervision hardly exists and in any case is badly organized. Infant mortality in the 38 creches, all forming part of orphanages, remains frighteningly high.[3]

An International Union for Child Welfare report from 1966 describes a visit to one of these orphanages:

> *Orphanage in the Province of Gia Dinh.* Passing through the main entrance of the staken enclosure there is a clearing, in the centre of which is a Catholic Church; to the right are several buildings: in one of them there were 150 children aged between 2 and 4 years who were crawling around in a room in which there was no play material, and only one woman in charge. In a covered space between this and an adjoining building were 30 girls from 6 to 10 years of age, running around quite happily, but there was no play material there either. Up on the first floor of this

building we found 90 girls between 10 and 15 years old, crowded into one large room; some of them were sewing but the majority were just standing around.

In a third building a little further away was another room where 50 handicapped children of different ages were lying in bed. Some of these youngsters were blind, or deaf-mutes and others were polio cases. The smell of urine and faeces was nauseating, and the woman in charge was playing with a baby who was lying on a mat on the floor. Outside, two women were washing sheets. No therapy was given, just physical care.

In another building, we saw 50 babies in one room where there were also some sick children; one nurse was feeding a baby with a bottle, and a girl was changing diapers.

The number of children with physical defects was relatively high: apart from those already mentioned there were youngsters with hare-lip, cleft palate, and so on. Incidentally, it is believed in Vietnam that such children bring bad luck and this is the reason why they are abandoned—to protect their families. Other "taboo" children are those who are born "on a day of ill-omen."

There were orphans, but they were outnumbered by half-orphans and social cases (children from too-large families who could not care for them because of lack of money and because of very overcrowded housing conditions).[4]

From a review of the available literature, it becomes apparent that the conditions described in this orphanage were quite typical of problems that faced most orphanages: absence of individualized care, inadequate facilities, and lack of adequate personnel, funding, and resources. While a government-run social welfare system existed in South Vietnam, it was in its early stages of development and, as with most nonmilitary government programs at that time, was given a low priority compared to the war effort. In addition, the care of orphans was not traditionally a government function, but was instead considered the responsibility of the extended family and the village.

Yearly, the number of children affected by the war swelled. By 1973, according to a report by Jean and John Thomas, South Vietnam's population included nine million children, of whom nearly one out of ten (880,000) had lost one or both parents.[5] While noting the destructive effects of the war and the massive social upheaval taking place, the Thomases observed that people depended most for assistance on the extended family, religious institutions, and the village, rather than central government services. These basic social support systems were still functioning even though the extended family had been placed under severe stresses, church affiliations had been disrupted, and whole villages eradicated. Of the 880,000 children who had lost one or both parents, 850,000 continued to live with their extended families. "[This gives] ample evidence of the residual strength of the family structure in spite of the battering it has suffered. Children are abandoned to institutions when the family cannot economically sustain them; often they are kept far too long and only brought to orphanages or hospitals when they are desperately ill."[6]

By 1973 20,000 children were in registered orphanages, and another 5,000 lived in unregistered orphanages or on the street.[7] One of the largest registered orphanages, Go Vap, contained more than 2,500 children. The number of orphanages had climbed to 130, and the government decreed that no more could be opened. It should be noted that these figures reflect an increase of 10,000 children in orphanages over the period 1966–1973. The actual increase was even larger, for these figures do not include the children who died in the institutions. According to some observers, sixty to ninety percent of the babies placed in orphanages died before their first birthday.[8]

The family circumstances of children in orphanages varied considerably, as did the reasons for their having been placed there. Some children had been accidentally separated from their families and taken to orphanages by police, soldiers, or others who found them. In some cases, both of a child's parents were dead; other children had been deserted. Some parents or other relatives left children in orphanages as a place of only temporary shelter or because they provided a safer place to live and a more dependable source of food than the family itself could provide at that moment. Agency workers also reported that some parents, after having made every attempt to care for a critically ill child, then left it at an orphanage, in a last attempt to save its life. No one will ever know how many parents planned to reclaim children and again assume repsonsibility for their care; nor will we learn how many were in fact abandoned. It is clear that in many cases the child's family initially intended the placement in an orphanage as a temporary arrangement. It was estimated that one-half of the 20,000 children in orphanages had living relatives who maintained some contact with them.[9]

Irrespective of the reasons for their children being placed in orphanages, however, the children typically received inadequate care. Despite the diligent efforts of committed staff members, the orphanages were overtaxed and unable to meet the needs of the children, a problem that increased in seriousness as the numbers of children admitted to an orphanage increased.

Throughout the war, many nongovernmental, bilateral, and international organizations were concerned about the well-being of children and the population at large. In the 1960s there were approximately thirty foreign voluntary agencies working in Vietnam with a total staff of about 400 foreign nationals. Like their Vietnamese counterparts, these persons often worked under extremely difficult circumstances, which included threats to their personal safety. Child welfare issues were also a concern at the international level as well, and a topic discussed at many meetings and conferences in the United States and Europe where participants demanded additional assistance and discussed policy and programs. While such national and international efforts undoubtedly contributed to the general welfare of children in need, they by no means fully met those needs.

All concerned agencies seemingly agreed that there were children in need, that conditions in the orphanages were poor, and that outside assistance was required. Yet there were major differences of opinion about how best to meet the needs of the children. Some workers believed that orphanages were intrinsically inadequate to the needs of children even under wartime conditions. It was noted, for example, that there were no orphanages in North Vietnam; orphans were placed with relatives or with other families through the village social welfare committees.[10] Similarly, some believed assistance should be directed to the prevention of family separations. To this end social welfare services were established, such as parent counseling to help mothers find alternatives to placing their children in institutions, and training and support programs were implemented to enable mothers to be economically self-sufficient. Day-care centers were established. Foster home care, although a new concept in Vietnam, was also introduced and proved to be feasible.[11] While such efforts no doubt contributed something, they were often "too little, too late."

Undoubtedly, the greatest controversy occurred over the question of the advisability of international adoption and the methods used by various adoption agencies. These agencies took the position that eligible children not receiving family care should be moved to other countries where such care could be provided. The social welfare services the agencies themselves provided to children and families varied considerably depending on experience, competence, and underlying philosophy. Some chose children very carefully, were thorough in their documentation of children considered for adoption, and were more careful in their choice of recipient families than were others. Some agencies, in addition to providing adoption services, also tried to arrange for local care and provided social services to families in an effort to prevent family separations. Other agencies focused only on the removal of children, used questionable practices to obtain them, inadequately documented their cases, and contributed little to the prevention of family separations or the development of local child welfare capabilities. The differences between the practices of different agencies was highlighted several years after the war when it was shown that out of seven authorized U.S. agencies, two were cited most frequently in connection with problems arising from their airlifting of infants for adoption in the United States.[12]

The degree to which opinions differed about the needed approach was illustrated in 1974 when a $7.5 million program was outlined by the U.S. State Department to improve the lot of orphaned or otherwise disadvantaged Vietnamese children. Included in the overall program was $470,000 to be paid to voluntary agencies to expand and improve intercountry adoption and related welfare activities. As reported by the *Village Voice,* American adoption agencies were said to have sought to channel most of the money into bringing Vietnamese children to the U.S. for adoption. Others objected to this emphasis, however. The American voluntary agency Clergy and Laity Concerned, for example, argued that the money should be spent on improving the lives of the children within Vietnam, so that they could grow up as Vietnamese in their own country. This group's efforts helped channel some of the aid money into Vietnamese programs, including one giving small grants to families so poor that they might have had to place their children in orphanages.[13]

Over the period of the war, a growing consensus developed among many international relief and child welfare agencies that international adoption was not the solution, for it affected only a comparatively small number of children at considerable cost while contributing little to the well-being of the majority of children in need. Opponents were critical of the fact that intercountry adoption alienated children from their heritage, culture, language, and religion, which they considered to be every child's birth right. The resulting alienation was believed to be psychologically damaging, and it was believed that children from other cultures might be at a social disadvantage. In addition, intercountry adoption was criticized for contributing little toward reducing the number of parent/child separations or toward developing local alternatives for care of children. Concern about the practice of intercountry adoption was justifiably encouraged by reports of gross abuses, such as the reported practice of one adoption agency which offered hospitalization for a sick child, then presented a hospital bill to the child's parents that they could never pay. The agency then agreed to forgive the bill in return for a signed consent for adoption.[14] International Social Service put an end to their own participation in the overseas adoption of Vietnamese children and began to state that there were, in fact, no eligible children in Asia for adoption. Even the Holt Foundation, which had established its offices to arrange intercountry adoption, began to offer broader social welfare services. It ceased to accept new children during the weeks when the fall of Saigon seemed imminent and processed only those children already selected for overseas adoption. Other agencies, however, continued (or even increased) their efforts to remove and place children internationally.

By the end of the war, the estimated number of children who had lost one or both parents had risen to 1,200,000; the number of children in registered orphanages was 17,055; and the number of children who were in nonregistered orphanages or considered "homeless" remained at 5,000.[15] Yearly, the number of children sent overseas for adoption increased: 200 in 1970–71; 485 in 1972; 682 in 1973; and 1,362 in 1974.[16]

In summary, the review of this situation confirms, as in other emergencies, that there were large numbers of unaccompanied children for whom special assistance was necessary. Social services were required both for the care of such children and, at an earlier stage, to provide assistance to families so that they need not separate from their children. No reliable studies were found that compared the care of children in orphanages with the care received in the homes of extended family members or in foster homes. In the opinion of this observer, however, the problems that arose in providing institutional care in this situation again suggest that the issue is not simply whether adequate facilities, staff, and money can be provided in order to improve institutional care. Instead, the issue is whether institutional care should be permitted at all, even in emergency situations. The problems of care within the institutions and of intercountry adoption generally are the result of inadequate care provided for the children within the traditional social system from the first moment that they are identified as unaccompanied. The social service efforts to support the traditional systems of caring for unaccompanied children and to develop alternative placement systems for children other than in institutions was "too little, too late," and may have been subverted by the efforts directed toward institutional care. Neither the government of Vietnam nor the international assistance it received protected the rights of the unaccompanied children nor provided assistance. Nor was the best interest of each child considered. The children were not provided care that met their developmental needs, their cases were not documented adequately, and insufficient measures were taken to identify the families and evaluate the potential of family reunion, or alternatively, to find other community-based family care systems.

OPERATION BABYLIFT

In early 1975, with the increasing threat that Saigon would fall, orphanages were under pressure to admit more and more children as ever increasing numbers of refugees flocked to the capital. Adoption agencies made every effort to speed up the processing and departure of children being sent overseas for adoption, spurred on by the concern of prospective adoptive parents and the fear, often fueled by rumor, as to the possible fate of children in institutions if the South Vietnamese government fell. In the first few months of 1975 some 400 children left Vietnam for adoption overseas and adoption agencies claimed that there were an additional 1,400 children eligible for adoption and who were being processed.

In March the adoption agencies, the United States government agencies including the Department of State, the Immigration and Naturalization Service (INS), and the Agency for International Development (AID) worked out accelerated procedures to move and admit into the United States all orphans who by April 10, 1975 were in the custody of agencies licensed in the U.S. to handle orphans (referred to in government jargon as "pipeline orphans"). Some $2.6 million was allocated to the project. AID was assigned responsibility for arranging transportation and care of the orphans en route. The adoption agencies maintained responsibility for selecting and documenting the children and retained custody of them during the trip until responsibility was transferred elsewhere. The Department of Defense authorized the use of military aircraft to fly the children from Saigon and designated three Air Force bases in the Pacific as intermediate stops during transit and three military bases on the West Coast as reception centers. The program was known as "Operation Babylift."

The South Vietnamese government was approached by the voluntary agencies and AID for permission to remove the children, and the minister of social welfare obtained the approval of the prime minister on April 2, 1975. According to an AID report on the project, adoption agencies were required to provide a list of the children, prior to their movement, to the Vietnamese minister of social welfare, "who after assuring himself that the children were adoptable (abandoned or formally released by the closest surviving relative or guardian),"[17] would not object to or intercede to prevent further movement. After "Operation Babylift" was approved evacuation programs by other groups were also mounted, such as an airlift of some 250 children to Australia and just over 100 children to Great Britain, sponsored by the *Daily Mail,* a British newspaper. Also, in some cases, the movement of children was sponsored and arranged by private individuals. Although the airlift was to be proven a success logistically, except for the tragic accident, the project was fraught with serious problems from the beginning.

The day after the Vietnamese government gave its permission for this operation, the first group of forty-five children were flown out of Vietnam on a privately arranged airlift, one of three, organized by the president of World Airways, an American charter airline company. The evacuation had not been authorized by the South Vietnamese government or the U.S. Government and most adoption agencies declined to send children on this unsanctioned flight.

The first authorized departure of children under "Operation Babylift" took place on April 4. Some 220 children were placed in a U.S. Air Force C-5A military transport plane, many strapped two to a seat, destined for the U.S. However, shortly after takeoff a lock system failed and a door blew off. Escorts attempted to pass the oxygen masks from one child to another since there were not enough to go around. Tragically, that plane crashed in a rice paddy while attempting to return to Saigon, killing seventy-eight of the children and many of their adult escorts. Many of the others were injured in the crash itself or suffered brain or other damage from the lack of oxygen. The accident did not deter further evacuation efforts, however. In fact, many of the children who had survived the crash were flown out the next day, an action that drew criticism for being unduly traumatizing to the children.

Within four days from the start of the airlifts more than 1,400 children, the initially targeted number, had left Vietnam. The government of Vietnam indicated that it would not obstruct the movement of additional groups of children and the evacuation of children continued throughout April. The total number of "orphans" who left for other countries is not known, but 2,547 children were evacuated as part of "Operation Babylift"; an additional 350 children were moved in three unauthorized World Airways flights to the U.S.; approximately 100 children were flown to England; and 250 children went to Australia. Of the children evacuated to the U.S. approximately 1,900 remained there, while some 650 were transferred to one of fourteen other countries. Ninety-one percent of the evacuated children were under the age of eight; fifty-one percent were under the age of two.[18]

On arrival in the United States most children were first taken to one of the reception centers set up in California and Washington where officials carried out preliminary immigration screening and the children were provided basic amenities and a health check by hastily organized civilian volunteers. Fifteen percent of the children were immediately hospitalized; nine children died during the evacuation or soon after arrival in the U.S. After completion of the reception procedures (a process that lasted anywhere from a few hours to as long as three days), the children were released to the adoption agencies that were given custody of them.

The arrival of planeloads of war "orphans" gave rise to massive public support in the United States for the project. Dignitaries, including the President, met the planes; the press gave the story full coverage; companies, as well as individuals, made contributions to cover immediate needs; and volunteers provided many of the required services. Associated agencies were inundated with inquiries and offers from the public to adopt the children.

The adoption agencies then flew children from the reception centers to wherever the children were to reside, namely, in forty-six states, the District of Columbia, and Guam. As "Operation Babylift" was organized for children expected to be adopted in the U.S., it naturally followed that the children were placed in prospective adoptive homes, either directly or after a brief interim placement in foster or group care. In England, however, the children were purposely placed in group homes rather than adoptive homes, a program approach that will be discussed in more detail later in this chapter.

Not everyone, however, approved of the airlifts or believed that they were necessarily in the best interests of the children. The International Committee of the Red Cross declared that foreign adoptions violated a Geneva Convention requirement that war orphans, whenever possible, must be educated within their own culture; and Caritas, the Vatican's relief organization, called the airlift "a deplorable and unjustified mistake."[19] The International Union of Child Welfare called the airlift "an error of judgment to be avoided." The airlift was described by others as kidnapping; as the taking of war souvenirs; as misdirected humanitarian concern motivated by guilt for the war; as a governmental manipulation to stimulate public support for the war; and as an unethical act that depleted Vietnam's needed human resources. Critics doubted that the children would be mistreated if they remained in Vietnam under a different government as was claimed by advocates of the airlift, and voiced concern about the trauma caused by moving children from one culture to another. Some considered the airlift to be only token assistance because it did not help to eliminate the cause of the problem, either by ending the war or by providing social services that would enable the children to be cared for within their own families or country.

There were many groups competing for the children. On arrival, a group of 196 children from An Loc Orphanage was caught in a dispute over placement authority between the voluntary agency that sponsored their evacuation and the authorized adoption agency that had arranged their departure in Vietnam and had custody. As part of a mediated outcome of this dispute the children were placed in foster care for approximately one month before being transferred to the care of an authorized adoption agency in another state, which then placed the children in prospective adoptive homes. Two hundred and fifteen Montagnard children were flown to the United States in the last of the unauthorized World Airways flights and then immediately transferred to flights for Denmark. After the children arrived in Denmark, however, the government of South Vietnam requested their return and the Danish government agreed. Their return was blocked by voluntary agencies which appealed to the European Court of Human Rights.

However, the debates and criticism for which the airlift is most remembered arose from the discovery by volunteers at the reception centers that some of the children were not orphans as claimed by the adoption agencies and that documentation on their cases was missing, inaccurate, or incomplete. Through discussions with the children it was learned that some had been purposely placed in orphanages by parents to avail the children of the opportunity of being evacuated, and others had been accidentally separated from their families. There was also uncertainty as to the validity of the claims made by adoption agencies that the children had been abandoned, and even more questions as to whether consent for adoption, where signed forms existed, had been given by parents knowingly and without duress.

On April 1975 a federal class action suit, *Nguyen Da Yen v. Kissinger,* was filed by the Center for Constitutional Rights (New York City) and the International Children's Fund (Berkeley) on behalf of three children brought to the United States who allegedly had parents living in Vietnam, and any other children in similar circumstances. The case claimed that the basic human rights of the children had been violated by their removal from the country without proper custody having been obtained, and by their continued detention in the United States in custody of parties other than their natural parents. The suit sought to force the government to make a determination of each child's adoptive status; institute procedures for tracing parents or relatives; prohibit adoption of the children until a search for parents or relatives failed; and immediately return any child found to have a living parent seeking its return. After various appeals, the court ordered the Immigration and Naturalization Service to review files and make plans for tracing overseas.

Approximately one year after the beginning of the law suit, the U.S. District Court, San Francisco, ordered a review of the information compiled by the INS. The court ruled that a majority of cases were in order and had adequate documentation and the case was dismissed. The INS then notified state and local authorities that the government was releasing most of the children for adoption. The plaintiffs charged that documentation on 978 children was too meager to prove that they were eligible and suggested that the 533 parental releases for adoption that were signed after March 15, 1975, be considered involuntary and invalid because they were signed in panic and confusion.[20]

Plaintiffs in the class action suit advocated active tracing for parents and relatives in refugee communities in the United States and in Vietnam. The International Union for Child Welfare and the International Committee of the Red Cross were contacted and agreed to assist. The U.S. government also considered a government-to-government tracing program, but political realities made this unrealistic. Opponents of tracing voiced concern that tracing might endanger parents, give rise to fictitious claims, and result in a "tug of love" between natural and adoptive families, which would not be beneficial to the children. Tracing was never systematically carried out.

The total number of Vietnamese parents and relatives who on their own traced and were reunited with children brought out on the Babylift is not known. Within the first ten months of the airlift the plaintiffs in the class action suit identified approximately a dozen parents and relatives of Babylift children among refugees in the United States. The number of parents and relatives who petitioned courts across the U.S. for the custody of Babylift children is not known because the decisions of local courts, whose decisions are unreported, are final unless the case is appealed to a higher court. The number of such cases is believed to be quite small, however. Over the next five years, several custody disputes were reported in the appeals courts, most involving relatives other than parents, seeking custody of children from adoptive parents. In most cases the court ruled in favor of natural parents, but if the dispute was between nonparent relatives and adoptive parents, the court usually sided with adoptive parents, stating that the decision was based on consideration of the "best interest of the children," a legal standard applicable in such cases[21] In Virginia, for example, the custody of a Vietnamese child living with a Vietnamese foster family for more than five years was sought by an uncle, a blood relative, also Vietnamese, who lived in California and was prospering in the United States. The court ruled in favor of the foster family.

IMMEDIATE POSTWAR EXODUS

The North Vietnamese took control of Saigon on April 30, 1975 and as the American military withdrew, tens of thousands of Vietnamese people in a vast flotilla of small boats followed the retreating Sixth Fleet into the South China Sea. They were taken to processing centers in the Pacific, then moved to one of four reception centers in the United States (in Arkansas, California, Florida, and Pennsylvania). When immigration proceedings were completed and sponsorships assured from American individuals or groups, they were released from the camps to go to the community of their sponsors. From April 1975 to December 31, 1975 some 138,000 Indochinese refugees entered the United States in this way, 800 of whom were officially recognized later as unaccompanied children.

The exact number of unaccompanied children in this influx is unknown, for some confusion existed as to who was to be considered as a member of this group. Vietnamese families traditionally would care for the children of relatives without legal formalities of adoption or custody. Sometimes at reception centers children cared for by members of their extended families were considered to be unaccompanied, while others considered to be accompanied were in fact servants.

Unaccompanied children were not officially recognized or provided special treatment in the transit camps in the Pacific or, initially, in the reception camps in the United States. When they were later identified in the camps, policies and programs were developed, but apparently without the benefit of experience gained from past influxes of children to the United States. Some of the difficulties are recorded in the observations of a team of psychiatric consultants working at one of the reception centers:

> Our recommendation regarding the desirability of sponsoring intact families despite their often large numbers was not followed. At times, legal problems worked against this goal. For example, some families included children of relatives or friends who remained in Vietnam. Immigration restrictions caused many of these children to be separated from their custodial families. Once these "unaccompanied" children were separated, camp administrators housed them in a special quonset hut. We made many strong recommendations that these children would do better in a receptive Vietnamese family than in a separate part of camp waiting for adoption by American parents. Nevertheless, county welfare agency personnel followed the standard adoption procedures for orphaned American children.[22]

The law did not permit the adoption of these children because most had families with whom they hoped to be reunited, and the Immigration and Naturalization Service would not release children without a legal custodian. Voluntary agencies were at the time not prepared to care for older accompanied minors as they were to be later. The release of the children from camps was, therefore, often delayed. It was a stressful situation for the children, some of whom were reported to be depressed and even suicidal.[23] By the time the camps closed, some 645 youth had been placed in foster care with American families or with distant relatives and a few with Indochinese families.[24] Most were adolescent boys.

THE "BOAT PEOPLE"

After the mass initial exodus in the first two months at the end of the war, a steady stream of people left Vietnam from the middle of 1975 to mid-1978 in what is sometimes referred to as the "second wave." The number of people leaving increased annually, with a drastic increase to some 163,000 in 1978–1979. Included in this "third wave" were the largest number of unaccompanied minors to leave Vietnam in any period. The number of Vietnamese to leave after 1979, in the "fourth wave," has decreased yearly, at least until 1984.

Each refugee left for his or her own reasons, but factors that are believed to have encouraged people to take the extreme measure of leaving home to risk great dangers at sea and an uncertain future included: new economic restrictions imposed by the Vietnamese government, relocation programs, conflict between Vietnam and China with an increase in harassment of ethnic Chinese, an increase in military conscription, and hardships caused by hard winters, flooding, and several destructive typhoons.

In the eight years following the war more than one-half million people left Vietnam. Their well-organized, secretly planned voyages took place in small boats, hence the widely-used label "boat people."

> Ages of the boat refugees from the south ranged widely, but most were under thirty-five. There were many women and children. Some young men said they left to avoid conscription. Some people said they were the victims of politically inspired harassment and persecution; others said they feared such treatment. Fear of what

might happen was a potent factor, ... A feeling of alienation from the new communist administration and identity with the old regime was common, often mixed with an economic motivation: a conviction that their livelihood was better before and could only get worse. They felt the future was bleak for themselves and for their children. Fear of 're-education' and of being sent to a 'new economic zone' were also pervasive.[25]

The exodus often occurred in the monsoon season in small, overcrowded, and sometimes unseaworthy craft. The sea voyage could last from a few days to as long as six weeks. The passengers did not know where wind, tide, and chance would take them and were in great danger of drowning or of being attacked by pirates, who often robbed, raped, and killed their victims. It has been estimated that as many as one-half of the boats never reached land. Often, a boat that set out for one country ended up in another that was neither the first nor the most expedient choice, because of weather conditions, high winds, and the passengers' poor navigation skills.

In spite of such dangers families paid their fares, usually in gold, and took the chance of reaching another country, where they sought a different political system, security, economic advantage, and the opportunity to secure a highly valued education for their children. On the average, about one-half of the passengers were children, women, and older people. When families could not afford the price of passage for all members, one or more children were sent instead. Children who were alone shared the fate of all in their party, often being abused or witnessing atrocities.

The number of recognized unaccompanied children to leave Vietnam between 1975 and 1978 was comparatively small. For example, approximately 800 unaccompanied children entered the United States during this period: 600 entered as part of the immediate postwar exodus, and another 200 entered over the next three years. Few were selected for admission to other countries. A report reviewing the experience of the Lutheran Immigration and Refugee Service, the largest American voluntary agency involved in providing services to Indochinese unaccompanied children entering the United States during this period, described the approximately 150 unaccompanied minors they assisted as being either from lower-middle-class ethnic Vietnamese families, from ethnic Chinese families of merchant class, or from Laos and Cambodia. The ages of the children ranged from two to eighteen years; most were older adolescents.[26] Eight-five percent were male. The same report noted that the unaccompanied minors who left Vietnam immediately at the close of the war were fleeing *from* a situation rather than fleeing to a selected refuge and thought at first in terms of temporary flight to safety. Only later did they realize flight had separated them from their parents and homeland. This was in contrast to the unaccompanied minors among the subsequent "waves," who believed they had no future in their homelands and therefore chose to leave or were sent by parents to rebuild their lives somewhere else.

THE CAMPS

If the boats were not rescued at sea, or did not capsize, they landed in Thailand, Singapore, Indonesia, Malaysia, Hong Kong, or the Philippines (known as first-asylum countries), where the refugees were usually permitted to stay temporarily as long as they were guaranteed relocation to other countries that would permit resettlement (known as "third" countries). Within each of the asylum countries refugees were provided shelter and basic amenities in camps that had, in some cases, been established by national organizations but more often by the United Nations High Commissioner for Refugees (UNHCR) and international voluntary agencies. Conditions varied greatly from camp to camp and from country to country in such crucial areas as treatment, personal safety and security, and in the availability of such essentials as water, food, and shelter.

During 1978 the number of recognized unaccompanied children in camps began to increase. Although unaccompanied children were registered with all other refugees, there was usually no further documentation of them after their arrival in the camps, however. In February 1979 the U.S. State Department estimated that there were between 500 and 600 unaccompanied children,[27] yet new boatloads of people were arriving daily with unaccompanied children among them and their number in camps increased drastically. By July their number in camps throughout Southeast Asia was estimated to be between 4,000 and 5,000.[28]

The year 1979 was one of appraisal and resettlement for unaccompanied children in these camps. In refugee camps throughout the region, UNHCR, encouraged by public concern for the children, organized assessments to determine the number and circumstances of unaccompanied children. Voluntary agencies such as Radda Barnen (Swedish Save the Children) and International Social Service seconded social workers to work for UNHCR in doing assessments of the camps to locate unaccompanied minors, survey their living conditions, establish a documentation system, identify and interview the children, identify special needs, suggest improvements in their living conditions, and assist in their resettlement.

The surveys confirmed that unaccompanied children existed in all camps. In mid-1979 the largest such number was in the camps in Malaysia where they constituted 4.6 percent of the refugee population for a total of some 2,500.[29] The age, sex distribution, and reasons for leaving were found to be similar among the unaccompanied children in the different refugee camps (with the exception that follows): Approximately four to six percent were under the age of twelve; most were adolescents in the age range of fifteen to seventeen (or claiming to be so when they were actually above eighteen); eighty to eighty-five percent were male. It is interesting to note that the statistics for unaccompanied children arriving in Hong Kong differed markedly from those for camps in other countries. There, out of a population of 56,000 refugees, only thirty-seven unaccompanied minors under eighteen years of age were identified in the assessment, a fact attributed to the different exodus patterns of the people who reached Hong Kong.[30]

Interviews with the children confirmed that most minors had left Vietnam with the consent and active cooperation of their parents. Older youths told of deciding to leave to avoid military conscription, to avoid being placed in the countryside to do agricultural work, or to further their studies. Parents, often choosing a favored child, had sent younger children with the intent of giving them a better life, many times with the expectation that they would be cared for by relatives or that the remainder of the family would join them later. The children's stories suggested other causes of separation as well, such as the following reasons: parents had difficulty earning enough to feed a large family and had therefore sent one or more children; parents and children had been separated on the way to the boats, with parents possibly arrested by the police; some children had been kidnapped by a neighbor or boat leader; some had been separated from their families for a considerable time and had been living as "street boys" or fishermen; some had joined the departing group without full understanding of where they were going; and others had lost parents at sea or in the camps.[31] A large number of children had relatives in another camp or in a third country. Younger children were not always certain what had happened to them, where their parents were, or why they had come.

The assessments themselves often stimulated action on behalf of the unaccompanied children as was indicated by one of the conclusions in the assessment of unaccompanied children in the camps in Hong Kong: " . . . these minors cannot be left in the present environment to earn their own living, or live off the charity of neighbors."[32]

Subsequent assistance provided to them varied from camp to camp. In a camp in Indonesia, for example, foster care was recommended for the younger children and the construction of an orphanage was discouraged. In other camps, special programs were often established to offer food and clothing, caretakers, educational opportunities, and special help in the resettlement process.

In attempting to define the problem of unaccompanied children through the UNHCR-sponsored assessments, much was learned about the assessment process itself. On the positive side, the experience confirmed the need for and usefulness of assessments to document the unique problems of unaccompanied children. Even though refugee program administrators often believed that unaccompanied children were not in difficulty, closer examination confirmed that there were children with particular needs who merited special assistance. Some children identified were in need of immediate material assistance. Camp records on the children were sometimes found to be totally lacking, incorrect, or incomplete. Problems often developed in the selection and resettlement process. The assessments confirmed that unaccompanied children were not being accepted for resettlement as quickly as were family units, and if selected, they often remained in the camps for many months longer than family units before resettlement countries facilitated their move.

On the negative side, the assessments did not always meet expections of program administrators. Many had been done hurriedly, with little standardization of definitions or methods. The interviews with children were usually very brief, without taking the time to establish a trusting relationship between interviewer and child (or between interviewer and youths, who stood to gain by not providing full or accurate details); they were often carried out through translators, or by people of a culture and language different from that of the children. For these reasons the assessments contained misinformation and uncertain interpretation of facts. The three most problematic issues were: (1) varying interpretations of what "unaccompaniedness" meant in the Vietnamese cultural context, for there were some children who were not related to those accompanying them, but were without question a part of the family, and others who had arrived with the extended family that did not, however, have a long-term commitment to the child; (2) difficulty in judging reliably the depth of the relationship of an unaccompanied child to the family or peers he or she was living with; and (3) difficulty in determining a child's correct age, for many of those over the age of eighteen claimed to be younger. Consequently, although the information gathered during the assessments was crucial, the files that were compiled in the process were later often found to be flawed. Furthermore, after the departure of the short-term consultants there was not necessarily any follow-up, nor were the same systems or techniques used again.

Documentation of children in many camps was never carried out. The only information on the unaccompanied children was often what existed on camp registration forms and what was written by embassy selection committees.

However, the assessments made of unaccompanied children in the camps in Southeast Asia in many ways represented a new beginning for the international response to unaccompanied children. Adequate policies and guidelines for action in the field relating to the care and protection of such children in refugee camps had until then not been available within such agencies as the UNHCR, UNICEF, the Red Cross organizations, or the voluntary agencies providing emergency services. New ground, or should we say "reinvented ground," was being broken when, as part of the assessments, the staff established definitions, developed forms, experimented with assessment techniques, and set policies. The process continues to this day.

With experience as the teacher and with more and more information about the children becoming available, staff were able to develop policies and procedures. In August 1979, UNHCR issued draft procedural guidelines for registration, tracing, and resettlement of unaccompanied children,[33] based on recommendations made by the American Council of Voluntary Agencies[34] and the International Council of Voluntary Agencies.[35] As part of the recommendations for placement it was established that adoption of unaccompanied Vietnamese children was to be avoided. In 1981 the Executive Committee of UNHCR adopted the statement that every effort should be made to trace the parents or other close relatives before minors were resettled and that tracing was to continue after settlement, in particular, before adoption was decided upon. In 1982 UNHCR issued more specific guidelines promoting durable solutions for unaccompanied children in Southeast Asia. These guidelines included criteria for defining the best interests of an unaccompanied minor, suggested at what age the wishes of a child should be considered in determining his or her best interests, and reconfirmed that unaccompanied children not of mature judgment could be repatriated upon the parent's request. Guidelines for care and protection were further defined in 1982 in the *UNHCR Handbook for Emergencies* and in 1984 in the *UNHCR Handbook for Social Services.*[36]

SELECTION FOR RESETTLEMENT

Between 1975 and 1978, admission of unaccompanied children to third countries occurred by and large on an ad hoc basis. There were very few voluntary agency programs for the international resettlement of children apart from those of adoption agencies. These agencies did not usually provide services to minors in camps, but were instrumental in initiating the resettlement of a small number of unaccompanied children to Canada and the U.S. in 1977.[37] Governments were not eager to accept youth whom they viewed, sometimes correctly, as "anchor" cases sent by parents to secure the family's right to immigrate at a later time. National resettlement services for adult refugees and their families were not generally interested in assuming the long-term legal, financial, and program responsibilities likely to be incurred in the settlement of children, and were not organized to provide long-term child welfare services. UNHCR at that time did not have systems to document and assess the needs of unaccompanied children among the refugee population, nor to provide special services.

The response to unaccompanied children quickly changed, however, during the massive exodus of people from Vietnam in 1978 and 1979. Vivid media descriptions of the hardships faced during the sea voyages and of life in the refugee camps generated strong public support for programs that would meet the refugees' needs. New initiatives were taken, primarily by voluntary agencies, revealing concern for the well-being of refugee children in particular. These initiatives were taken primarily by voluntary agencies to resettle unaccompanied children from asylum camps to third countries. Obstacles that had previously impeded resettlement of unaccompanied children and other refugees were quickly overcome, and the number of unaccompanied minors resettled increased dramatically in 1979, when, for example, some 900 unaccompanied children were taken to France, as compared to 250 in 1978; 1,000 were taken to the United States, as compared to about 200 in 1978; and about 500 were taken to Australia. Generally, these minors were given preferential treatment both in gaining admission to a country and in receiving services once admitted.

Such preferential treatment by U.S. authorities is a case in point. While maintaining their usual admission criteria based on refugee status and/or humanitarian concern, the officials considered these children a "high-risk" population and therefore agreed to take those not admitted to other resettlement countries, processing their cases more quickly than those of other refugees. In easily granting refugee status, the officials gave the children the benefit of the doubt as to their articulated fear of persecution. Moreover, the federal government agreed to reimburse agencies completely for the expenses incurred in resettling unaccompanied children and agreed to provide child welfare services or maintenance subsidy payments until they reached their eighteenth birthday, thereby resolving uncertainties for the voluntary agencies that resettled children as to who would pay for the long-term services required. In most countries, this preferential admission and support service diminished or ceased about 1982.

The process by which unaccompanied Vietnamese children finally reached a resettlement country was extremely varied; it was certainly more than a matter of the minors' choice. Those who were rescued at sea were usually offered sanctuary by the country from which the ship was registered. Many of the unaccompanied children who eventually got to Norway, Holland, and England were part of such rescues. From first-asylum camps, the minors were usually offered the chance to go to countries where family members awaited them. Other minors, when it appeared that they might not be selected to go to the country of their first choice, chose or at least agreed to go wherever they would be accepted even if it meant going to a place where they had no family members and about which they knew nothing. Fast decisions were often necessary, for immigration offers were quickly made and quotas quickly filled. Sometimes unaccompanied children agreed to go to one country hoping that it would be a stepping stone to another.

Requirements for immigration to each country differed and their procedures varied. Some countries accepted only children who were to be reunited with family members, while others only accepted young children. Some sponsoring agencies accepted whatever child was recommended by UNHCR, while others had field teams that participated in the selection process and even attempted to match sponsor and child while the child was still in the camp. At a minimum, leaving the camps, in most cases, was dependent upon approval by immigration officials and upon the child's having an individual or group sponsor in the host country who was willing to assume responsibility for his or her care and placement. Leaving the camps usually involved cooperation among officials of the country of first asylum, UNHCR (which often recommended which children were to be resettled), the sponsoring agency (which may have selected the children to be resettled and made arrangements for their placement), and embassy immigration officials, who screened the children to ensure that they were unaccompanied, under age, and met other admissions criteria. Problems in this collaborative process often delayed settlement and influenced the choice of the country to which a child finally immigrated.[38] For example, in the time required for one selection team to complete the admission procedures, the selected child may have been identified and removed by a selection team to another country.

Countries of first asylum and UNHCR often permitted the varous embassy selection teams direct access to the unaccompanied children. From a child welfare perspective this proved to be problematic at times. Selection teams and immigration officials were often required to make judgments as to the relationship of unaccompanied children to the people caring for them. In the absence of someone very familiar with the child's situation, decisions were made on the basis of brief interviews and with little background information or supportive documentation. This sometimes led to separation of siblings or other family members. In camps in Malaysia, for example, it was found that selection missions from some embassies were wrongly assuming that the unaccompanied children were orphans and available for adoption. Selection teams sometimes caused separations of related children by selecting only the younger of siblings. Although this did not play a major role, there was some competition among agencies and selection teams for the youngest children, particularly girls. Children were also sometimes separated from relatives or other individuals with whom they had a permanent living arrangement because of selection criteria that rigidly admitted nuclear family units, but not children from the larger extended family.

In Thailand, UNHCR attempted to use its own social workers to document the minors and participate in placement planning before the children were considered by various agency and country selection teams. (Indeed, it can be argued that only when UNHCR carries out such screening is it fulfilling its mandated protective role for the children.) Unfortunately, in the early stages the embassy workers often processed their cases faster than did UNHCR. By the time information on family circumstances had been obtained, the child may already have been accepted by and on its way to a third country. In 1981, UNHCR attempted to impose a moratorium on the removal of Vietnamese children until full documentation of them had been completed. This resulted only in creating a backlog of cases and longer waiting times for children in the undesirable conditions of the camps. In 1982, UNHCR recommended the establishment of a screening board in each asylum country to review the circumstances of unaccompanied children as part of the process of placement planning. Such a screening board is not known to have been established except in Thailand.

In most countries, it was necessary for officials to control access to the camps and to unaccompanied children to prevent individuals and agencies from privately spiriting them away. Usually such individuals and agencies, although undoubtedly well-intentioned, appeared to have very little knowledge about the children's situations, the relevant legal framework concerning child placement, or the child welfare practices commonly accepted as necessary to protect the rights of all parties and assure that children are cared for adequately.

RESETTLEMENT

Between 1979 and 1985 unaccompanied Vietnamese minors, identified among refugees rescued at sea and in the refugee camps throughout Southeast Asia, were resettled to Australia, Austria, Belgium, Canada, Denmark, France, French Guyana, the Federal Republic of Germany, Great Britain, Italy, Japan, Netherlands, New Zealand, Norway, Switzerland, and the United States. The countries admitting the largest number of children were the United States, France, Australia, Germany, and Canada, with the United States accepting the majority.

The majority of unaccompanied children were resettled to third countries between 1979 and 1982 after which there was a slowdown in resettlement. In 1982 the Orderly Departure Program, which enabled people to migrate directly from Vietnam to resettlement countries, began to function more effectively; the mass exodus of "boat" refugees from Vietnam began to wane; small voluntary agencies were fully occupied by the responsibilities of the first groups of unaccompanied children they had sponsored, and many countries became less willing to admit additional unaccompanied minors. UNHCR redirected its efforts from an emphasis on resettlement, to the provision of services more appropriate for long-term stays in the camps.

It should be remembered, also, that the full story of the exodus of the Vietnamese unaccompanied minors is, at the time of this writing, still in progress. It is still developing month by month, with a continuing exodus of people from Vietnam by boat and with unaccompanied children continuing to arrive in first-asylum camps, some of whom are then being resettled to third countries. Acceptable alternatives to resettlement do not yet exist, so that by 1985, resettlement continued to be considered the only "durable solution" for unaccompanied Vietnamese minors arriving in refugee camps. Most minors are resettled to third countries to rejoin family members, and the United States continues to accept approximately fifty percent of the minors without relatives in third countries; few are presently accepted by other countries. The length of stay in camps is increasing, but UNHCR statistics suggest that the number of unaccompanied Vietnamese minors in camps of first asylum in Southeast Asia is declining: 2,800 in July 1984; 1,750 in September 1984; and 1,300 in March 1985.[39]

The total number of unaccompanied minors to leave Vietnam from 1975 to the present is unknown. Suggesting that the figures are conservative, Elizabeth Lloyd estimated that 30,000 to 40,000 unaccompanied minors have landed in countries of first asylum between 1975 and 1983.[40] Lloyd based her estimate on analysis of UNHCR statistics that indicated that about seven percent of Vietnamese boat people arriving in countries of first asylum were identified as unaccompanied minors.[41] Of 43,800 arrivals in the Southeast Asian region in 1982, for example, 3,500 were children not in the care of a close adult relative.[42] In addition to the seven to eight percent who were identified as not in the care of close adult refugees, Lloyd estimates that an additional seven to eight percent of the refugee caseload were minors who arrived with relatives (rather than parents) who should also be included in the statistics on unaccompanied children. She suggests that approximately sixteen percent of the total Indo-chinese refugee caseload are minors not with their parents.[43] In most countries, statistics are not available to verify this estimation. In the Netherlands, where such statistics do exist, unaccompanied minors make up seventeen percent of the total Southeast Asian refugee population.[44]

Accepting all the uncertainties of the available statistics, the number of unaccompanied minors who received services from voluntary agencies within all resettlement countries between 1975 and 1984 was approximately 22,000.[45] Unaccompanied children who had neither parents nor relatives to care for them made up approximately one percent of the refugee population admitted to France,[46] the United States,[47] Great Britain,[48] and Australia.[49] This figure does not appear to be constant, however, for in 1985, based on UNHCR statistics of "boat refugee" camps in Malaysia, Indonesia, and Thailand, unaccompanied minors were four to six percent of the refugee camp populations.[50]

On their arrival in a resettlement country, unaccompanied minors were usually received in a manner similar to that for other refugees. In Australia, unaccompanied minors went first to migration centers for one to two years with other refugees. In Canada (in Ontario), newly arriving unaccompanied minors remained at a "staging" center from two to four weeks and awaited the results of medical tests. In England, they were initially cared for in refugee reception centers. In France, unaccompanied children stayed for about two weeks in a reception center with other refugees and then were transferred to a special orientation center for a two-month observation period before being placed elsewhere. In the Netherlands, unaccompanied minors went with other refugees for ten days to a "first-stop" refugee reception center, then to a refugee reception facility for six to nine months, and from there to a supervised halfway house before longer-term accommodations were arranged. In Germany, the children admitted in 1979 were first placed in youth villages, children's hospitals, transit hotels, and homes, for a period of from two to twelve months.

Although fewer than ten percent of unaccompanied children arriving in the U.S. were placed in special reception programs, various program models have been tried. Youth admitted to a reception program in New Jersey, for example, spent two to twelve weeks in an orientation program before placement arrangements were finalized.[51] Some reception programs have used group homes, and in the last several years, United States Catholic Conference (USCC) has increasingly used "reception foster families," which are ethnically similar or American foster families in which at least one other Vietnamese minor is already living who is considered to have successfully integrated into the home. Most unaccompanied children going to the United States, however, were immediately taken to the home of their preselected foster family after being met at the airport by a bilingual representative of the sponsoring agency and the family with which he or she would initially live. Many foster home programs in the U.S. perceive the initial accommodation with families as a reception program, with foster parents having commitments or expectations only to assist the unaccompanied minor to establish independence on his or her arrival. In such programs flexibility is often stressed.

In all countries these reception programs were established in 1979, in haste and under pressure because of the sudden admissions of large numbers of refugees, including children. Most reception programs have been continually experimented with and revised repeatedly. No fixed or standard reception program has evolved, although there seems to be consensus among child welfare agencies in all countries that the reception is least stressful if newly arriving adolescent unaccompanied children have the opportunity to stay first in a "neutral" reception arrangement where they can begin the adaptation process into their new environment with the help of acquaintances and peers of the same ethnic background and language. The "neutral" reception period also provides social workers with the opportunity to make their own determinations of the minors' personal characteristics, which is often helpful in arranging the most suitable placement and providing the most effective support services. In addition to child welfare issues raised by reception programs, in all countries, one of the determining factors of the form of reception programs has been the availability of funding. In most cases the programs have represented a compromise between what social workers felt would have been ideal and what was possible within financial constraints.

The question as to the most appropriate mode of care for unaccompanied Vietnamese children after resettlement to a third country was an issue from the onset of the programs. In early July 1979, UNHCR circulated a copy of "Recommendations Regarding Unaccompanied Children," which had been drawn up in 1977 by an ad hoc group of the American Council of Voluntary Agencies (ACVA). Citing as the basis for their recommendations the experience gained in working with unaccompanied children during World War II—the Hungarian exodus, and the Cuban exodus—the group recommended the following:

> 3. The overriding principle in any sound planning for minors dictates that there be opportunity for an individualized assessment of each young person's situation prior to the formulation of any plan for resettlement, and that planning be based on a variety of alternatives.
>
> 4. These assessments should first and foremost consider the possibility of preserving existing familial or personal ties which the unaccompanied minor may already have to other refugees—whether such refugees are still in South-East Asia, in the process of movement to or already resettled in, another country. This alternative would require that the families or persons with whom the minor is to be resettled be involved in the planning process, with due consideration given to the practical realities of their assuming the social and legal responsibilities involved.
>
> 5. For those minors who are truly without ties, the most appropriate alternative would most likely be some form of foster care. The experience of child welfare agencies in the U.S. suggests that in developing foster care plans for minors, a variety of choices should be considered, including:
> —Placement, where possible, with families of similar cultural background to the minor involved.
> —Placement with families not of the same cultural background who have special sensitivity to cultural differences and the ability to respond to the needs of uprooted young people.
> —Some form of group placement. Group placements have proven effective in the U.S. in planning for older adolescents, especially those who may have difficulties in adjusting to a family foster home.[52]

The recommendations also stated that a competent professional team carry out an in-depth assessment of the children in the camps, so that when resettled to another country the children should "become eligible for the same services and protection that are available to all children and youth in their place of residence"; and that adoption would likely be appropriate only in rare and isolated situations due to the older age of the youth, the difficulty of adopting older children, the legal and documentary requirements required for adoptive planning, and the questionable severing of legal ties to parents or other relatives. They urged that children not be separated from their families or relatives.

At the end of July 1979, nongovernmental agencies meeting under the auspices of the Geneva-based International Council of Voluntary Agencies (ICVA) restated and reaffirmed the principles recommended by the ACVA ad hoc committee cited above. They stressed the needs for a dossier on each child as part of the documentation, continued tracing of parents or other family members, and preserving the possibility of family reunion.[53]

The ACVA and ICVA recommendations reflected a certain consensus among voluntary agencies as to the general principles for resettling unaccompanied children. However, in the ensuing years, debate among the many concerned voluntary agencies has been considerable as to mode of placement—in particular, as to the relative merits of "group care" and "foster care" for adolescent minors. The debates are often couched in terms of the "wrongness" (culturally biased) or "rightness" of either form of placement. Studies and experiential descriptions of the resettlement of unaccompanied minors from Southeast Asia in Australia[54], Canada[55], Federal Republic of Germany[56], France[57], Great Britain[58], the Netherlands[59], and the United States[60] have supported the need for various placement options to satisfy the needs of unaccompanied minors of different ages, in different circumstances, with different expectations, in different host cultures. These reports consistently emphasized the importance of preserving the cultural and ethnic ties of arriving minors. There continues to be differences of opinion as to how these goals are best achieved. The relative merits of each placement type and the cultural considerations in placement are discussed more fully in Part II.

Virtually all observers of the resettlement of unaccompanied minors agree that the process of resettlement is stressful, and is even more stressful for older adolescents than for younger. Whether it is more stressful for youth than for older adults is not certain. A study of the mental health problems among the Indochinese refugee population in the U.S. found that the persons most at risk were nineteen to thirty-five years old.[61] In Holland, twenty to twenty-five percent of the unaccompanied minors were reported to have difficulties or problems, which were categorized according to levels of seriousness: minor resettlement problems, openly expressed migration problems, concealed migration problems, and problems arising from premigation difficulties.[62] Minors with the latter problem were found to be a "high-risk" group for which special psychosocial assistance was more likely to be needed. Similarly, a U.S. study of the psychiatric problems among adolescent Southeast Asian refugees confirmed that adolescents identified as psychotic had been psychotic or had evidenced serious personality problems prior to migration and that following migration their problems had worsened.[63] Various authors have described resettlement for unaccompanied minors as an adjustment process. An Australian author, under the heading "Resettlement: A Hard Time for All Concerned" described the adjustment as a staged process that progressed from excitement and exhilaration, to grief on realization of losses, to grief resolution and new attempts to integrate, to the finding of avenues to satisfy expressed needs.[64]

The resettlement of Vietnamese unaccompanied children required social welfare systems that could accommodate unaccompanied minors in widely varying circumstances. Some minors expected upon arrival in a resettlement country to reunite with parents or relatives already settled, only to discover after arrival that reunification was not possible. Others were admitted based on the claim that they had no relatives in third countries, and then after arrival were found to have relatives, as was estimated to have been the case for as many as ten percent of the children admitted to Ontario, Canada.[65] Some unaccompanied children admitted to a resettlement country with relatives were separated from them after arrival. In Australia, sixty percent of the children arriving with persons other than parents separated from the relatives, most within six months of arrival.[66] The separation of unaccompanied children from relatives is considered to be a problem of some magnitude in the United States and other resettlement countries, but the incidence of separation has not been determined.

Inadequate documentation for arriving children proved a serious complication for programs receiving unaccompanied children in countries of resettlement, just as it had "haunted" care and resettlement efforts in the camps. Even when documentation had been completed in the camps, records were not released by UNHCR for fear that the documents would be misused to the detriment of the child or the family. From the perspective of the agency receiving the children, however, full documentation on each child was extremely important in the planning of the child's arrival and in the arrangement of appropriate placement and necessary support after arrival.

Discrepancies existing between the claimed and the actual age of minors proved to be even more serious a problem in resettlement countries than it posed in the camps. Although the claimed age of arriving children often corresponded roughly to the age shown on accompanying documents, the actual age was often two to eight years more than claimed. The number of unaccompanied minors with age disparity is unknown although thought to be fairly high. Age disparity was a particular problem in Australia, Canada, and the United States, where minors were often placed in foster care. Every effort was made to place the child according to the age which the family and social welfare agency thought was most appropriate for the home situation. The arrival of an individual much older than the claimed age predictably resulted in the child wanting the treatment and the independence more appropriate to his actual age. The age discrepancies in Ontario, Canada were said to have often caused serious dislocation in the household and were rated as a significant probable factor in "breakdown" of foster family arrangements.[67]

While the selection and international resettlement system that has operated since 1979 has proven to be overwhelmingly successful in providing for unaccompanied children in refugee camps the opportunity to begin new lives in third countries, it has failed, first, to document adequately the situation of each child; second, to establish a social welfare system by which the parents' wishes in Vietnam can be safely confirmed; and, third, to establish a system by which children can be returned to Vietnam if the parents wish them to return and if they themselves wish to return. This is only to suggest that repatriation must be an option for those children whose best interests would be served in returning. Experience confirms that most unaccompanied children left Vietnam willingly and with the support of their families. However, there have been cases of children who were accidentally taken to sea, children orphaned during the voyage, and children who for other reasons may have wished to return to family members remaining in Vietnam. That no unaccompanied children out of more than 20,000 resettled since 1975 have been returned to their families in Vietnam would suggest that adequate policy and social welfare programs are not in place to ensure that the best interests of the children are always considered.

There are various reasons why no unaccompanied children have been repatriated. Most important, policy makers have feared a public outcry from Western countries if children were to be returned to a communist country. Other reasons include fear of endangering the families who chose to send their children away, fear that the children might be endangered upon their return, and fear that the return of children might become a political tug-of-war between countries with the children in the middle. Whether these fears are justified and whether social serivce systems could be developed by which children could be returned without endangering them or their parents has not yet been demonstrated. In all past emergencies, including the Hungarian exodus of 1956, some unaccompanied children returned to the care of their families in their home countries when all political obstacles were removed and guarantees were satisfactorily given that the child and parents would not be endangered.

Few issues are more sensitive than the question of whether the continued resettlement of unaccompanied minors from the refugee camps in Southeast Asia has acted as a "magnet" that encourages others to leave or be sent from their country and families. Lloyd, for example, argues that the special resettlement of unaccompanied minors has indeed encouraged other children to leave or be sent out of Vietnam. She therefore recommended in 1983 that the Australian Government encourage UNHCR to make greater efforts to repatriate unaccompanied minors, cease to process "anchor" cases, ensure that unaccompanied minors receive a minimum of publicity so as not to encourage others to leave, and be prepared to defend longer stays in the camps for the minors. Social welfare agencies providing direct services to unaccompanied children argue the opposite—that letting the minors stay longer in the camps does not deter them from leaving Vietnam, and on humanitarian grounds, the length of time in the camps should be as short as possible and every effort should be made to provide the minors with a new beginning.

9

Cambodian Crisis

On Wednesday, October 24, 1979, eight thousand Khmer men, women, and children were moved by buses and trucks from encampments on the Thailand-Kampuchean border to a barbed wire enclosed field some sixty kilometers away in Thailand. Thousands more were moved each day so that within eight days 28,000 Khmer people were on a thirty-three acre site named Sakaeo Displaced Persons Camp. In subsequent days, the population of the camp reached 31,000 people. Many of the arriving people were in very poor health, ravaged in particular by famine and a fatal strain of malaria. Although the trip from the border to the camp was short and over good roads, some people died in transit, and during the first week of the camp, at least thirty-five to forty people died each day. Unaccompanied children were among the first persons in the arriving population.

BACKGROUND

Although Cambodia claimed neutrality in the Vietnam war, it became a victim of the war by virtue of its having a common border with Vietnam. The fact that Cambodia's eastern border area and port were increasingly used by the North Vietnamese as a supply route precipitated retaliatory attacks by the South Vietnamese and the United States military. The stability which Cambodia had experienced for decades was threatened further as national groups, particularly those of the Khmer Rouge forces, vied for power and control. In 1970, Prince Sihanouk, who had ruled the country for thirty years, was deposed by Marshall Lon Nol in collaboration with the Khmer Rouge. From 1970 to 1975, the Pol Pot forces expanded their domination over larger and larger sections of the country.

As the Vietnam war progressed in the early 1970s, the country suffered increasing war damage. In addition to its destabilizing effects on the Cambodian political structure, the Vietnam war also caused some 500,000 casualities in a population of approximately seven million and massive dislocation, affecting an estimated 3.3 million refugees.[1] It also separated families and caused inflation of over 200 percent per year, which made life quite difficult for the average family. In April 1975, the Khmer Rouge toppled the Lon Nol government and immediately forced the evacuation of the three million inhabitants of the capital, Phnom Penh; changed the name of the country from

Cambodia to Kampuchea; and began a forced restructuring of Khmer society, as described in the paper, *Human Rights, Wars, and Mass Exodus:*

> ... The subjection of the country to a programme of radical political and social change in which the upper strata of the former society were eliminated and the denial of the most basic freedoms caused severe hardship to the great mass of the population, has been widely reported. The Khmer people, expected to work from dawn to dusk with inadequate supplies of food or medicine and with the constant threat of summary execution, fell prey to widespread epidemics or simply succumbed to exhaustion and hunger ...

> The remaining Vietnamese and Chinese who had constituted the commercial backbone of the former society were driven out of the country or forced into agricultural co-operatives being established as the declared first priority of restoring the national economy. From the cities, people were moved at gunpoint to rural areas. Collectivization became the government's second most important policy priority after national security. With abolition of a currency or banking system, the co-operatives were designed to be as self-sufficient as possible. All property—even personal possessions—was deemed to be held in common. When a constitution was ratified in January 1976, it abolished the monarchy, terminated Buddhism as the official state religion, expropriated all private economic enterprises and established peasants, workers and members of the army as the ruling proletariat.[2]

For three years the Pol Pot regime brutally ruled Kampuchea, causing the deaths of millions of people by execution or starvation and related diseases. Finally, in December 1978, the Khmer Rouge were defeated and driven out of Phnom Penh by Vietnamese forces. During the first half of 1979, as the Khmer Rouge forces retreated across Kampuchea and into the mountainous region of western Kampuchea, tens of thousands of men, women, and children were forced to accompany them. The year 1979 proved to be difficult for other Khmers as well, since widespread food shortages and famine prevailed, culminating several years of bad crops, disruption of planting caused by fighting between the Vietnamese and the retreating Khmer Rouge forces, and the collapse of the collective systems as hundreds of thousands of people attempted to return to their home villages. By mid-1979, with the memory of the atrocities perpetrated by the Khmer Rouge forces still fresh in mind, and in the face of food shortages and continued fighting, some 600,000 Cambodians trekked from within Kampuchea to the Thai-Kampuchea border area north of the Thai border town of Arayaprathet, where village-like agglomerations were established. Most of these people strongly opposed the Pol Pot regime and were known as Khmer Serei—"free Khmer." Continued fighting and food scarcity also forced the Khmer Rouge forces and the civilians they had forced to move with them from the mountainous regions to the Thai-Kampuchean border. However, they remained separate from the Khmer Serei in encampments south of Arayaprathet. The 31,000 Khmer taken to Sakaeo camp in October 1979 were soldiers and civilians from the Khmer Rouge group.

UNACCOMPANIED CHILDREN IN CAMBODIA

A 1974 United States Agency for International Development (USAID) report from Phnom Penh about orphans in Cambodia indicated that before the war there were no unaccompanied children to speak of because it was customary for relatives to take children whose parents were dead into their homes and raise them. "For a Khmer to refuse to take in the orphaned children of a relative was considered a denial of family responsibility and would more than likely bring down the criticism of the whole community, especially in the closely-knit rural areas. Sometimes, however, when there were no relatives to raise an orphaned child, the child would be turned over to the pagoda for the monks to raise."[3] In this way, the issue of orphaned children was resolved at the family or village level, and the government played only a minor role. Orphanages were rare. Adoptions were not uncommon but were usually arranged by relatives and members of the family with the government merely recording the arrangement for legality.

The orphan problem in Cambodia became more serious with the involvement of Cambodia in the Vietnam war. With more and more parents killed, more children had to be assimilated into the extended family structure. The economic hardships and displacements of the times also caused some parents to abandon their children and this limited the ability of relatives to care for additional children. The pagodas continued to care for children but were often full and unable to meet the full need. Some orphans joined the military. The *New York Times* in 1973 reported that an entire company of orphans existed in the 13th Brigade.[4] The USAID report also cites the forced separation of families under the Khmer Rouge as another cause of unaccompanied children:

> The Khmer communists have, since the beginning of hostilities, actively pursued a policy of sealing off the areas under their control and of carrying off large numbers of rural Khmer to the forests or mountains for indoctrination and economic exploitation. One side result of this has been the appearance of many children who have been in effect orphaned, not because their parents are dead, but because they have been completely cut off from their parents who are in Khmer communist areas. Takeo offers an example of this type of problem where there are 368 young Khmers, mostly students, whose parents are in Khmer insurgent controlled areas and have dropped completely from sight.[5]

The total number of orphans in Cambodia in the 1970s is unknown. Some estimates were as high as 250,000 but accurate statistics were virtually nonexistent and, in any event, in Khmer usage, the term "orphan" referred to children who had lost a single parent as well as to those who had lost both parents. In 1974, the Ministry of Social Action and Labor estimated that there were 3,000 to 4,000 "genuine" orphans who were not being properly cared for: most living with relatives or in the provinces, 100 were estimated to be in refugee camps, 200 to 300 in recognized orphanages, and 500 in pagodas.[6] Faced with increasing numbers of "orphans," the government and several national and international humanitarian organizations instituted relief programs, primarily by building more orphanages. In 1974 the number of orphanages had risen from one in 1970 to five in Phnom Penh and three in Battambong, Cambodia's two major cities.

In reviewing the plight of orphans in Cambodia in 1974, the author of the USAID report concluded, however, that "there is no distinct and separate orphan problem in the Khmer Republic. True, there are orphans who are real orphans and who are completely alone and have no one to care for them. But the number of such orphans is very small. The great bulk of all the orphans in both Phnom Penh and the provinces do have somebody to take care of them."[7] To substantiate this claim he cited a survey

of 6,000 chldren by the Save the Children Fund which indicated that there were practically no children without someone to care for them. Similarly, a Phnom Penh police
survey of sixty street children revealed that almost all of them were living with relatives who were trying to care for them.[8]

> This is not to say that most orphans are happy and well taken care of. Quite the
> contrary, many are undernourished, in very bad health, and in desperate need of
> assistance. But it is not because they have no one to care for them; it is because
> those people who are caring for them just do not have the resources to raise them
> properly. The orphans in the Khmer Republic are not suffering because they are
> alone. They are suffering because they have been caught up in the poverty which
> has enveloped a large section of the general population in the Khmer Republic.[9]

This report highlighted the fact that Khmer families were in fact caring for most
orphans and recommended that orphanages "under any guise" be limited. It suggested
that relief assistance was indeed needed, but instead of orphanages the assistance
should be directed to monthly monetary grants to Khmer families caring for orphans.
When Pol Pot came to power in 1975, all such programs were terminated. All foreigners were immediately expelled and the people of the cities evacuated to the
countryside.

Family life under the Pol Pot regime is described in the paper, *Human Rights,
Wars, and Mass Exodus* as follows:

> As a further step towards levelling society, private households were practically abol-
> ished in 1977. All meals had to be taken in communal kitchens, thus virtually elim-
> inating family life. Indeed, families were frequently split, the children being taken
> away from their parents' influence. Schools had in any case been closed, most teach-
> ers liquidated . . . [10]

In addition, some children were accidentally separated from their families during
mass population movements, as during the evacuation of the cities, and by the deaths
of parents. The full extent of family separations is not known. At least one scholar
suggests that while it was the policy of the Pol Pot government to break up extended
families into nuclear families, there is little evidence to suggest that it was their policy
to purposefully separate preadolescent children from their parents.[11] The same author
admits, however, that in actuality, there are reports of young children being taken
from parents to be placed in children's centers. The separation of the greatest number
of children from their families came as a result of a governmental policy which forced
adolescents to leave home to work in mobile work brigades to build irrigation systems,
clear forests, and the like. Often they lived under extremely harsh conditions and did
not see their families for weeks or months. Adolescents were generally treated
as adults under the Pol Pot regime—as they tended to be in pre-Pol Pot peasant
society—marrying soon after puberty, and often leaving home during their teenage
years.

ARRIVAL IN THAILAND

From 1975 to 1978, during the Pol Pot years, a small but steady stream of people
managed to escape from Kampuchea across the Thai border. Although it is estimated

that only one in ten were successful in their bid to escape, some 35,000 reached Thailand and were then resettled by the UNHCR to third countries. The number of unaccompanied children in this group is not known.

In 1979, the number of Khmer to enter Thailand increased dramatically. By May some 80,000 to 90,000 had sought sanctuary there. They fled famine, war, and life under the Khmer Rouge and the Vietnamese. The Thai government considered the new arrivals to be "illegal entrants" rather than "refugees" and UNHCR was not given access to these people. By October there were an estimated 200,000 Kampucheans in temporary village encampments scattered over a 100-kilometer stretch of the border. With evidence of massive starvation and sickness among the Kampuchean groups south of the town of Arayaprathet, and after international commitments were made to provide the needed assistance, Thailand declared an "Open-Door" policy on 19 October 1979. The incoming Kampucheans were to be sheltered in six camps, called "holding centers" which were to be built on sites designated by the Royal Thai Government.

UNHCR was notified of the Thai decision to transport thousands of severely ill Kampucheans from the border to a site in Thailand on October 21. With only two days to prepare, preliminary steps were taken to organize basic goods and services, but essential services were still being established as the arriving Kampucheans were assisted off the buses. In every way the situation was an emergency: food had to be acquired and distributed; water had to be trucked into the camp; a latrine system had to be built; a temporary hospital had to be set up with medical personnel recruited and medicines acquired; plastic and bamboo had to be provided to enable people to build their own shelters; and so on. To make the situation more difficult, the monsoon rains turned the crowded camp into a field of mud and lying water. Although the hospital was established by the fourth day, it was months before some services, such as sanitation, were satisfactorily organized.

Within the first twenty-four hours, in the midst of famine and the monumental efforts required to establish essential services, relief workers were faced with the question of how to provide for children who were alone. In comparison with the massive problems concerning the camp population at large, the issue of unaccompanied children was indeed small but, nonetheless, one that required and deserved action. During the first several days of the camp, children who were alone and in need of shelter and food were taken to an intensive feeding center operated by Catholic Relief Services (CRS), a United States based relief agency. However, the intensive feeding facilities were urgently needed for the large number of severely undernourished persons throughout the camp. So on the fourth day, the unaccompanied children were moved from the intensive feeding center to donated tents set up especially for them. The children were referred to as "orphans" although it was unknown whether their parents were deceased, in the camp, in the hospital, on the border, or in Kampuchea.

The number of unaccompanied children quickly increased. At the end of the first week approximately 400 children called "orphans" were being cared for by foreign volunteers and a Khmer staff they hired. In addition to the children who were identified as being alone, there were other children throughout the camp who were living with families other than their own—just how many was unknown, and they were at that time provided no special attention.[12]

UNHCR social workers registered the unaccompanied children and documented 340 of them. Approximately 200 were under thirteen years old; few were infants. The remaining 140 were between fourteen and eighteen years of age. On the basis of stories given by the children, UNHCR social workers identified eighty-five children as probable genuine orphans.[13]

Within a month the children were moved again, this time from the tents to newly constructed barrack-like buildings known as the "orphanage," which was located in the hospital compound. While the barracks were an improvement over the tents, the arrangements still proved unsatisfactory, for the children were separated from the remainder of the camp and the barracks seemed dark, overcrowded, and chaotic. UNHCR delegated the responsibility for the care of the children to various voluntary agencies, an arrangement which did not work well. A Protestant evangelical organization was given responsibility for child care, but there was an immediate clash of personalities and differences in cultural and religious views relating to the care of the unaccompanied children between the designated agency staff and persons from other agencies also concerned about the well-being of the children. The person in charge of the orphanage had little training or experience in child welfare, and relief workers disagreed with the style and institutional form of the care being provided. Also, many considered the Christian proselytizing of the Buddhist children in the orphanage to be disturbing because of the potential for coercion when children were in such need and so dependent upon the orphanage services.

As would be expected, almost every agency and individual involved with the Khmer population was particularly interested in ensuring that the children were cared for, and especially that those children without families were provided care and protection. The condition of the unaccompanied children was therefore a matter of great concern to many relief workers. While each interested party had his own idea of what was best for the children, there were essentially three opposing positions: that unaccompanied children should be (1) immediately removed from the camp and resettled to a third country; (2) removed from the camp but provided services in Thailand; and (3) left in the camp until their situation was fully evaluated.

The arguments most often expressed for the immediate removal of the children from the camp were that they needed immediate family care, better physical conditions, and a safer environment should there be fighting or repatriation of the camp population. One vocal advocate for immediate resettlement of the unaccompanied children stated:

> The orphans' needs are clearly for love and personal attention, the comfort and security that only a family can give them; to be removed from the squalor and despair of the refugee camps and the uncertainty of life in the border settlements, and into decent, loving homes before the dehumanization they already have suffered spreads and destroys them like a cancer; to be given a future free from the constant threat of death by disease, starvation and war, a future that is the birthright of every child in every country.[14]

The arguments most commonly expressed for keeping the children in the camp, at least temporarily, were that the children were not necessarily orphans and that their precipitous removal without full documentation would diminish the chances of family reunion, particularly if the children were placed in foreign foster or adoptive homes. This argument was supported by letters sent to UNHCR asserting that relatives and parents had not always been reunited with Vietnamese Babylift children once the children had been placed in foster or adoptive homes. The argument was strengthened by a well-publicized case of a Khmer mother searching for her child who had been taken to Switzerland by a Swiss doctor who, as a volunteer in the orphanage, had assumed that the child was an orphan. The supporters of this argument thought the children should be cared for by people of their own culture and that adequate services for the children could and should be established within the camps.

In response to vivid media presentations, the public in many countries contacted voluntary agencies and government officials to offer homes for the Kampuchean "orphans." Encouraged by public demand, agencies and officials lobbied the UNHCR for immediate action and sent representatives to the camps to determine the situation firsthand. Offers to evacuate the orphans en masse came from many lands. Visitors to the camp who were interested in the children were plentiful: curiosity-seekers, representatives of professional relief and child welfare agencies, people who came to search for desired children or to photograph them for fund-raising efforts, workers who promised to facilitate their resettlement, and so on. In order to control abuses and protect the children, UNHCR was obliged to impose strict visitation limits. It expelled at least one relief worker caught photographing individual unaccompanied children and circulating the pictures to prospective sponsors in another country.

In the absence of a clear official policy, a competent agency to delegate the responsibility for the care of the minors, and the establishment of a satisfactory program, all interested parties felt compelled to act with determination for what they believed was best for the unaccompanied children. Volunteers working in Sakaeo camp, particularly during the disorganized first month or so, took almost any action they felt was needed, including in at least one case, the removal of unaccompanied children from the camp. Those who believed the children should be resettled lobbied officials within their home countries, the Thai authorities, and the staff of UNHCR. Several agencies carefully engineered press briefings in Switzerland and the U.S. to put public pressure on the UNHCR to release children. Relief workers in the Sakaeo camp who opposed the removal of the children took whatever steps they could to ensure that the children remained. So, for example, when UNHCR approved the movement of a group of unaccompanied children from Khao I Dang camp to a camp called Khao Larn, administered under the patronage of the Thai Royal family, the selected unaccompanied children were hidden in homes throughout the camp by Khmer house parents.

The evacuation of unaccompanied children to France in December 1979 was also embroiled in debate. As in many countries, extensive media presentations of the poor conditions in the camp and the plight of the children in the orphanage generated strong public pressure to bring the children to France. French officials, including French President Valery Giscard, lobbied hard for their release and promised the public that the children would arrive by Christmas. UNHCR acquiesced, agreeing to the resettlement of about 105 children with the understanding that the children were not to be adopted and that France would accept the family members if they were identified. News of the planned resettlement was not well received by relief workers in the camp who believed that many of the selected children were probably not orphans and were being removed too hastily.

In an effort to prevent the resettlement of the children, a petition protesting their removal was drawn up in the camp and signed by some 115 persons. One woman threatened to kill herself in protest, and some volunteers attempted physically to block the gates to the camp. The children were nonetheless removed and arrived in France on Christmas Eve. Differences of opinion about the care of the children existed in

France also. To prevent the arriving Khmer children from being admitted into the institutional-type receiving center used for most incoming refugee children, a French voluntary agency, having been tipped off as to the flight schedule of the arriving Khmers, managed to be the first to meet the children at the airport and immediately admitted them to a hospital for observation. From there they were placed in foster homes.

The only unaccompanied Khmer children removed from the camps in the latter part of 1979 were the 105 children moved to France; 127 children resettled to Germany by the German Red Cross without the agreement of UNHCR and, a few who left with relief workers who managed to arrange private authorization or were smuggled out of the camp.

KHAO I DANG

On 18 November 1979, less than a month after the first refugees arrived in Sakaeo camp, Thai officials informed UNHCR that, for security reasons, it was believed necessary to move as many as 200,000 of the Khmer people on the Thai-Kampuchean border to a site in Thailand, ten kilometers inland. The move was expected to begin three days later and the designated spot to which the people would be taken was a three-square kilometer brush-covered field at the foot of a small range of hills. It was labeled an "emergency reception center" and given the name Khao I Dang.

The UNHCR, associated voluntary agencies, and the Royal Thai government, again on an emergency footing, prepared for the sudden influx of tens of thousands of people in only a few days. Personnel and resources were effectively mobilized in a monumental effort, and before the first new Khmer residents arrived by bus, roads had been constructed, latrines dug, water tanks installed, a bamboo hospital built, and food and housing materials acquired. The camp was designed and built incrementally, section by section, with each section able to accommodate 10,000 to 12,000 people. Each section had its own core community services, such as water and food distribution points, a clinic, feeding center, welfare office, administrative office, and recreation field. The Sakaeo and the Khao I Dang camp proved to be quite different. Many adults in the Sakaeo group were associated with the Khmer Rouge forces, while almost all of the people arriving at Khao I Dang were people very much opposed to the Pol Pot regime. Although the major causes of death were the same in both camps—fever/ malaria, undernourishment, pneumonia, and diarrhea—the people arriving in Khao I Dang were not in such a severely deteriorated physical state as those of Sakaeo. Death and hospital admission rates in Khao I Dang never reached the level they did in Sakaeo. Another important difference between the camps was that while Sakaeo, for the first several months, was excessively overcrowded and plagued with sanitation problems, Khao I Dang was more spacious and clean and had a more healthy atmosphere.

During the very first days of the Khao I Dang camp, any children without adult care were taken to the hospital where they were provided with shelter and food. Such children were usually identified in one of three ways: by the ICRC registration and medical screening process which all arriving Khmers underwent, by one of the Khmer leaders who would typically indicate that a family no longer wished to care for a child they had been assisting, or by the identification of unaccompanied children on the border prior to their transfer to the camp. As the number of unaccompanied children began to increase, six tents were erected for them next to the hospital and Food for the Hungry, a voluntary agency based in the United States, was given responsibility for their care. Just as in Sakaeo, the children were immediately referred to as "orphans" and tents became known as the "orphanage." Within five days 250 unaccompanied children were placed in the orphanage. Some 150 of them had been living as a group on the border under the care of Khmer women who had on their own initiative collected the children, provided for them, and then moved with them to the camp.

Few of the staff members of the voluntary agency with responsibility for the orphanage had had experience in providing child care services. Most were young foreigners who could not speak Khmer and who worked in the camp for quite short periods before returning to their home countries. Although the standards in the orphanage were quite minimal—including tents, simple cots, and a single elevated water tank—the orphanage was perceived by others in the camp as a place that enjoyed certain advantages: special foods, special clothes, access to a soccer ball, games organized by young foreign volunteers, many visitors, and the hope of resettlement. A fence was built around the orphanage and it was common to see children from the camp lined up at the fence watching the children in the orphanage; children who until a few days previously had been peers in the border camps and who now lived quite separate and privileged lives.

Even within the first few days of observing the operation of the orphanage and the community's response to it, camp administrators became increasingly convinced that children in the orphanage should be cared for by Khmers rather than foreigners, and that the children should live as part of the community rather than in separate facilities with goods and services different from those provided for other children in the camp. On 25 November, four days after the first refugees entered Khao I Dang camp, the decentralization of the orphanage was discussed at the camp coordination meeting where it was noted that of an estimated 7,000 children said to be "orphans" living in the border camps, only 300 were in institutions. This suggested that the Khmer community had been providing for the children. The matter was referred to the Khmer leaders within the camp who agreed to the establishment of a children's center in each section. The design of the first two children centers was based on the recommendations of the Khmer housemothers who had cared for the children on the border. Four barrack-like buildings, each housing twenty-five children, were grouped around a central courtyard.

POLICY DEVELOPMENT

The program that was developed for unaccompanied Kampuchean children is understood best in the context of how assistance was generally provided to refugees and displaced persons in Thailand in the previous five years. The Royal Thai Government maintained final authority on all matters concerning refugees and displaced persons living on its territory, and between 1975 and 1979 provided and coordinated assistance to refugees through its Ministry of the Interior. During that same period,

UNHCR, on invitation from the Royal Thai Government and in accord with its international mandate, assumed responsibility for the protection of refugees in Thailand and maintained a staff for this purpose. UNHCR also provided finanacial support to the Royal Thai government for the cost of food, water, shelter, and other essential services provided to refugees. While committed to ensuring that refugees had access to essential services, UNHCR remained essentially nonoperational, and to a large extent, carried out its assistance responsibilities by supporting the day-to-day work of other agencies. International nongovernmental organizations for their part, were allowed by the Royal Thai Government to provide assistance in the refugee camps, and contributed substantially by providing both personnel and finances for many program activities.

In response to the massive influx of persons into Thailand from Kampuchea in 1979, however, the former arrangement between UNHCR and the Royal Thai Government was modified and, henceforth, UNHCR was given primary responsibility for the provision of all services to Kampuchean displaced persons except for the security of the Kampucheans which remained the responsibility of the Supreme Command of the Royal Thai military. UNHCR, in turn, in an attempt to maintain its nonoperational mode of operations, relied primarily on nongovernmental agencies for the development and implementation of in-camp services. In the first few months of the emergency, agencies were permitted relative freedom to develop programs along whatever lines they chose. Within six months, however, UNHCR had developed a much firmer program management system and took a more directive role in setting policy. They decided which agencies would be responsible for the various services, insisted that most programs be carried out under written contract between the agency and UNHCR, and provided either partial or total financing for most of the services implemented by the nongovernmental agencies. This mode of operating was very much reflected in the way programs for unaccompanied children were carried out.

There were exceptions to this general mode of providing assistance to displaced persons. A camp known as Khao Larn, for example, was operated independently under the patronage of the Queen of Thailand. Another exception was a "camp establishment team" organized under UNHCR auspices in the fall of 1979 for the physical development of the six new camps for arriving Kampucheans and for the establishment of essential services within those camps. The camp establishment team was an exception because it was very much operational rather than advisory.

While unaccompanied children identified among the initial arriving groups of Kampucheans from the border were provided immediate services by relief workers in the camps, the issue of unaccompanied children was dealt with at a planning and policy level within UNHCR, largely in response to the criticisms and debates generated by voluntary agency personnel within the camps about how and where unaccompanied children should be cared for, and by the international lobbying efforts of agencies and politicians wishing to resettle unaccompanied children to third countries. During November 1979 UNHCR received many such criticisms and demands.

In determining what action to take UNHCR emergency operations staff in Thailand solicited the advice and opinions of persons working with unaccompanied children in the camps and consulted lawyers and social workers at UNHCR's Geneva headquarters. UNHCR staff in Geneva, in turn, sought the advice of persons outside the organization who were known to have dealt with the problem of unaccompanied children in previous emergencies. Little printed information was available about the care and protection of unaccompanied children; most of the advice was based on opinion and personal experience, some of which was contradictory. However, on the basis of these consultations, three cardinal decisions were taken by the UNHCR Regional Co-ordinator for South East Asia: Kampuchean unaccompanied children should not be moved from the camps before their circumstances were individually determined; the circumstances of each should be individually assessed; and all should be provided with adequate care during whatever period they were in the camps. A temporary moratorium was imposed on the resettlement of unaccompanied Khmer children until each case had been fully documented and attempts had been made to locate parents or relatives. These decisions were reflected in the UNHCR position defined in a public statement issued on 5 December 1979 by the UNHCR High Commissioner.[15]

The UNHCR statement of December 5 cautioned interested parties against the precipitious removal of children before the children's circumstances had been properly ascertained through family tracing procedures. It reaffirmed that efforts were to be taken within the camps to ensure that the needs of unaccompanied children were met. It conveyed UNHCR's insistence that family members be given the right to reunite with unaccompanied children whenever they could be located, and that the Khmer children should not be adopted when the whereabouts of their parents was unknown. The statement also cautioned against the removal of children from their ethnic and cultural milieu. The December 5 clarification of policy was much needed and was a bold statement based on principle—as the UNHCR authors then saw the issues—rather than a position based on political expediency. Considerations of political expediency would have been more likely to result in permission being given for the immediate removal of unaccompanied children from the camps to third countries in deference to the international pressures for such action.

In conjunction with its statement of policy UNHCR also unveiled a $9.6 million dollar program, entitled "Special aid to Kampuchean refugee children" in which UNHCR proposed to provide medical care, supplemental feeding, education, and recreation to all of the estimated 60,000 Khmer children in the displaced persons camps, and to establish a special program for the care of unaccompanied children. The planned program was welcomed by parties concerned about the care of unaccompanied children—even some of the most vocal critics—and bilateral donors committed the funds without delay. UNHCR's position that unaccompanied children should be provided for as part of, rather than separate from, other Kampuchean children was reflected in the decision to present the special program for unaccompanied children as a part of the more comprehensive program for all Khmer children in displaced persons camps. To coordinate the children's project activities a special staff position— Children's Programme Officer—was created at the UNHCR regional headquarters.

While the general principles for the services to unaccompanied children were defined in the UNHCR December 5 statement, the actual development of the services was to a large extent a pioneering effort on the part of those responsible for the programs. No UNHCR officer, and very few voluntary agency staff members working at the regional or field level in Thailand, had previous experience in the development of services for unaccompanied children in an emergency, and many had little or no previous child welfare experience. Even with concerted effort, the programs for unaccompanied children developed slowly, often without clear understanding of what final options would be available to the children.

INTERIM CARE

The presence of unaccompanied children in each new camp meant that provision had to be made for their care. Placement in camp hospitals was obviously inappropriate for more than a few days. Within a short period of time it was also obvious that the "orphanages"—the first form of interim care established in several camps—were untenable as a method of care. In November 1979, as an alternative to care in orphanages, UNHCR decided that the unaccompanied children were to be cared for in community-based "children's centers" during the time when the circumstances of the children were being documentated and their family traced. Each children's center was to be composed of eighty to one hundred children. The children were to be cared for in family-like units of up to ten children per houseparent, and children under five years were each to have a single houseparent. The physical layout of the centers consisted of long barrack-like structures or individual housing units situated around a common courtyard. The intent of this program was that unaccompanied children would be provided a living environment that provided stability, consistency, and structure, and that met their physical, mental, and developmental needs in a way that preserved their cultural heritage. Program planners believed that through the children's center concept the isolation and separateness observed in orphanage programs could be avoided and that the small family-like units would lend themselves to a stable living environment and allow for better individual care. Such children's centers were established in most of the camps for Kampuchean displaced persons.

From the establishment of the first children's center in December 1979, the number of Khmer chldren in children's centers rose quickly to 2,163 in April 1980. Because of spontaneous family reunions, tracing successes, and repatriation and resettlement, the number dropped to 612 by November 1981. By October 1982 the number had increased to 726, and then declined to 266 in October 1983, 70 in October 1984, and 65 by October 1985.[16]

Program planners insisted that every effort was to be made to ensure that the children would remain integrated in the community. Children's center facilities were to be constructed of the same building materials and with furnishings similar to the homes lived in by other children. Unaccompanied children were to use the same services used by others—hospitals and clinics, supplemental feeding centers, and schools—and were to receive the same foods and clothing as all children in the camps. One voluntary agency in each camp was given responsibility for the children's centers, which included choosing and training Khmer houseparents. The agencies which were given this responsibility included Catholic Relief Services, Holt Shahathai, the International Rescue Committee, and Terre des Hommes-Germany.

Virtually all observers agreed that the children's centers were successful in meeting the physical and social needs of the children. However, although they were welcomed as a model of emergency care during the first months, controversy over the appropriateness of the children's centers began to grow over time. For example, one of the social workers who interviewed the unaccompanied children for documentation spoke out forcefully against the centers as they were being run, complaining that they were so influenced by foreign agency administration and staff that the program was not a Khmer program as had been envisioned. The children's centers were seen as unduly influencing what unaccompanied children believed to be in their own best

interests, particularly with regards to resettlement. The pervasive orientation to resettlement to a third country by unaccompanied children in the Khao I Dang camp was,
unfairly, attributed by critics to the influence of foreigners providing services to the
children and to the selection of Khmer houseparents who were themselves committed
to resettlement. Another criticism was based on the observation that because of the
higher standards and services which they offered, the centers tended to encourage families to give up the children they had been caring for. In spite of the intent of the
organizers to maintain equality between unaccompanied children in the centers and
children living with their families, the children's centers were perceived throughout
the camps as a place of privilege—those there enjoyed priority consideration for resettlement, language training, access to foreign volunteers, gifts, special foods, and so on.
Adolescents often wished to join the children's centers and some parents and relatives
purposefully placed children in the centers under the category of "unaccompanied"
to enable the child to take advantage of perceived benefits. In one camp, Khmer leaders estimated that as many as fifty percent of the children in the children's centers had
parents or relatives in the camp or on the border. It was also found that many individuals older than age eighteen claimed to be younger in order to qualify as a minor.
One survey of just over 400 children in the unaccompanied children's centers confirmed that over one-fourth were actually eighteen years or older.

A most serious charge was that while physical needs were being met the psychological and developmental needs of children were not. Even a Children's Center Coordinator in the Khao I Dang camp took this position, noting that there was little stability or permanence in the relationships between houseparents and children. He
confirmed that the turnover of houseparents was high and that many children had
had from two to as many as ten different houseparents.[17] This problem confirmed the
concerns of critics that the Children's Center concept did not give adequate consideration to the need for consistency of caretakers, especially for the youngest children.
This concern was finally acknowledged by UNHCR in mid-1980 when in the midst
of efforts to interview all unaccompanied children, the documentation team was redirected, as a priority, to write final recommendations on all children under five years
of age.

An assessment of the children's centers program which was carried out by social
workers intimately familiar with the children and the program, attempted to examine
both the advantages and disadvantages of the program being implemented. In citing
the advantages of children's centers, they noted that for many children there seemed
to be no available alternative. Moreover, the centers helped to establish bonds
between small groups of children, offered a secure environment in which the children
appeared happy and well cared for, facilitated the tracing and reunification process,
and provided a setting in which special attention could be provided to parentless children. Against these advantages were disadvantages such as that children in children's
centers were competitive and manipulative for opportunities to go abroad, rejected
the idea of family life and many aspects of Khmer culture, were perceived as living
separate from the rest of the community and, in the case of youths older than sixteen,
were believed to be avoiding the responsibility of caring for themselves.[18] Some of
these disadvantages came to light when Khmer leaders in one of the camps were asked
for their opinions. They suggested that the children's centers were propagating a
"Western lifestyle" while the rest of the camp upheld Khmer traditions. They felt that
the children were inadequately supervised and were developing behavior character-

istics considered "bad" or "impolite" in Khmer tradition. One additional disadvantage came to light. It became increasingly evident that with the organization and administration of the special program for unaccompanied children by "outsiders," the Khmer community did not take responsibility for the children once they were admitted to the children's centers, but rather assumed them to be the responsibility of the United Nations.

In response to the perceived weaknesses, there was a constant attempt to modify the children's center programs. The practice of building special buildings for children's centers was stopped in favor of using identical buildings to those used by families throughout the camps. Billboards were removed to make the children's centers less obvious, and fences were discouraged so as to minimize the isolation they tended to create. Independent living arrangements were established in the camps for the oldest minors. Khmer leaders and other concerned persons were encouraged to take a more direct role in planning and implementing programs for the children, and attempts were made to identify houseparents less likely to be resettled and therefore able to remain longer with the children. Efforts were made to reduce and limit the number of foreign visitors to the centers and the number of expatriates working directly with the children. Even at the end of 1980, various social workers advocated a year-long moratorium on the resettlement of unaccompanied children so as to promote the opportunity of determining the children's full circumstances without the distortion caused by the enticement of resettlement opportunities, but such a moratorium was never implemented. These corrective measures were considered helpful but did not fully resolve the uncertainty that remained as to whether the children's centers were necessarily the best interim care approach, or whether the children could have been cared for equally well in some other way, notably by families within the Khmer community.

Foster care was the principal alternative to the children's centers for providing care for unaccompanied children in the camps. At the beginning of the emergency, some relief workers claimed that foster care was alien to Khmer tradition. It was also claimed that children who were fostered were not likely to be afforded equal standing with other children in the adoptive home, and might be used as servants, or even be physically mistreated. However, various assessments confirmed that many children not with their parents were being cared for by other families, and while many unsatisfactory foster home arrangements were identified, many others appeared to be successful and happy. Unaccompanied children often reported to the documentation teams that they arrived with families whom they had met and lived with in Kampuchea or had joined while living on the border, and arrangements had been made spontaneously between the child and the family. The stability of the relationships was reflected in the fact that over thirty-two percent of the children had lived with these families for two or more years and seventy-three percent had stayed with only one family—facts identified in a survey of the foster homes carried out by the Swedish documentation team.[19] An assessment by social workers from the International Rescue Committee found that spontaneous foster family arrangements had occurred because the families had taken pity on an unaccompanied child in need of assistance whom they had come to know. That assessment revealed that the majority of the children being fostered were fourteen years old or older, and foster families were often recently married couples, widows, or single persons. These findings were somewhat surprising as they were different from Western assumptions about the willingness of families to foster adolescent children and the types of families likely to be foster families.[20] It was also noted that while deliberate placements often failed, spontaneous arrangements were quite successful.

Both observation and assessment of existing foster arrangements confirmed that some Khmer families, although often in very difficult circumstances themselves, were still willing and able to provide for children who were in need of assistance. However, it also became evident that successful foster care depended upon the arrangement being voluntary both on the part of the family and the child, and on being free of outside pressure, moral or otherwise. The extent to which foster families might have been found for children in the child care centers would only be a point of conjecture, and widely differing opinions were given by social workers in the camps with responsibility for the children. One agency advocated that children should not be fostered by Khmer families, while another placed almost all the unaccompanied children in their charge in Khmer foster homes and claimed the children were well cared for.

However, the question that remains is not simply whether fostering arrangements could have been found for all the children in the children's centers, but to what extent the existence of the children's centers disrupted the foster family process and encouraged the separation of children from families who had been caring for them. We also need to consider how permanent the foster arrangements were.

Many, if not most, unaccompanied children arrived in the camps with a family. While foster family arrangements were believed to be distinctly advantageous for those children who had benefited from a long-term commitment from a family, for many children the foster care provided by a family was decidedly a short-term arrangement. Breakdowns are believed to have occurred for many reasons such as: the perceived advantage to the child of being in the children's centers, abandonment by foster families, relationship breakdowns, the child's acquaintance with another family he wished to stay with, the foster family's perception as to whether the child in their care enhanced or diminished their resettlement opportunities, or whether the foster family wished to return to Kampuchea unencumbered. Foster family breakdowns also sometimes occurred prior to arrival in the camps, within the camps, or even after resettlement.[21]

An ideal solution other than children being well cared for by their own families did not exist—each alternative had advantages and disadvantages. The ambiguity was reflected in a UNHCR evaluation of services provided to the Kampuchean displaced persons in Thailand:

> In spite of the experience gained over the past two years project staff have not been able to determine whether unaccompanied minors should be placed in child care centres or foster homes. Assessments of the situation have never conclusively indicated the most advantageous approach. Unaccompanied children in child care centres are known to be physically healthier than camp children in general but serious doubts remain regarding the extent to which the centres are able to meet the children's psychological and developmental needs. Foster homes better meet the children's emotional needs but resettlement has caused a constant turnover of foster parents that has often disrupted this arrangement. In some instances minors have had as many as four or five foster parents thus creating an unstable and unsatisfactory home environment. There have also been numerous complaints of foster parents neglecting their wards and as a result special monitoring of foster homes has had to be undertaken.[22]

REGISTRATION, DOCUMENTATION, AND TRACING

In accordance with the UNHCR's policy to individually assess all unaccompanied children, it was considered essential that such children be identified—"registered" so as not to be overlooked—to be fully "documented" when time permitted, and their families "traced." A distinction was commonly made between the terms "registration," "documentation," and "tracing." Registration generally referred to a brief listing of personal facts about a child; documentation referred to the process by which more in-depth information about the child's past and present circumstances were ascertained and assessed by social workers through interviews with the child, his caretakers, and his acquaintances. Tracing in this situation generally referred to the special efforts carried out to locate parents and family members other than through interviewing the child or his acquaintances.

As part of the admission procedures to the Khao I Dang camp, all families and individuals were registered by ICRC. Similar registrations were also completed throughout the other Kampuchean displaced persons camps after people had already arrived—a process that required months of work. The information recorded on this registration included the reported relationship of people arriving and living together, and by obtaining this information, 3,692 children who were not living with their parents or close relatives were identified during 1980, although the maximum number of children who were residents in children's centers at any one time was about 2,100 children.

While this registration information continued over time to serve as important reference material to personnel involved in documentation and tracing of unaccompanied children, its usefulness was limited by the fact that a coordinated information system did not initially exist between services within the camps, and because the information was initially considered confidential and not shared. Therefore during the initial months of the emergency each agency which provided services to unaccompanied children—the hospital, "orphanages" (when they existed), and children's centers—developed independent registrations forms, if they registered the children at all. The lack of a standardized information-sharing system for children resulted in no registration for some children and reinterviewing for others.

The seriousness of the registration issue was demonstrated by how easily a young child's identification could easily be lost, or how he could be separated from existing family members, simply by the lack of adequate records. A relief worker involved in the tracing of unaccompanied children recorded examples of the consequences of inadequate hospital records.[23] It should be noted that the lack of death certificates and of family details in death records made it impossible to determine whether or not the parent of an unaccompanied child had died in the camp hospital or whether there might be surviving family members. Lack of information about mothers admitted to the maternity ward made it impossible to relocate those who left children in the hospital, as it sometimes happened, particularly in the maternity wards. Inadequate documentation of children who were moved from one location to another resulted in children being brought to the camp hospital admissions ward from the border area with no more documentation than a verbal statement from the ambulance driver and a label on his admission record that said "Border." Another way in which children were separated from parents occurred when very young children were transferred from ward to ward within the camp hospital, as from obstetrics to pediatrics, without accompanying documentation as to the whereabouts of their parents.[24]

With the recognized need to document the circumstances of each unaccompanied child came the responsibility of finding an appropriate agency to assume that task. Voluntary agencies which were involved in the international movement of children were not considered appropriate because it was feared they would interpret what was best for the children only considering their own program approach. Radda Barnen (Swedish Save the Children Fund) was invited to carry out the documentation of unaccompanied Khmer children and in January 1980 sent a social worker for a preliminary exploration of the camps. In March a team of three Swedish social workers arrived and began by developing, testing, and revising a documentation form. They carried out the interviewing of all unaccompanied children over the next nine months after having increased the size of their team by adding additional Swedish interviewers and by working in collaboration with UNHCR social welfare personnel and children's programme personnel.

During the first year of the documentation efforts, interviews were conducted in English by the Swedish social workers and translated to Khmer by a Khmer translator. After the Khmer translators had gained some experience and had the trust of the social workers they also carried out interviews independently. Each interview with a child required approximately one and one-half hours. Special guidelines for the interview of children of less than three years old and for children from fourteen to seventeen years old were developed to accord with the different interviewing techniques and the somewhat different information required of different aged children. In general, the information that was sought included a chart of relatives and their previous addresses; details about when the child last had contact with them; a short history of the child's life experiences; the amount of schooling obtained; their general health condition; and any information provided by adults who were caring for the child. Relatives and acquaintances known to be living in the camps were also interviewed to acquire information about the location of parents or relatives and to explore the possibility of the child being cared for by that acquaintance or relative.

Initially, the Radda Barnen social workers were to compile a dossier of the facts about each of 1,200 children then in children's centers in the Khao I Dang camp. Later, the social workers were asked to interview children in children's centers in all camps and children in foster care as well—in total more than 2,600 children. To avoid the need to reinterview the children or recopy information from the children's dossiers, the social workers were also requested to complete special forms for tracing as part of their responsibilities. Their workload was further expanded in mid-1980 when requested by UNHCR to offer final placement recommendations on each child. This request arose from the realization that even though the final decision as to whether a child should remain in Thailand, or should be resettled, or should be repatriated was still to be made by an interagency review committee convened by UNHCR, nevertheless, a recommendation from the social workers who had talked with each child, his caretakers, and his acquaintances was also essential.

Expansion of the documentation responsibilities of the social workers reflected, to a large extent, the need for information about the children for different purposes—needs which became evident or which changed over time. Information about unaccompanied children was needed to formulate general policy and program guidelines; to identify individual needs; to ensure that the available care adequately met the needs of the children; to identify information by which family members of the children might be located and the family situation assessed; and to provide information on which long-term plans for the children could be adequately made. When information was collected for only one of these purposes, rather than for comprehensive assessment, new forms had to be developed for each purpose and children reinterviewed.

The documentation of the children was not without difficulties. Many inconsistencies were found between what was written on the registration form, what was known about the children by the staff of the children's centers, what information provided by the children to the social workers, and what information was believed to be true by other persons familiar with the children. The discrepancies included location and existence of the parents, the number and names of relatives, and the age of the minor. To give one example: a boy had reported to a staff member after completing an earlier registration form that his father had resettled in the United States, but later the same boy reported that his father died of famine in 1977.[25] Interviewers found that children were not always forthcoming with the desired information. Sometimes they refused to tell about siblings or relatives inside the camp. Other children sometimes provided more names of relatives than they had initially listed. Questions were also raised during the documentation process as to the most effective techniques for interviewing children, the possible distortion of the information by interpreters, the possible misinterpretation of information by interviewers from a culture different from that of the child.

Uncertainty about the credibility of the information given in the documentation process plagued the program. In the largest camp some sixty-three percent of the unaccompanied children claimed in the documentation interviews that both parents were dead and thirteen percent claimed to have no living relatives.[26] While some children were truly orphaned and without living relatives, such claims often proved to be suspect. This was soundly established by the fact that of the first 735 parents whose existence was confirmed through tracing, 109 (19.5 percent) had been reported as dead by their unaccompanied children.[27] The reasons for the discrepancies were not always clear, but the major factor was considered to be the desire of the children—often with prompting from parents, relatives or friends—to report whatever would be most likely to assist them in achieving their goals, particularly where there was interest in being resettled to a third country. Other suggested reasons for giving inaccurate information included: the cultural practice of giving the answer which the interviewee understood was desired by the interviewer, information based on incorrect rumors, psychological factors through which children might fantasize or repress information, and the possibility of children learning new information from acquaintances about their families, or remembering things they had forgotten. Interviewers had particular difficulty in determining the actual age of the children. Many children did not know their exact age, as birthdays are not typically celebrated in the Khmer culture. More problematically, individuals older than eighteen often claimed to be younger so as to be eligible for the special services available to minors. Uncertainty about the accuracy of the information given by the children was countered by checking and rechecking the information given against information given by other persons who knew the child and possibly his family.

The documentation process showed that approximately two-thirds of the unaccompanied children were boys and most unaccompanied children were between twelve and seventeen years old. Less than five percent were under four years old.

Approximately half of the children were accompanied by brothers or sisters. Whereas early in the emergency some relief workers had assumed that unaccompanied children were children without living parents or relatives, review of the circumstances of separation suggested that the separation had often occurred for reasons other than the deaths of parents. For example, many children were separated when forced to join mobile work teams; others were separated during the evacuation of Phnom Penh; still others were accidentally separated during the fighting. The following case history gives an account of an unaccompanied child considered by the documentation team to be "typical":

> A boy, now aged thirteen, used to live together with his father and mother and four siblings in Battambang City. His father was a merchant and his mother sold vegetables at the market. In late 1975 the family was forced to the countryside to work in the fields. He was then totally separated from the rest of his family and lived in a group of boys under the age of twelve. He worked in a mobile team of more than 100 boys very hard from early morning till late at night. He had just started school in 1974 but was not yet able to read or write Khmer and there was no possibility for education while in the mobile team. He stayed in the mobile team for almost four years, during that time he had no chance to see anyone from his family.
>
> When the Vietnamese attacked Kampuchea in January 1979 he ran away from the mobile team (Kong Chalat) and went back to his village to look for his family. A neighbour told him his parents had died from disease and hard work in the fields. He was not able to find out where his older siblings were living. He stayed with his neighbor for some months and hoped to meet his relatives. But no one showed up. The shortage of food was severe and he decided he had to leave the neighbour-family because there was not enough food for everyone. Rumours told him that there was a chance to get food and even a possibility to be resettled in a third country and even get an education if he headed for Thailand and the refugee camps at the border. So in August 1979 he left Battambang City with a group of persons he didn't know. When they reached the Thai border he received food. He stayed with two persons from the group in 007 camp for sometime till there was fighting between different groups within the camps. At that time he again was separated and picked up by relief workers who put him on a bus which brought him to Khao I Dang.
>
> He was screened and then placed into an orphanage where he now has been living for four months. He claims he has no relatives abroad, but he wants to study and be resettled to a third country.[28]

In addition to interviews with the children with the purpose of receiving information from the children as to where parents or other family members might be, independent tracing efforts were organized as well. Bulletin boards, to which the pictures of unaccompanied children were attached, were erected in all Kampuchean displaced persons camps and in camps on the Thai-Kampuchean border and were manned by personnel of the International Rescue Committee. These bulletin boards proved to be a comparatively simple and effective system of tracing.

The second tracing method was the circulation of "tracing books." Using a photo-off-set printing process for good quality reproduction of the pictures, these books were made up of the tracing forms which had been completed by the documentation team for approximately 3,200 children. Redd Barna (Norwegian Save the Children) published 350 copies of each volume and, with the assistance of more than 200 selected and trained Khmers, circulated these books household by household to some 20,000 to 30,000 families throughout the camps. At each house, page by page, data sheets of the children were reviewed, a process which required about two and one-half hours. From the circulation of these volumes, the bulletin boards and word of mouth, more than 2,000 people identified a child either as their own child, a relative, or an acquaintance. (Identification did not necessarily mean that a child would join the person who identified him.)

Whenever a child was identified, an interview was carried out to record as much information about the child as possible. The interview report was then given to the International Rescue Committee documentation team who reinterviewed the informant for more in-depth information. While many of the claimants were not willing or able to care for the child, they did provide valuable information about the circumstances of the child and his family. By mid-1982, at least 2,100 children were reunited with families through the combined tracing efforts of all agencies.

Photographs proved to be the single most useful aid in the documentation and tracing processes. Pictures were taken of each unaccompanied child with his or her name and tracing number included in the photo. Because of the frequent need for duplicates, black-and-white negative film was essential. Pictures were attached to the dossier, and to the tracing forms, and to tracing billboards. Polaroid pictures proved useful in the verification process.

The registration, documentation, and tracing process can be summarized as follows: (1) On admission to a camp, the unaccompanied child was registered and assigned a tracing card number. (2) An in-depth interview as then carried out by a social worker and information was collected which comprised the main dossier of the child. In addition, a special tracing form was completed and a photograph was taken of the child. (3) The dossier and tracing form were then sent to a documentation and tracing coordination office where a file (consisting of the initial registration form, the dossier, and the tracing form) was established for each child. (4) Pictures were then attached to both the dossier and tracing form and the original file was returned to the camp. (5) Photos were attached to tracing bulletin boards and tracing forms were compiled into tracing books then circulated, and any information collected through the tracing process was added to the file. (6) Each child was given a medical checkup and the medical report was added to the files. (7) After a tracing period of a few months, the child was reinterviewed and his file reexamined to assess any information found through tracing. If possible, a recommendation was then written by the social workers which included the recommended long-term plan for the child. (8) The completed file was then passed to the UNHCR-convened interagency Review Panel for review and final decision. Many times a third interview or more was required before the full circumstances of the child were available and for some children a psychological evaluation was added. (9) Case monitors ensured that the decision made by the Review Panel about each child was appropriately carried out.

In spite of the difficulties the individual documentation and tracing of such a large group of chidlren in a comparatively short period of time was an outstanding example of a collaborative interagency effort for the benefit of the children. Radda Barnen's documentation efforts were completed by the end of 1980. The circulation of the Tracing Books by Redd Barna was completed in early 1981. The International Rescue Committee continued to document unaccompanied children, compile the files of unaccompanied children for a decision by the UNHCR review committee, assist in the reunification of families within the camps, and to maintain the tracing bulletin boards throughout the camps and on the border until these functions were assumed by UNHCR at the end of 1983.

LONG-TERM CARE

In instructing the documentation team and the review committee as to alternatives to be considered in recommending long-term care arrangements for unaccompanied children, the Bangkok office of UNHCR identified four options in their order of desired preference: family reunion, foster care in the holding centers camps, remaining in the children's centers, and resettlement to a third country.

By June 1982, 1700 children had been reunited with family members within the camps and an additional 438 had departed Thailand for reunion with family members in third countries. Family reunion referred to children rejoining biological parents, primary family members, or other family members such as siblings, aunts and cousins. Reunion was complicated by the fact that family members were located in any of at least four locations: in Kampuchea, on the border, in one of the camps within Thailand, or in some other country around the world. Uniquely different tracing and reunification efforts were required to facilitate the reunion of unaccompanied children with family members in each of these different locations.

One of the most dramatic reunification programs occurred at tracing stations set up on the Thai-Kampuchean border where the International Rescue Committee had erected tracing bulletin boards and tracing centers. Family members of some unaccompanied children were living within the border area itself, but many more were living in villages within Kampuchea. Response to the border tracing service was gradual but several months after the tracing centers were established many families traveled long distances to the border from within Kampuchea in search of a missing child. Policy makers were concerned that this service might act as an incentive to encourage additional people to leave Kampuchea, but persons involved in the tracing efforts attested to the fact that claimants for the children did not come to the border with expectations to join the children in the relief camps but sought their return to Kampuchea. The difficulty of the journey was itself evidence of the sincerity with which families sought their children, for travel was often difficult and dangerous, particularly in the rainy season, and travelers were often required to pay travel fees enroute. In an analysis of the first forty-two children reunited with relatives at the border, twenty-five reunited with parents, eleven with siblings, and six with relatives or other people. Verification techniques, as employed in the camps, included the use of photographs and the comparison of personal and family details given by the claimants with those given by the unaccompanied children. Passing cassette recordings between the claimants and the children was found to be very helpful in reestablishing rapport and in answering questions raised by each of the parties. Before the border tracing was terminated in 1983, 189 children were reunited with relatives on the border and returned to Kampuchea with them. Workers who participated in the border tracing operations were convinced that many more children could have been reunited with family members in this way had the border tracing efforts continued longer and if tracing within Kampuchea had been facilitated. An analysis of the reunifications that occurred at the border concluded that "separation times of three and more years [did] not indicate reunification to be unlikely."[29]

While the border tracing and reunification program of the International Rescue Committee and UNHCR provided an essential service for the families served, even more family members would almost certainly have been identified if an effective family tracing program had been organized within Kampuchea. Similarly, more children would have been reunited if an acceptable system had been established to facilitate the reunion of unaccompanied Khmer children in Thailand with their families within Kampuchea when both parties desired such. A Red Barna sponsored analysis of just over 2,000 files—containing documentation and tracing information—of the unaccompanied Khmer children in Thailand was carried out at the end of 1980. Evidence from the information gathered about the children suggested that as many as fifty-three percent of the unaccompanied children could have parents alive, and that eighty-nine percent of the children had reported siblings and relatives to be alive in Kampuchea. The analysis also indicated that fifty percent of the children came from three provinces, reflecting where tracing was most likely to be beneficial.[30] On the basis of this analysis, tracing within Kampuchea was recommended by the Red Barna paper. Various tracing techniques were formulated for use within Kampuchea. One proposed effort would have sent a form letter seeking information about the family to the leader of each home village. In another, pictures of all unaccompanied children were printed in sheets in a newspaper format with the intent that the "newspaper" might be widely circulated within Kampuchea. Neither of these approaches were permitted.

Official efforts were made by ICRC, UNHCR, and other interested parties to establish a tracing program within Kampuchea to search for family members of unaccompanied children in the displaced persons camps in Thailand. Kampuchean officials authorized the search for children younger than fifteen years. However, many difficulties obstructed the program including interagency differences. At least seventy files were submitted to the Kampuchean Red Cross for tracing and relatives of at least thirteen children were found. The initial claims proved the potential of family tracing and by June 1981 some 400 additional files had been selected, screened, and prepared for transfer to the Kampuchean Red Cross. These efforts were futile, however. Reunification proved to be an impossibility because of a lack of agreement about how the relationship of the claimant to the child was to be verified, and because the return of any children was made a political issue relating to the recognition of the Kampuchean government. ICRC terminated its attempt to organize a tracing effort in June of 1983.

Repatriation was considered by many persons to be the most desirable solution, for it would have allowed the unaccompanied children to reconnect with their families and remain within their culture. Despite UNHCR's repeated attempts to facilitate a voluntary repatriation program that guaranteed the safety of the Kampucheans and that was politically acceptable to all parties, a voluntary repatriation program was never a real option. A major consideration was that many of the Kampucheans did not wish to return to their homeland and had journeyed to the camps with the intent to resettle in another country. Others did, however, plan to return when they believed it was safe to do so. Surveys of the wishes of the unaccompanied children mirrored the different opinions of adults—some wished to return to their home villages, while the majority, particularly those from urban areas, desired to be resettled.

UNHCR's position remained that resettlement to a third country was to be considered only after a guarantee was given by the resettlement country that children would not be adopted and that the option for family reunion would be preserved. Planners expected tht after being resettled unaccompanied children would be cared for by a Khmer foster family, a non-Khmer foster family, or in a group home.[31] The moratorium on the resettlement of unaccompanied Khmer children that had been imposed at the end of 1979 remained in effect until mid-1980 when UNHCR decided that resettlement was the most viable option for unaccompanied children who had not been reunited with their families through tracing. The first children to be offered to the embassies for resettlement were thirty-seven infants and toddlers who it was believed needed to be in a family environment as soon as possible and for whom too little information was available for tracing and verifying claims. The resettlement of older children began in early 1981, if the family of each had been searched for.

The Radda Barnen social workers, who had the responsibility of documenting the unaccompanied children, drew up in collaboration with UNHCR age-related criteria for making recommendations for the long-term placement of the children, then they scheduled the interviewing according to categories of perceived vulnerabilities: medical emergencies first, then children with handicaps, then children birth to five years, then adolescents from fourteen to eighteen years, then finally, all others. Following are the factors they considered in making their recommendations for the younger children:

- mental and physical health conditions of each child;
- living conditions in the camp;
- detrimental effects of long-term residence in the holding centers;
- vulnerability, particularly of the youngest children
- the minor's own opinion (if possible, depending on the age of the minor);
- the relationship between the housemother and child, e.g., period of contact, the strength of the bonds between them, the housemother's capability and willingness to care for the minor, the opinion of the housemother in charge of the children's center;
- the number of previous separations from guardians;
- the value of further tracing.[32]

For unaccompanied children between the ages of fourteen and eighteen, the following factors were to be considered:

- the wishes of the minor;
- the mental and physical health conditions of the minor;
- the child's social patterns and ties with the children's centre;
- the total situation and life history of the child including: experiences during the last five years in Kampuchea; location of relatives in third countries, Thai camps, or Kampuchea; years of education; job skills, if any;
- the opinion of the houseparents;
- the value of further tracing.[33]

Nine additional points were listed as guiding principles for social workers in offering recommendations on all children:

1. Under no circumstances should siblings be separated.
2. The minor should always be placed together with another Khmer child with

similar background and experiences, preferably with a close friend from the same camp (if placed with an indigenous family), or in a group home in a third country.

3. Minors with psychological and physical handicaps should be resettled in a third country if proper treatment is not available in the holding centres but can be guaranteed in the resettlement countries.

4. Mentally retarded children should normally remain in their culture of origin if taken care of by a sensitive, caring adult.

5. Further tracing will continue for all cases with a view to family reunification.

6. Repatriation for those minors with strong desires for returning to Kampuchea in the future if this is possible, accompanied by a responsible adult unless the minor is very mature and independent.

7. Resettlement for those minors who have a realistic picture of life in a third country and have a strong wish to go there.

8. Resettlement for those minors who have relatives in a third country with whom they wish to live.

9. Detrimental effects of long-term residence in the holding centres in Thailand.[34]

During the period from 1975 to 1984, at least 1,700 children were reunited with family members in the camps in Thailand, 120 or so were reunited with family members on the Thai-Kampuchean border and then returned to Kampuchea, and some 2,023 unaccompanied Khmer children were resettled from Thailand to other countries including: 897 to the U.S., 638 to France, 143 to West Germany, 93 to Canada, 83 to Australia, 58 to Switzerland, and 55 to Belgium.[35]

10

Analysis of the Problem

Past experience substantiates that unaccompanied children exist in virtually every war, famine, refugee situation, and natural disaster. As discussed in Chapter 1, millions of children have been separated from their families in emergency situations over the past fifty years. Each child, each separation of a child from a family, and each emergency is unique. Yet comparison of emergency situations shows that there are many similarities including, for example, the ways in which parents and children separate, the reasons for the separations, the needs of unaccompanied children, and the problems in providing services. Thus, children were separated from parents for many of the same reasons during the Spanish Civil War as during the Vietnam War; similar legal issues arose for unaccompanied children who crossed national boundries after the Hungarian uprising as for those who entered Thailand from Kampuchea; and similar questions about the placement of unaccompanied children arose after World War II as were faced by relief agencies in recent emergencies. The recurrence of similar issues and problems in past and present emergencies affirms the usefulness of examining past experience as a basis for improving the assistance provided to unaccompanied children in present and future emergencies.

Various terms have been adopted in past emergencies to describe children not in the care of their parents or guardians. In addition to "unaccompanied children," which has been used since World War II, other terms have included: "abandoned children," "beggar children," "detached children," "foundlings," "homeless children," "orphans," "separated children," "street children," "vagabond children," and "waifs." The most commonly used term is "orphan," and although the term implies a parentless child it has often been erroneously used to refer to any child not accompanied by his parents and to children who have lost one parent. This has frequently caused misunderstandings and has led to mistaken assumptions about the existence of parents and about services required for the children. Sometimes the term is qualified as "half-orhpan," "orphaned of father," or "orphaned of mother." The authors recommend that the word "orphan" be reserved to only refer to a child who has lost both parents. Experience has shown that it should not be assumed that a child's parents are dead simply because a child in not in their company because they may be separated for other reasons. The term "unaccompanied child" is preferable to the other terms because it avoids unsubstantiated implications about the cause of separation or the existence or intent of the parents.

Table 10-1. Stages in the Experience of an Unaccompanied Child

1. Pre-separation:	when the child is in the care of the parent before separation.
2. Pre-identification:	the interval before parent/child separation and identification of the child as unaccompanied.
3. Emergency care:	the interval immediately after identification, measured in days.
4. Interim care:	the interval between identification and when long-term care is provided, measured in weeks to months.
5. Long-term care:	when a child is in a care situation considered to be permanent, measured in years

Table 10-2. Overview—Unaccompanied Children in Emergencies

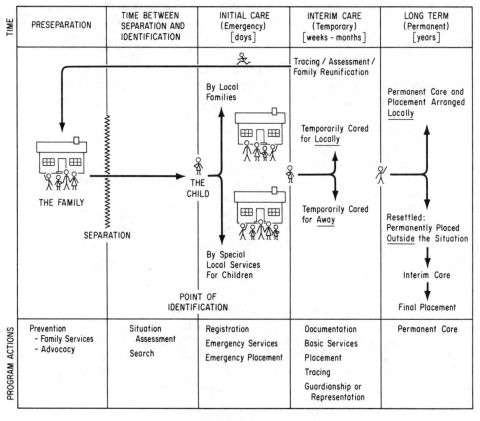

The experience of an unaccompanied child as well as the actions taken on his behalf falls into five categories: preseparation, preidentification, emergency care, interim care, and long-term care. (See Table 10-1.)

Within these five periods assessments are required to determine whether assistance is needed. If so, decisions must be made and actions taken to provide care and protection most suitable for each child. The assessments upon which decisions and actions are based include analysis of the cause of the parent/child separation; the psychological, social, and legal situation of the child; and the assistance provided by local and outside intervenors. Experience confirms that all of the above factors are significantly interrelated. For example, the cause of separation may influence the placement of a child; the type of emergency care provided may affect the long-term care received; and the availability of tracing services may determine whether family reunification is possible. Therefore, services to unaccompanied children are best provided as a program that integrates long-term goals with emergency actions. Table 10-2 illustrates the relationship between the periods, decisions about the kind of assistance commonly provided, assessment considerations, and general program responses.

The separation of children from parents can be divided into two broad types of separations: involuntary and voluntary. These two types of separation focus on the intent of the parents and child and can apply both to the time of separation and the time of potential reunion. This intent is important because it bears directly on the issue of when parent/child separation can and cannot be prevented, and the issue of whether or not reunion is desirable or possible. Within these two groups there are nine categories of parent/child separation based on intent and circumstances: abducted, lost, orphaned, runaway, removed, abandoned, entrusted, surrendered, and independent. (See Table 10-3.) Obviously, both intent and circumstances are subject to change over time.

Table 10-3. Categories of Parent/Child Separations

Involuntary Separation: Against the Will of the Parents.

1. Abducted:	a child involuntarily taken from parent(s)
2. Lost:	a child accidentally separated from parents.
3. Orphaned:	a child whose parents are both dead.
4. Runaway:	a child who intentionally leaves parents without their consent.
5. Removed:	a child removed from the parents as a result of the loss or suspension of parental rights.

Voluntary Separation: With the Parent's Consent.

6. Abandoned:	a child whose parent(s) has deserted him with no intention of reunion.
7. Entrusted:	a child voluntarily placed in the care of another adult, or in an institution, by parents who intend to reclaim him.
8. Surrendered:	a child whose parents have permanently given up their parental rights.
9. Independent:	a child living apart from parents with parental consent.

Ressler and Steinbock

• *Abducted children.* Children have been abducted in various emergencies, particularly in wartime. In World War II, children between the ages of two and twelve were kidnapped for a secret Nazi adoption program and adolescents were abducted as slave laborers. In the Greek Civil War, insurgents abducted thousands of Greek children and took them to neighboring Eastern European countries. The Khmer Rouge from 1975 to 1979, after taking children from their families, placed them in mobile work teams. In Zimbabwe during the war for independence children were kidnapped from homes and schools to serve in guerrilla movements. Such kidnappings are sometimes used as a means of coercing parents.

• *Lost children.* Accidental family separations have been recorded in most emergencies although they seem more likely to occur in wars, refugee situations, or famines with large-scale population movements and emergency evacuations. In the confusion of a mass exodus and during bombing raids for example, children are sometimes inadvertently separated from parents.

• *Orphaned children.* Some children are left without adult care because both parents die. This type of separation occurs in all emergencies. Experience suggests, however, that only a minority of the unaccompanied children are orphans, contrary to commonly held assumptions. An exhaustive tracing program after the Nigerian Civil War, revealed that only 327 children out of 3,922 unaccompanied children were found to be orphans. Similarly, an analysis of the records of some 2,500 unaccompanied Kampuchean children in the displaced persons centers in Thailand in 1980 indicated that, while a majority of the children did not know their parents' whereabouts, many of these parents were possibly still alive.[1]

• *Runaway children.* Some unaccompanied children leave their homes against the wishes of their parents. Runaways were found, for example, among unaccompanied Lao children in the refugee camps in Thailand, among the street children in Addis Ababa during the Sahel drought in the early 1970s, and in most emergencies.

• *Placed children.* Some children who are unaccompanied in emergency situations have been removed from their homes earlier by the authorities because of either the child's or the parent's behavior. For example, some of the unaccompanied Hungarian youth who fled after the 1956 revolt and some of the unaccompanied Cuban youth who entered the United States in 1980 came from jails and mental institutions. During emergency situations, as in nonemergency times, children may also be taken from their parents because of abuse or neglect on the parent's part.

• *Abandoned children.* In virtually every war, refugee situation, and famine, some children are abandoned. A child may be abandoned at any age, although it occurs more often in infancy. Over a fifteen-year period after the Korean war more than 80,000 children were abandoned in Korea. More recently, in a camp having an estimated population of about 20,000 persons on the Thai-Kampuchean border during 1980 to 1981, at least seventy children were abandoned over a six-month period, that is, about one every three days. Forty-four of the children were less than three months old. Some were abandoned in the maternity ward of the hospital or at the food distribution center; others were found elsewhere in the camp or in the forest nearby. The mothers of seven of eleven children abandoned in the maternity ward were unmarried. Some of their stated reasons included; "[I am] too poor"; "I prefer a boy [or girl]"; "I'm afraid of having to run with a burden in case [there is] fighting"; "[I am] very busy"; "I have no husband."[2]

• *Entrusted children.* In difficult times parents frequently ask family friends and neighbors to care for their children. Although no statistics exist, most separations occur in this way during wars, refugee situations, famines, and natural disasters.

Families have also often entrusted children to child-welfare institutions, orphanages, and boarding schools during emergencies.

● *Surrendered children.* Sometimes a child is unwanted, or wanted but parents are unable to keep the child for such reasons as destitution or illegitimacy, so parents consign the total responsibility for the child to another person or party and terminate their own rights and obligations as parents. Parental rights are terminated to facilitate adoption, for example. In some situations less formal means of surrender have been used by parents and children have been given or sold to persons or agencies.

● *Independent Children.* Some children, particularly older adolescents, are likely to be living independently through their own choice and that of their parents. This is true for many unaccompanied minors who have migrated in times of war, as part of a mass population movement, and in famines. Thousands of unaccompanied Vietnamese adolescents, for example, left their homes and country with the consent of parents. Usually they have maintained personal communications and family members have continued to influence decisions made by the absent child.

In every war, refugee situation, or famine, and in many natural disasters, there are likely to be unaccompanied children in all or most of the nine categories mentioned above. In the refugee and displaced persons camps in Thailand in the early 1980s, for example, there were unaccompanied children who had been abducted from their families by Pol Pot forces. Some had been accidentally separated from their families during bombings and emergency evacuations. Others had been abandoned. Some children were living with other families with parental consent (for example, an infant being cared for by the friend of a hospitalized mother). Older adolescents were sometimes living in the camp independently, although child and parent(s) knew of the others' whereabouts. Some children were orphaned, as when a mother had died during childbirth. Some children had left home against their parents' wishes.

Several important implications arise from recognition that family separations may occur in quite differing circumstances and with various possible intentions on the part of parents and children. First, it challenges the common assumption that unaccompanied children are generally orphaned or abandoned. Second, because individual circumstances are likely to be so varied, the plight of each unaccompanied child must be individually assessed. Third, assessment must include a determination of the intent of both parents and child regarding separation and reunion. Fourth, the circumstances at separation, the intent of both parents and children, and the potential for reunion can only be determined if the parents are located and consulted. Tracing is therefore likely to be required.

CAUSES OF PARENT/CHILD SEPARATIONS

The underlying causes of parent/child separations are specific to each family and situation and are extremely varied. Common causes may, however, be grouped into seven categories: social, psychological, cultural factors; parental inability to provide care; emergency-related circumstances; perceived opportunities; military/government policies; relief interventions; and child's initiatives.

• *Social, psychological, cultural factors.* In considering the problems that cause parent-child separations in emergencies, it should be remembered that even in "normal" times large numbers of children throughout the world live in foster and adoptive homes, in group homes and institutions, and as street children. The same social, psychological, and cultural problems that ordinarily cause separations are exacerbated by emergencies—as children may be rejected by step-parents, abandoned because of physical handicaps, given away because of illegitimacy, deserted as a result of an unwanted pregnancy, or neglected because they are of the wrong sex. Cultural mores and religious values also play an important role. For example, women abandoned illegitimate children in Korea in part because such births resulted in social ostracism and job loss.

• *Parental inability to provide care.* Separations sometimes happen because parents are unable to care for their children. Parents may be absent because of work, imprisonment, or conscription for example. Poverty and single-parent status are the two most commonly identified characteristics of the families of unaccompanied children. Abject poverty is an important cause of parent-child separations, and single-parent families are often the poorest of the affected population. Separation may also arise from family relationship problems between parents or between a child and parents.

• *Emergency-related circumstances.* Hostilities, famine, and other life-threatening circumstances may cause additional separations. Sometimes the separation is accidental, as may occur in bombings and mass population movements. Separations may also be purposeful. Families living in danger have often entrusted children to the care of others as a preventive measure. Sometimes the separation is a life-saving measure. During the Nigerian Civil War and during the Ethiopian famine in the early 1970s, starving children were left in feeding centers by parents who had no food and who believed that such centers would provide for their children. During the Vietnam War, orphanage personnel often received critically ill children after the parents had apparently exhausted all other means to care for them.

• *Perceived opportunities.* Many parent-child separations during emergencies have occurred for reasons not directly related to threats to survival, but for perceived opportunity or benefit. Sometimes children have been sent away by parents so that they may be spared hardship or removed from environments the parents believed to be morally harmful. For example, some children were sent overseas from Germany before World War II and from Cuba after the revolution by parents who opposed state indoctrination of their children with ideas contrary to personal beliefs. In past emergencies, separations often occurred because parents wanted to provide children with opportunities, goods, or services that were not locally available or that parents could not afford, such as educational opportunity. In the years after the Vietnam war, thousands of children left their families to travel at great risk to seek opportunities in resettlement countries. In virtually all disasters, stories are told of children who were intentionally placed by parents in orphanages and children's programs for the children's benefit. When such programs restricted services to children without families, parents often temporarily concealed their own existence. During the Ethiopian drought of the early 1970s and in the Kampuchean displaced persons camps in Thailand, parents whose whereabouts had been unknown sometimes appeared unexpectedly to claim their children when it was rumored that children's centers were to be moved.

• *Military and government policies.* In wartime and refugee situations, malevolent policies are sometimes purposefully adopted to separate children from families, as was the Nazi policy forbidding foreign slave laborers from caring for their own children. The abduction of children by military forces has occurred in many emergencies. Drafting or inducement of children into the military is also a phenom-enon in many countries and a cause of separation. Sometimes government policies purposefully cause separation, as did the decision of the Greek Government to evacuate children from contested areas during the Greek Civil War. Policies can also inadvertently cause separation. For example, immigration and refugee reset-tlement policies that provide to unaccompanied children resettlement possibilities not available to the larger population may encourage parents to separate from their children in order to give them special advantages.

• *Intervention.* Sometimes the separations are the unplanned result of the way relief assistance is provided. For example, relief workers have sometimes removed children from a dangerous area or to a medical facility without notifying the family or others in the vicinity. In the Nigerian Civil War, as in many other emergency situations, the lack of proper records or adequate documentation for children in hospitals, orphanages, and other facilities made it impossible to establish their identities, particularly for the very young children. Parents in relief centers were sometimes encouraged to place their children in special facilities where food and education were provided. In various emergencies, children with handicaps and spe-cial medical problems are sometimes separated from their families for treatment in foreign countries. Where the amount of relief goods and services provided to unac-companied children in separate facilities is far greater than that available to other children, family members may separate to allow the child access to these services. This paradox haunts child welfare efforts in virtually every emergency: the very existence of special programs for unaccompanied children may have a "magnet effect" which actually causes more separations. Often by wanting to do what seems best for the child, the parent separates from it purposefully.

• *Child's initiatives.* Parent-child separations also occur because of a child's own circumstances and wishes, with parental consent or involvement only sometimes at issue. After the Hungarian revolt, some older children fled the country fearing persecution because they opposed the political system. In that case as well as in others, some children took the opportunity to leave bad homes, to seek fortune or opportunity in other countries, or simply to follow their friends. Many times older adolescents have left home to avoid being drafted into the military, particularly when the conflict is unpopular.

Three program implications arise from review of the reasons why parents and children separate in emergency situations. First, the diversity of reasons affirm the unique situation of each unaccompanied child and reconfirms the need to document and assess the specific circumstances of each. Second, assessment is therefore required in each situation to determine the causes of separation. Third, as most of the causes of separation are related to resolvable problems, many parent-child separations can be prevented with appropriate intervention.

Advocacy or social welfare services may help prevent separations. Advocacy includes public and private efforts to change policy or administrative procedure in order to defend or protect the interests of unaccompanied children. Advocacy, for example, may be required when the causes of parent-child separations are military actions, such as abduction, government policies, or inadequate relief policies. Many of the difficulties which cause family separations are familial problems which can be influenced by social welfare services. Social welfare services include any assistance which helps the individual or family cope with the difficulties faced. Social welfare services, for example, may prevent parent-child separations by enabling destitute families to develop a means of income, by ensuring that single-parent families have necessary economic and social support, and by ensuring that various common social welfare services are available to assist individuals and families with the many problems that can lead to family disintegration.

WHEN PARENT-CHILD SEPARATIONS OCCUR

Parent-child separations occur as a continuing phenomenon in most war, famine, and refugee situations. Some unaccompanied children identified during an emergency are likely to have separated from their families prior to the emergency and others separate during an emergency. In Ethiopia during the drought of the early 1970s, children separated from families because of the famine were found together with unaccompanied children who had been living in orphanages or on the streets before the famine. Children were separated from families over the six-year duration of World War II and during the entire Nigerian Civil War. In the refugee and displaced persons camps in Thailand between 1975 and 1984, the United Nations High Commissioner for Refugees (UNHCR) repeatedly attempted to move all known unaccompanied children from the camps, but new unaccompanied children continued to appear. Separation of children also continues after an emergency, as when a war has ended, refugees have settled, and the situation appears stable. The number of institutionalized children increased for more than fifteen years after the Korean War. After World War II, programs for unaccompanied children in the displaced persons camps were required for at least seven years—new children continued to be found by the tracing programs, families continued to separate, and unwanted children continued to be born. The program implication is that services for unaccompanied children are likely to be needed over the duration of and after an emergency. This challenges the assumption of relief agencies which assume that the issue of unaccompanied children is a fleeting problem which is quickly resolved.

THE CHILDREN

Unaccompanied children in past emergencies have included both boys and girls, and have been children of all ages, from infancy to the age of majority. Yet some broad trends exist as to age group and gender. Abandoned children are often infants. Abducted children may be of any age. In World War II, for example, young children were abducted for adoption and older children for work programs. Children who have been orphaned or accidentally separated from parents, and those who have been entrusted by parents to the care of others, fall into all age groups. Most runaways are adolescents as are those youth who are living on their own with parental consent. A majority of the unaccompanied Hungarian, Cuban, Lao, Cambodian, and Vietnamese children who crossed national borders as refugees were adolescent and male. Abandoned infants after the Korean war were predominately female. The age span of unaccompanied children in emergencies confirms that a range of services may be required to meet the needs of children of differing ages. As is well known, children of different ages have different needs. The needs of children, based on age-related developmental differences, are elaborated in Part II.

LOCATING THE CHILDREN

If a search for unaccompanied children were to be carried out in an emergency situation, where would they be found? Experience confirms that many, a majority in most situations, are likely to be living with other families—in the homes of brothers and sisters, grandparents, aunts and uncles, and other relatives; in the homes of neighbors and family friends; and in the homes of new acquaintances. Some, however, are found in hospitals where they are brought for treatment, abandoned or surrendered at birth, or left alone by the death of the parent who accompanied them there. They are also found in any facilities providing residential care for children, such as orphanages, children's centers, group homes, children's villages, and boarding schools. They may be at any other facilities where food and/or temporary shelter are provided, such as relief and feeding centers, welfare offices, and religious establishments. Sometimes unaccompanied children have been found in jails, prisons, detention centers, and around military camps. They may also be found in public places, along roads, in markets, in forests, or on the streets in larger cities. Two implications arise from the fact that unaccompanied children in an emergency may exist in any or all of these circumstances. First, a determination of the existence of unaccompanied children must include surveys of all the above-mentioned places and any others where they might have taken refuge or have been placed. Second, all services that may assist unaccompanied children should be provided with technical support and training to ensure that the special needs of the children are adequately met. For example, in the Korean War, as in many other emergencies, active family tracing was not provided for children in many institutions.

UNACCOMPANIED BUT NOT ALONE

In past emergencies children, at the time they are identified as being unaccompanied, had been separated from parents for greatly varying lengths of time—some only for days; others for many years. Some, such as street children, maintained contact with their parents who lived nearby; others may have been separated before the onset of the emergency and may have had no contact with their family for periods longer than the emergency itself. Some unaccompanied children identified after World War II had been separated for seven years or more. Interviews with unaccompanied Khmer children in the camps in Thailand in 1980 revealed that they had been separated from parents for periods ranging from a few days to more than five years. When unaccompanied by parent or guardian, children typically establish relationships with other individuals. In every emergency examples abound of unaccompanied children who, without outside help, integrated into new homes and established intimate long-term relationships with nurturing caretakers, who may have been relatives, neighbors, old friends, or new acquaintances. Strong peer relationships also developed. After World

War II, for example, groups of unaccompanied children who had survived extreme hardship through mutual support were identified, and subsequently resisted every effort to separate them from each other. Irrespective of the length of time a child might have been separated from parents or guardian, documentation, family tracing, and assessment of the potential for family reunification is required. However, the longer the separation from the family, the more likely that a new relationship will have developed which, from a psychological perspective, becomes the primary criterion for determining family reunification or alternative care.

UNACCOMPANIED CHILDREN AT RISK

Left on their own, unaccompanied children have usually survived by taking advantage of new opportunities. As noted above, they have often been taken into homes or services. Many, however, were not so fortunate as to find circumstances which met their needs and protected their rights as individuals. They lived in situations in which they were very unhappy—in displaced persons camps or centers, in homes willing to provide only temporary accommodation, and in extreme situations, have been abused or exploited for the labor they could provide. Some unwanted children have been passed from one family to another and others have lacked such basics as food, clothing, shelter, medical care, and education.

Unaccompanied children are uniquely at risk during emergency situations. Malnutrition and preventable diseases have caused the deaths of millions of children in past emergencies—the unaccompanied child is the most vulnerable. A study of abandoned infants less than three months of age at a Khmer refugee border camp in Thailand found an unusually high mortality rate, more that twice that in any developing country.[3] Survival also depends upon adult protection in bombings, emergency evacuations, and famines. Sometimes protection is required to prevent children from being abducted or forced to participate in military actions. In Thailand in 1981, UNHCR had to move unaccompanied children from one displaced persons center to another to protect them from being recruited forceably to fight with the guerrillas. In some situations unaccompanied children have been violently treated or sexually abused. After the 1980 Cuban Mariel exodus to the United States, some unaccompanied Cuban minors lacked adequate protection and were raped in detention centers where they were being held with Cuban criminals. Children must also be protected against actions taken presumably on their behalf that may, in fact, violate their basic rights and interests and those of other family members. In many past emergencies unaccompanied children required protection when well-intentioned intervenors, acting only on their own authority and without having made an adequate assessment of each child's circumstances, attempted to carry the children away. After World War II intervention by the United Nations Relief and Rehabilitation Agency (UNRRA) was necessary to prevent independent agencies from moving unaccompanied children to other countries until the circumstances of each child had been determined. During the Nigerian Civil War, some children were placed on planes and sent out of the country without parental approval. In the United States during 1975, people reportedly went into the reception center for Vietnamese refugees at Camp Pendelton and simply removed unprotected, unaccompanied children. In the same way, in the first weeks of the Kampuchean crisis, people visited the Sakaeo camp in Thailand and, without authority or an assessment of the children's eligibility for adoption, removed children assumed to be orphans. In other situations, children have been released to foster families without consideration of the children's wishes, without verification of the claimant's identity, and without home studies done to determine the suitability of the placement. These are but a few of many experiences that substantiate the need for special protection of unaccompanied children in emergencies.

There are two implications arising from the chance arrangements unaccompanied children typically exist in. Experience has shown that it should not be assumed that unaccompanied children have by chance found situations that meet their needs or protect their rights. These vulnerable children should be identified and their situations assessed whether they live alone or with families or in institutions. Any necessary intervention must consider the psychological relationship that may have developed between a child and his new caretakers. Such psychological considerations are discussed in more detail in Part II of this book.

THE EVACUATION OF CHILDREN FROM EMERGENCY SITUATIONS

Organized programs to move children away from emergency situations have caused many parent-child separations in the past. Such children can be described as *entrusted*, usually to authorities or agencies. Selected examples of evacuations are given in Table 10-4. Most of the children evacuated were between the ages of five and fourteen, although there have been some as young as one year and others as old as eighteen years. Evacuations have taken place for at least three reasons: as a preventive measure, before there was any immediate threat to the children's lives; for rescue, when the lives of the children were endangered; and for support, when the lives of the children were not actually threatened but to provide the children with special assistance.

Table 10-4. Examples of Past Evacuation Programs

	Number of Children Evacuated	Age Range	Estimated Number Reunited	Estimated Length of Separation
Spanish Civil War to other countries	23,000	5–12	18,000	months—permanent
World War II within Great Britain	730,000	5–14	730,000	months—7 years
English children to other countries	13,000	5–14	13,000	5–6 years
Finnish children to Sweden	69,000	1–14	52,000	4–6 years permanent
European children to Palestine	15,000	6–18	unknown	permanent
Greek Civil War within Greece	14,500	3–14	11,500	months—3 years
Nigerian Civil War	5,000	1–7	4,500	1–3 years
Cuban Refugees to the U.S. (1960–67)	15,000	6–18	unknown	1–7 years
Vietnam War (Babylift) to the U.S. and England	2,500	1–12	less than 10	N/A

(These may be classified as preventive evacuation, rescue evacuation, and supportive evacuation.)

Evacuation has also been used as an emergency measure to protect children already separated from parents and indeed is recommended for unaccompanied children in article 24 of the Fourth Geneva Convention, as discussed more fully in Part III. Children evacuated from Nigeria and Vietnam during the wars in those countries, for example, came predominately from orphanages, hospitals, and feeding centers.

Nine observations emerge from a review of these evacuations. First, the evacuation of children differs from other forms of separation in that it is usually organized by a well-intended third party. In most of the situations mentioned in Table 10-4, parents (or guardians) voluntarily agreed to the evacuation of their children and signed consent forms. Although voluntary, these separations were often promoted by the organizers. During the Spanish Civil War evacuation organizers produced a film and extensively used the media to generate public support. The Greek and English evacuations were organized and encouraged by the respective governments. The initial impetus for the evacuation of the Spanish and Finnish children came from small groups of concerned foreigners who first mobilized public support in their own countries, and then gained support from officials of the host government, who in turn, encouraged parents to consider evacuation. Therefore, while parental concern for the well-being of the children has been decisive in evacuating children, evacuation organizers and their enticements and encouragements to parents to send their children often played an important role.

Second, past evacuations have often been initiated and carried out by organizations without child welfare or placement experience. In the Spanish Civil War, a variety of agencies, including short-term voluntary agencies, churches, political parties, and trade unions evacuated children and cared for them. In Great Britain in World War II, civil defense planners organized the evacuation of children. The evacuation of children within Greece during the Greek Civil War was a government/military operation in which a newly established social welfare organization cared for the children. In the Nigerian Civil War, the evacuation of children was implemented by relief agencies. A licensed child welfare agency was involved in the evacuation of the Cuban children and in the Vietnamese babylift operation. Common child care and protection services normally provided to children away from the care of their families were many times not provided. Organizers thought of an evacuation as if it were only a logistical operation, a civil defense exercise, a medical operation, or a summer holiday trip for children. Child care issues were usually not adequately considered.

Third, the length of separation was much longer than parents, children, or organizers expected. Most evacuations were planned as short excursions for children and the separation was expected to end within several months. Some children were in fact reunited with families soon after separation, but in at least five different evacuations the general length of separation is estimated to have been between two and five years. (See Table 10-4) Approximately twenty percent of the Spanish and Finnish evacuees never returned.

Fourth, past evacuations have also consistently confirmed the difficulty of ensuring adequate care for children once they are separated from their families. Parents assumed that evacuation organizers would take good care of the children but this did not always happen. For example, the Spanish children evacuated to Mexico suffered many hardships in a poorly planned and executed program, although well-intentioned and labeled a "model" by organizers. In England, the initial reception program and at least one institutional placement proved unsatisfactory. Review of the placement of the Basque children, as with other evacuated children, confirms that some children were very unhappy in the placements provided, whether they were in foster care, group care, or institutions. In the Spanish Civil War and in others, the care of evacuated children was usually planned and carried out on the assumption that the children would within several months be returned to their families; therefore only short-term arrangements were necessary. As the period of care was much longer than expected, the initial programs often proved inadequate for longer-term care.

The children's reactions to their experience depended on the kind of care and placement they received. From the information that is available, most children apparently look back on their experience positively. For some, however, as for those evacuated in the Spanish Civil War, the evacuation was an unhappy experience and the lack of loving care was frequently recalled.

> The youngest children sent into Belgium and Mexico tended to suffer such trauma, commenting that they almost 'died of fear,' or suffered from a lifelong lack of confidence, or cried for weeks, or still have a feeling of insecurity, or vomited for years, or lacked parental counsel when entering young manhood or womanhood. Some few still live at home or with siblings, being unable to live independently.[4]

The separation of siblings usually has had a devastating effect on children. Lagaretta also noted that the initial reception was important to the children because it conveyed a lasting impression as to their welcome. It was also helpful to the children if attention was given to maintaining the exiled children's fluency in their native languages, and supporting their cultural traditions and folklore, and also continuing their traditional religious practices. Receiving familiar foods contributed to a sense of well-being.

The limited financial resources of the organizing agencies often limited services provided to the children. Program funding was usually based on the assumption that services required for the evacuated children would be only temporary. As the period of care extended, many agencies had difficulty finding additional sources of funds and had to reduce services.

Fifth, the evacuation of a child sometimes led to changes in the relationship between parent and child. Some parents lost interest in the child, or were happy for the child to remain in the care of the evacuating organization or with the foster family. On their part, children established new relationships with their caretakers and sometimes had no desire to return home. Even though Swedish foster parents had signed forms stating that the evacuated Finnish children could not be adopted, some filed petitions for adoptions after the children had integrated into the family, and in several cases the courts honored the petition to protect the relationship that had formed. Others who returned home found adjustment difficult. Some had become accustomed to different foods and better living conditions during evacuation than they were to receive at home. They had not shared the common difficulties of other family members and found their place in the family altered by siblings born in their absence. Such changes make reintegration difficult for some children.

Sixth, although evacuation organizers usually tried to establish records and maintain communications between children and their families, they could not always do so. Logistical requirements were often underestimated and many unanticipated problems arose, such as: inadequate documentation of the children, confusion when responsibility for the children was transferred from one agency or family to another, loss of records, change of placement of the child after separation, change of location of the parents, and disrupted communications. In Nigeria, as in other emergencies, young children were evacuated without proper documentation. In the Vietnamese babylift there were reports of missing records, loss of identity tags, and confusion as to children's identities.

Seventh, unexpected political complications have prejudiced the outcome of evacuations. Children evacuated to Mexico during the Spanish Civil War could not return to Spain because Mexico did not recognize the new Franco Government. Children who returned to Spain as adults after having been evacuated to Russia were considered politically suspect and were denied equal opportunity and treatment as citizens. In many evacuations, whether the children were invited into a country, and when and if they were returned home, became national political issues.

Eighth, evacuation sometimes led to changes in ethnic identity, religion, and language. Evacuated children, as might be expected, adapted to their new environment by learning the language, accepting the social customs, and even adopting the religion of their caretakers. The children's age, length of stay, and living conditions influenced such changes. As in most evacuations, some children sent to neighboring countries in the Nigerian Civil War returned home unable to speak their native language because their caretakers had belonged to a different ethnic group.

For many reasons the separation of children from their families for evacuation can NOT be recommended. However, although evacuation may be traumatic for a child, it may not necessarily result in disability. Lagaretta in her study of children evacuated in the Spanish Civil War concluded that a stay of from three to six months had no more effect on the adult lives of the children than that of a strange holiday.[5] For those who had been evacuated longer, she found no obvious traumatization for the majority of those interviewed as adults. This was generally found to be true of the English and Finnish children who were evacuated.

INTERVENORS: THOSE WHO MAY ASSIST

In the absence of the family, the care and protection of unaccompanied children becomes a public responsibility by conscience and law in most societies. Many benefactors intercede on behalf of the children in emergencies, both from within and outside the affected community. In emergency situations outside intervenors often assume that local intervenors will be overwhelmed by the crisis and thus be unable to provide services to unaccompanied children. This has not, however, been substantiated in past emergencies. In all emergencies reviewed most of the assistance provided to unaccompanied children has come from local and national intervention. Outside intervenors, for their part, have also contributed substantially. In virtually all major emergencies international assistance for unaccompanied children has arrived in the form of money, relief supplies, and personnel.

INTERVENORS FROM WITHIN THE LOCAL COMMUNITY

Family and Relatives

Experience has shown that when children are separated from their parents they are most likely to be cared for by other family members—adult brothers and sisters, grandparents, and aunts and uncles. In Guatemala, for example, a sampling of eighty-eight children who remained in their home villages after being orphaned by the conflict of the 1980s revealed that ten (11%) were cared for by older siblings, forty-two (48%) by grandparents, fourteen (16%) by uncles, and the remaining 25% were cared for by neighbors or left the community.[6] Families often consider themselves responsible for unaccompanied children. Some legal systems and religious traditions require relatives to take responsibility. In the Guatemala highlands orphaned children traditionally go immediately to their grandparents' home. On the nineth day after the parental funeral, the remaining family members meet to decide with whom the children should live.

Unrelated Individuals and Families

In virtually all emergencies individuals and families have spontaneously provided assistance to children other than their own, even in the face of danger, scarcity, and risk. In Holland during World War II, Dutch families hid and sheltered at least 15,000 Jewish children. In the mass evacuation of northern France in 1940, individuals who were themselves fleeing, searched for and assisted unaccompanied children. In refugee and displaced persons camps, families have adopted and cared for unaccompanied children. In the holding centers in Thailand, a study of the Kampuchean families who were caring for children other than their own showed that the majority of foster care arrangements had taken place spontaneously during the crisis. Of the eighty-two families interviewed in that study, fifty-one percent had not known the child before initiating foster care.[7] Even in countries where foster care is not customary, such as Nigeria, families voluntarily assumed responsibility for children when foster care programs were organized after the civil war.

Social Welfare and Court Services

Community social welfare and court services usually have legal authority for programs for unaccompanied children. When they have not themselves provided direct services, agencies have delegated the responsibility while usually maintaining at least a supervisory role. In the Spanish Civil War, the social welfare department of the Republican government established "children's colonies," assisted needy families, and encouraged the evacuation of children. During the Vietnam War, the Republic of Vietnam operated most of the orphanages in the country. After the war in Nigeria, the national social welfare department carried out a large-scale tracing and family reunification program.

Independent Group Efforts

Privately organized efforts have played major roles in assisting unaccompanied children in past emergencies. Such assistance has often come from private hospitals, schools, child welfare institutions and agencies, churches and temples, and social and relief organizations, such as the National Red Cross societies. Some provided assistance as an extension of already existing child welfare services, while others hastily established services expressly for unaccompanied children. Throughout Europe during and after World War II, and similarly in other emergencies, private initiative led to establishment of programs, children's centers, orphanages, and schools to provide for children generally and unaccompanied children specifically.

INTERVENORS FROM OUTSIDE THE LOCAL COMMUNITY

In most large-scale emergencies, individuals, independent agencies, and international organizations from outside the affected community have also helped provide services for unaccompanied children. Sometimes outside intervenors have been called in by local or national authorities, but most often the agencies initiate assistance. A review of past experience confirms that various types of intervenors from outside the victim community usually offer assistance.

Individuals

Individuals acting on their own initiative often go to areas affected by emergencies and offer to assist. They usually give constructive help to established organizations and programs. There are, however, in most emergencies, examples of individuals working independently of all established programs, and contrary to child welfare principles.

Independent Agencies

The number of private international voluntary agencies providing assistance in emergencies is steadily increasing. At the time of the Russian Revolution and subsequent famine, for example, very few international agencies existed. By the end of World War II, there were approximately fourteen private international organizations working with unaccompanied children in the displaced persons camps in Europe. More than 60 private voluntary agencies have assisted in each major emergency since 1970. Most do not generally provide services explicitly for unaccompanied children, although many may have done so in one or more emergencies.

International Organizations

The United Nations relief organizations were established after World War II. The United Nations Relief and Rehabilitation Agency (UNRRA), was founded to assist after the war with repatriation, relief, and rehabilitation and was the first U.N. agency to offer assistance to unaccompanied children in those years. By 1948 the problem of unaccompanied children was well recognized and was specifically mentioned in the mandate of the International Relief Organization (IRO), successor to UNRRA. In 1952 the United Nations High Commission for Refugees (UNHCR) replaced the IRO, and although unaccompanied children are not singled out in the UNHCR mandate, they were mentioned in its founding conference. The organization has interpreted its general protective responsibilities for refugee populations to include unaccompanied children. The United Nations Children's Emergency Fund (UNICEF) was created in 1946. UNICEF's founding statutes also do not specifically mention assistance to unaccompanied children, but such aid is considered within the general mandate of the organization's work with children. By and large, UNICEF has restricted its efforts to the medical, nutritional, and educational areas and has not, until recently, provided special services to unaccompanied children. (The legal bases for UNHCR and UNICEF actions for unaccompanied children are discussed in Part III.) Other UN bodies, such as the United Nations Educational and Cultural Organization (UNESCO) and the World Health Organization (WHO) have, since World War II, occasionally provided specialized assistance to unaccompanied children.

The International Committee of the Red Cross (ICRC) has throughout its history many times given special assistance to unaccompanied children by providing material goods, initiating protective actions in dangerous situations, and tracing and facilitating family reunification. For example, in the Spanish Civil War, ICRC arranged for the family reunions of children who were stranded by the sudden beginning of the war. After World War II it helped trace missing families or children in many countries. After the Greek Civil War the United Nations General Assembly asked the ICRC in association with the League of Red Cross Societies to help return abducted Greek children to their homes. During the Nigerian Civil War, it was responsible for the coordination of all relief including assistance to children. Although it has provided specialized assistance in some emergencies, by mandate ICRC assists only in armed conflicts, usually limiting its services to the emergency period (except for central tracing services). ICRC has not provided specialized assistance to unaccompanied children in the way of care or placement, nor like UNHCR, UNICEF, or other UN agencies, does it assume legal responsibility for unaccompanied children.

PROBLEMS WITH ASSISTANCE

The substantial efforts made in past emergencies must be examined in the context of actual needs. In many emergencies timely assistance has been provided, but in most, the assistance has been inadequate, late, and sometimes misguided. Many unaccompanied children have received basic assistance, but tens of thousands have died for lack of care, food, clothing, or shelter. Many unaccompanied children have been reunited with their families, but they are only a small percentage of those who could have been reunited. Some have received interim or long-term alternative care that has met their needs, but hundreds of thousands of others have lived as street children or have been placed in institutions or otherwise never received care adequate for their developmental needs. Sometimes efforts to provide care have only increased the trauma, as when children have been held in interim situations for extended periods, or passed from one care-giver to another, or separated from the few people to whom they had been attached. Some children have been provided with assistance according to an established legal and child welfare system, but in many situations, existing laws relating to the care and protection of children have not been observed and the rights of the children and families have not been respected.

There are unaccompanied children in nonemergency times as well as during emergencies and all nations have formal and informal legal and social welfare systems to provide for children separated from their parents. As mentioned above, local, national, and international emergency relief assistance is usually provided to children during and after emergencies. In light of the existing structures and substantial efforts made to assist unaccompanied children, why then have the needs so often not been met?

To begin with, in many past emergencies the local social welfare systems have proven ineffective. In some situations the exigencies of a large-scale emergency exceeded the resources of the established services, particularly when the number of unaccompanied children increased by thousands in a very short time. Sometimes local systems were not effective because other pressing needs were given priority over those of unaccompanied children. In other situations, the social welfare and legal procedures that would otherwise have been used to evaluate the circumstances of each child and assess the interests of all parties were disregarded as superfluous. In wars, the legal and social welfare services often gradually disintegrate, particularly on the losing side.

Secondly, in refugee situations unaccompanied children are away from usual community services. Legal and social welfare structures in the communities from which the refugees came do not ordinarily survive in refugee camps and temporary settlements. Displaced persons therefore have depended upon others for such services or for opportunity to recreate their own. When unaccompanied children have crossed national borders, the host country determines services for their care and protection. When the host country has opposed their arrival, alien children have often been denied access to local child welfare services, and consequently, routine child welfare procedures for care and protection have not been followed. This has been and continues to be a problem in many first-asylum countries. Thus, unaccompanied refugee children often live in a care and protection vacuum: separated from home area services, but not eligible for or unprovided with local services.

Thirdly, because systematic searches for unaccompanied children have not been routinely carried out in emergency situations and because the children's circumstances have not been documented and evaluated, programs have been based largely on conjecture and assumptions, often fallacious. Relief services have not anticipated large numbers of unaccompanied children in emergencies and have therefore often been late in providing aid. The needs of the children were often assumed to have been met locally by existing relief measures with the result that assistance to the affected community has often been inadequate. The unaccompanied children have been commonly assumed to be orphans or abandoned; as a result, efforts to find and reunite the children with their parents have been lacking. The general public, sometimes misinformed by the news media, has assumed that unaccompanied children are infants, and relief agencies have sometimes assumed them to be self-sufficient adolescents. In both cases, actions inappropriate to children's ages have followed. Programs for unaccompanied children have been established on the assumptions of recent and traumatic separations from parents. Intervenors have not realized that some may already have been apart from parents for years because of reasons unrelated to the emergency. They have also assumed that required assistance should be short-term rather than a continuing social service.

Fourthly, in many emergency situations the assistance to unaccompanied children is inadequate because there is no program for meeting the developmental needs of the children. Children receive food, clothing, and shelter, but not the equally essential care and nurturance necessary for normal growth and development.

Fifth, because a systematic and coordinated approach to the care and protection of unaccompanied children has been lacking, programs have developed largely by chance. For example, if someone within the situation was committed to family tracing and worked to establish it, then families were traced. On the other hand, if the interest of the intervenor was in establishing orphanages, than tracing may not have been attempted. In the complex and uncertain environment of large-scale emergencies responsibility for the care of unaccompanied children is often ambiguous. To a large extent, programs have been undertaken by whatever agencies or persons happen to have been involved in the emergency, not necessarily by those with experience and expertise in the care of children.

Sixth, usefulness of outside assistance has been limited by the lack of programmatic guidelines for international organizations and voluntary agencies. As a result, these organizations have not provided systematic assessments, guidance, and coordination to solve the particular problems of care and protection for unaccompanied children. There are noteworthy exceptions in which major efforts were made on behalf of unaccompanied children, such as UNHCR's assistance after the Hungarian uprising in 1956 and its activities during the Kampuchean crisis in the early 1980s. In many situations, however, assistance has been less than adequate, and coordination and continuity have faltered. There is presently no international organization to act for unaccompanied children in all emergencies: UNHCR provides assistance only in refugee situations; UNICEF has until recently largely restricted itself to providing food, clothing, and medical care; and ICRC is limited to conflict situations and does not provide child welfare services. In some regional emergencies this means that international support for unaccompanied children has existed on one side of a border but not the other. Sometimes international organizations have actually contributed to the institutionalization of children by funding only orphanages or children's centers while withholding support for other forms of care.

The failure to satisfy the needs of unaccompanied children is due, in part, to the staffing and orientation of the intervening agencies themselves, both internationally and nationally. Some agencies working with unaccompanied children have neither emergency nor child welfare experience. Although others have had emergency experience, they lacked the orientation and background of working with children in general and unaccompanied children in particular. Some agencies with adequate child welfare expertise were unfamiliar with the special problems of assisting children in emergencies. Furthermore, other groups offered only narrowly defined, predetermined services such as intercountry adoption or the establishment of children's villages or orphanages. Many programs for unaccompanied children lack such essential services as tracing.

The absence of experienced staff and the lack of special training also limits the effectiveness of programs for unaccompanied children, despite the good intentions of the agencies. Agencies often claim to have worked with unaccompanied children for years, yet individual staff members may lack personal experience. Few agencies employ regular child welfare specialists and even fewer have provided preparatory training or technical support in child welfare issues to field staff. Agencies are inadequately prepared to establish emergency care procedures, assessment techniques, tracing methodologies, and verification procedures, nor are they prepared to establish more general social services. Seldom have emergency plans included contingency measures for the protection of unaccompanied children. Where agencies ahve provided exemplary services, it has often been because particular staff members were committed to and knowledgeable about unaccompanied children. However, unless these individuals are present in later emergencies, the agency's expertise is often lost.

Decisions about unaccompanied children are many times shaped as much by extraneous influences as by the needs of children. The actions of intervenors often reflect different philosophies, political preferences, religious beliefs, military priorities, or cultural biases. Rather than focusing on the genuine needs of unaccompanied children, programs are sometimes based on donor and agency interests. For example, orphanages are sometimes built because donors have contributed money for this purpose, not because orphanages best meet the needs of children. In the past, short-term crisis intervention measures have often become permanent because of the self-perpetuating nature of organizations themselves. This was clearly demonstrated after the Korean and Nigerian civil wars when orphanage directors opposed the closing of their institutions. Political factors sometimes prevent the reunion of unaccompanied children and their families. After World War II, for example, the U.S.S.R. refused to permit UNRRA to search for unaccompanied children in the Russian zones of Austria and Germany. After the Greek Civil War, Eastern European countries refused for years to permit the return of children confirmed by the Red Cross to have families in Greece. The establishment of effective programs for tracing the families of unaccompanied children has not been permitted in Kampuchea, and political considerations have prevented the reunification of children claimed by parents.

Four recent international developments likely to contribute to an improvement in the care of unaccompanied children deserve mention. In 1982 UNHCR proposed guidelines for the care of unaccompanied children in refugee situations in the "UNHCR Handbook for Emergencies." These guidelines were amplified further in the "UNHCR Handbook for Social Services" published in 1984. In 1986 UNICEF published "Assisting in Emergencies: A Resource Handbook for UNICEF Field Staff," which includes guidelines for the care and protection of unaccompanied children. Also in 1986, the UNICEF Executive Board adopted resolutions which sanction UNICEF program support for "children in especially difficult circumstances" and recognized unaccompanied children as one such group.

II

UNACCOMPANIED CHILDREN FROM A PSYCHOLOGICAL PERSPECTIVE

Central to a discussion of unaccompanied children in emergencies are issues arising from family separation and loss. Desertion, loss, and separation in infancy, childhood, and adolescence constitute universal themes for subsequent grief, despair, and bereavement. Yet, separation or loss never occurs in isolation of other factors, especially in emergencies, where events of extraordinary intensity—violence, hunger, destruction, deprivation, and enormous social change—can compound the initial trauma of family separation or loss and lead to far more suffering. Conversely, during emergencies, there are also factors which can protect children from developmental harm; chief among these is the presence of family, other trusted adults, and to a lesser extent, peers.

Thus, while family separation or loss is indeed the central issue, any assessment of psychological risk needs to be multidetermined. First, one must begin with the children themselves as previous family backgrounds, cultural experiences, temperamental and personality characteristics, and most importantly, age at the time of separation or loss will determine to varying degrees how a particular child is affected by, understands and responds to, the stress of family separation or loss.

Secondly, close attention must be paid to the kinds, numbers, and length (acute-sustained) of adversities the child is subjected to after the initial separation or loss and before reaching the "safety" of the relief community. In some cases, unaccompanied children have immediately received adequate adult care in reasonably stable communities. Here, disruptions were kept to a minimum and children's outcomes were generally favorable. In other cases, however, family separation and loss has occurred against a backdrop of concurrent stress, trauma, and deprivation. The far greater adversities these children were subjected to resulted in more serious adjustment problems and often in long-term psychological disorders.

Finally, an analysis of the relief community's response to unaccompanied children must be included. As detailed in Part I, initial assistance for unaccompanied children has been both absent and present, and when present, effective as well as ineffective. Further, unaccompanied children have been cared for in foster and adoptive families, group homes and institutions, which have been located both within their original communities as well as within different communities and nations. The kind, quality, and cultural setting of these placements significantly affects the adjustment of each child.

With this basic outline in mind, Part II of this book is divided into five chapters:

Chapter 11 provides a broad overview of children's psychological and social development as it is shaped by family and cultural influences. It is written primarily for individuals who work directly with unaccompanied children, or who formulate policy on their behalf, and yet hold expertise in areas other than child welfare. This overview serves as a baseline for understanding the different psychosocial needs, capacities, and limitations of children of various ages as well as for the psychological struggles of unaccompanied children in emergencies.

Children's responses to natural disasters, wars, and refugee movements are examined in Chapter 12. This review indicates that children are usually able to endure the emotional stress and physical hardships of most emergencies provided they remain with their families and their parents are able to continue caring for them in their accustomed manner. Often, too, the community itself can afford children the kind of familiarity and predictability essential in helping them cope with the misfortunes that arise during emergencies. All emergencies, however, become enormously significant the moment separations occur and children's primary attachments to the family and, in some cases, community group are disrupted.

Chapter 13 focuses specifically on unaccompanied children during war and refugee situations. Distinctions are made between unaccompanied children who immediately received adequate adult care in reasonably stable communities, and those whose separation experiences were compounded by other stresses and adversities. Often it is this distinction, more than the initial separation or loss *per se,* that determines the subsequent psychological fate of an unaccompanied child and, correspondingly, should be a central consideration in formulating intervention strategies and selecting substitute placements.

Chapter 14 looks at family reunion from what is referred to as a "child's sense of time." This review suggests that the significance of separation, from a child's perspective, will often depend upon its duration as well as the developmental stage during which it occurs. In general, the younger the child the more quickly separation will be experienced as a permanent loss accompanied by profound feelings of deprivation. In the same way, the length of time the child is away from his natural parents and in the care of another adult can significantly affect family reunion efforts. In addition to time, family reunion may be further complicated by sociocultural factors, poverty, and changes in family membership.

In the past, unaccompanied children have been placed cross-culturally, usually through international adoption and refugee resettlement programs for displaced children. Chapter 15 thus begins by reviewing literature on transcultural adoption, attempting to distinguish findings on children's adjustments generated from more systematic research from ideological arguments for and against this practice. It concludes with a review of recent resettlement programs for unaccompanied children from Southeast Asia. Much of this discussion is based on two background papers prepared for the larger study by Jan Linowitz, who remains the lead author of this chapter.

11

Normal Patterns of Psychological and Social Development

Children are different. From birth onward, they interact with their environment on the basis of innate individual differences which lead to countless variations in character structure and personality development. Yet, whatever these differences, children of similar ages are also very much alike. The concept of "developmental lines" is based on the understanding that human development unfolds according to a series of internally structured stages which are related to age.[1] Maturation from infancy, through childhood and adolescence, brings with it many predictable changes and interrelated attainments. As children grow up, they generally become physically stronger and more adept, are better able to tolerate frustrations and understand events, and are more competent in dealing with people and things.

Any forward movement along the developmental continuum towards maturity, however, depends on a number of favorable "external influences."[2] All children born with normal potentials need sufficient bodily care, human affection, and intellectual stimulation on an ongoing basis if these potentials are to be realized. In most societies, the family is recognized as being the fundamental unit most capable of providing for children's physical and psychological needs on a long-term basis.[3]

Families, both in function and composition, differ between cultures and according to the demands of the particular environment. Families, for instance, may be nuclear or extended and consist of blood-related or adoptive members. Nonetheless, in all societies the adult-child relation within the family is distinguishable in at least two fundamental ways.[4] First, adults nurture and safeguard their children, command and discipline their behavior, and serve as essential emotional attachments for them. Secondly, and this is partly connected to the first statement, adults hold the position of social authority in this relationship and assume responsibility for the transmission of social rules and values to their children.

While there are basic commonalities that characterize family relationships all over the world, there are also significant differences. Normally, parents guide their children's behavior in a range circumscribed by their community, society, heritage, and physical environment.[5] Parents, in other words, raise their children in a manner that ensures their eventual adaptation to the demands of a given society. These large-scale, systematic differences in child-rearing and socialization patterns lead to noticeable differences in children's behavior.

Corresponding to the theme of developmental lines, this chapter is divided into four main subsections—infancy, early childhood, later childhood, and adolescence—which represent rough approximations of what can be considered "normal" or "expected" stages of psychological and social growth during these years. Each subsection begins with a brief examination of the major psychosocial accomplishments of the particular developmental stage being described and then moves to a discussion of how these accomplishments are linked to family and cultural influences. Again, as noted earlier, this chapter is illustrative rather than exhaustive; it is an effort to provide a basic understanding of age-specific needs, capacities, and limitations which is essential for individuals and agencies assuming responsibility for children unaccompanied during an emergency.

INFANCY: FAMILY ATTACHMENTS—A SECURE FOUNDATION

An infant is both dependent as well as potentially capable. He is dependent in the sense that he is unable to feed himself or care for himself in any effective manner whatsoever. Yet he is capable in the sense that, given a responsive family setting, he can and does perform actions such as crying, sucking, grasping, and reaching that elicit nurturing responses in his family members, thereby ensuring his growth and survival. Babies are also capable in the sense that they are born with differing innate characteristics which help shape how others respond to them. A parent, for instance, may respond quite differently to an infant who, for largely biological reasons, is active and cries more than an infant who, in comparison, is passive, quiet, and more content. Thus, while "parenting" behavior is in part determined by the parent's personality, past experiences as a parent, and cultural heritage, it is also in part determined by the unique characteristics of the child himself. It is in this sense, then, that one speaks of the "give-and-take" or reciprocal nature of the parent-infant relationship.

This early reciprocity, however, needs to be put into proper perspective. While the infant does make a contribution to the relationship, the adults, obviously, exert far more influence and bear far greater responsibility than the child. It is highly unlikely, for example, that a neonate possesses conscious awareness that his actions will lead to an intended goal. Rather, these earliest actions are largely *reflective* as opposed to *intentional* behaviors, and it is the well-timed, carefully coordinated responses of the parents that provide the social context through which an infant is eventually able to realize his own intentionality. Thus, while an infant is indeed capable and active in one sense, he cannot, as Winnicott suggests, "exist alone, but is essentially part of a relationship."[6]

From the child's perspective, neither blood ties nor the mere provisions of food, shelter, or physical protection are, in and of themselves, sufficient fuel for adequate psychosocial growth. Rather, a child's emotional and mental development is inevitably linked to the quality of adult care and support he receives. Indeed, Erikson suggests that the most critical accomplishment to emerge during infancy is what he refers to as a "sense of basic trust"; a general feeling that one's needs will be met and that the world is a safe place to be.[7] A child's sense of trust, or lack of it, is essentially the result of absorbing and integrating into the infant's own rudimentary sense of self the expressive responses and caring attitudes of his parents during the first year-and-one-half to two years of life. Here, Erikson stresses the unconscious, affective aspects of the adult-child interaction. The more the child is responded to in a loving and affectionate manner with concern and commitment for his own individual well-being, the more likely he will enter the following stages of childhood with a fundamental sense of security and hope. In contrast, a child who has not had the advantages of warm, devoted parents, and has not internalized a sense of being loved, valued, and wanted, tends to emerge from infancy feeling insecure and pessimistic about what the future will bring.

Rohner's cross-cultural study has documented some of the advantages of a child gains from an internalized sense of being loved and wanted.[8] Rohner found that children who are accepted by their parents were reasonably self-assured, self-reliant, and were often compassionate and responsible towards adults and peers. In contrast, children who were not accepted by their parents tended to be critical of themselves and others, more aggressive in their social relationships, and appeared to struggle over feelings of wanting but being unable to reach out emotionally towards others.

"Attachment" is a term many social scientists use when referring to the psychological bond between a child and his parents, while "attachment behavior" is often used to describe how this bond evolves. Attachment behavior begins at birth as a "dialogue," so to speak, between an infant and his parents. As an infant signals his needs to be fed, held, comforted, and stimulated, and his parents respond accordingly, the infant slowly begins to make a connection between his signals and his parents' responses. Long before this connection is possible, however, the infant must have had experiences in which his needs have been consistently met. Since there is no single act that guarantees an attachment will be made, the minimum requirement for its development is prolonged intimacy with at least one nurturing person.[9]

Obviously, the parents' interpretation of the infant's signals will not always be the correct one. The critical point, however, is that they have responded to the infant's signs in an individualized way. In doing so, the parents provide both the meaning for what would otherwise remain merely random behaviors as well as the motivation for further and eventually more selective signals. It is, thus, over time and through intimacy that an infant's instinctual drive for bodily satisfaction is drawn, for the first time, towards another human being. What the infant now brings to this relationship is not solely his need for physical care but his emotional demands for affection as well.

Normally, infants make their first clear-cut attachment sometime around the sixth or seventh month of life. There appear to be two basic characteristics that distinguish this attachment from earlier responsive behaviors. First, attachments are age-specific. While younger infants are able to direct their vision, reach out, and smile towards others, it is not until around the sixth month that they are able to clearly distinguish one person from another. Consequently, before this time, infants respond to their parents largely on a biological or need-fulfilling basis. Once, however, an infant has clearly discriminated his parents from others and has formed an attachment, his psychological as well as physical well-being depend in many essential ways on close and continuous contact with them. Secondly, attachments are person-specific. Once they are made, an infant tends to express positive attitudes towards and becomes more expressive with his parents than with others, and may react with fear or protest when he is approached too suddenly by a stranger. Children between the ages of six and twenty-four months, moreover, find it difficult to tolerate separations from their parents, even the brief interruptions that are frequent in family living.

Indeed, children of this age expend a great deal of energy, especially when they begin to crawl and walk, attempting to remain close to their parents, and may greet their impending departures with cries, clinging, or other expressions of alarm.

In many cases, an infant's first attachment will be to his mother. However, since attachment behavior involves a psychological interaction superimposed onto bodily care, the infant will attach himself to that person who, on a hour-to-hour basis, provides for his ever-present nutritional and physical needs. In extended families in which child-care duties are shared, the infant will benefit from more than one primary care-giver so long as stable relationships and responsive social care are offered by each. And while an infant's attachment to his primary care-giver(s) remains paramount during the first two years of life (that is, it is more intense and not interchangeable), other family members, in both direct and indirect ways, are essential to the child's well being. In many societies, not only does a father, for example, have his own important relationship with the child, but by providing emotional and material support he also has an influence on the mother-child relation. Other adult relatives as well as children who interact with the child and help the primary care-giver in various ways also enhance the well-being of both. In this sense, the family is a system of individual as well as interpersonal relationships in which each member exerts an influence on the others.

These tenuous first attachments have critical immediate and long-range developmental implications for the child. Not only are family care-givers the principal source of an infant's nutritional well-being, but they provide the graduated stimulation so necessary for early motor, sensory, and cognitive growth. A secure attachment, for instance, increases the infant's propensity for exploring his environment, including the people in it.[10] The prelinguistic exchanges between parents and an infant also form the conceptual base upon which formal language is later built.[11] Moreover, these earliest attachments are the necessary precursors of all subsequent social relationships. The securely attached child tends to be more self-assured, competent, curious, and sympathetic in his relationships with both peers and adults than the child who has not had the opportunity to form solid early attachments.[12]

Children first experience society through their bodies. Essential physical contacts during infancy become the child's first social events and these form the beginning of the psychological patterns of his later social development. A society, subgroup of society, culture, or ethnic group's basic ways of organizing experiences are transmitted to the child's early body experiences and tie that child to his native milieu. Childrearing patterns not only ensure the child's own growth and development but also guarantee the continuation and preservation of a culture's unique qualities and tradition.

Attachments can form with a number of people who care for the infant on an everyday basis. Early on, Mead found that infants within extended families often form more than one primary attachment.[13] She suggested that several responsible adult care-givers may be better able to provide stability than a single one by lessening the stress of separation from the mother. More recently, Ainsworth, in her study of African Baganda infants, also observed that when the household included a grandmother or the mother's cousins who shared in childrearing responsibilities, the infant's attachment to the mother was often less exclusive.[14]

The ways in which parents interact with their infants may also affect behavior and subsequent psychological development. Comparative studies, for instance, point out that in some tropical parts of the world it is not uncommon for infants to sleep in their mother's bed. Mothers, too, often transport their babies by carrying them in their arms, on their hip, or by binding them with cloth to their stomach or back. However, in other areas of the world, particularly where the temperature drops below freezing, infants are often put to sleep in beds, cribs, or cradles and in rooms separate from their parents. Because of the colder climate infants are more heavily swaddled or clothed and are often transported in small carriages or cradle boards.

As a result of the "closer contact" engendered in some tropical areas, infants are less often separated from their mothers and tend to develop somewhat earlier attachments and more dependent types of behavior. In contrast, in colder climates, there is less body contact and parents may exert earlier pressure on their infants to become more self-sufficient at night. It is believed that this type of parent-infant interaction can result in a more "ambivalent" dependency which fosters earlier independence.[15]

EARLY CHILDHOOD: SOCIALIZATION AND IDENTIFICATION

During early childhood, from roughly eighteen months to the fifth year of life, children continue to undergo a rapid acceleration of physical and mental maturation. Gross motor skills become progressively well coordinated to the extent that reaching, holding, walking, and running are no longer activities in themselves but become important means to new discoveries and achievements. Fine motor control abilities evolve to the point that children are able to draw straight and curved lines, and by the age of five, are usually able to create in drawings a recognizable person and house, although certain details are still meager and the proportion of their figures remains somewhat poor. Language is more impressive during these years than at any other time in development. In all cultures, children, at about the same time they begin walking, begin uttering their first recognizable words. By age of three or so most children possess a functionable vocabulary of over one thousand words, are able to talk in sentences, and have mastered the basics of grammar. Increases in children's ability to use and command language leads to increases in their understanding of their environment, themselves, and what others expect of them.

For three-, four-, and five-year-olds, there is also a rapid increase in their use of imaginative or pretend play. Children often create imaginary playmates or act out various roles such as being a "mother" or a "father" or "nurse" or "doctor." Pretend or make-believe play has an important role in children's psychosocial development. It becomes a way of preparing for or anticipating various roles and exploring or mastering fears and anxieties by reenacting in safety some event that has been frightening. Imaginative play often assumes a kind of wishful thinking; children often reenact in play not what actually happened but what they would like to have happened. Fears, too, also change somewhat during these years. Earlier, children are frightened by actual occurrences such as loud, unexpected noises, unfamiliar people, or strange animals. Now children are also frightened by imaginary dangers such as ghosts, witches, etc., with nightmares becoming increasingly more common around the fourth or fifth year. Children of this age are often apprehensive of new or unfamiliar settings, although they usually manage quite well when accompanied by their parents.

The developmental accomplishments of this age do not occur in a social void but are linked to family attachments in a number of important ways. As Erikson suggests, a child's emerging sense of self autonomy and initiative is dependent upon stable adult relationships and that it is the "tolerant firmness" of the parents that provides both the motivation for the child's explorations as well as the necessary restraints which ensure these actions will remain within the bounds of safety and the child's still limited capacities.[16] By both granting and withholding bodily satisfaction, parents teach the child that not all of his wishes can be met all of the time. This increases the child's capacity to tolerate immediate frustration and provides the child his first opportunity to exercise self-control. Parents, through the judicious use of praise and criticism, also guide their child's behavior towards socially acceptable standards. Through sharing objects and adults affection with brothers and sisters, the child also learns of the need for fairness and mutual cooperation in relationships with others. And most importantly, stable parental ties provide the child with an opportunity to receive and return love, and in the process, come to realize his own self-worth and individual competency.

A very young child has no innate sense of right and wrong and no prior knowledge of what is permissible and what is prohibited. Rather, a knowledge of right and wrong is acquired through interaction with parents and other family members. In general terms, *socialization* is the process by which children are helped to become responsible members of their society. The essential ingredients of socialization are *self-control* and *social judgment,* and these ingredients are acquired in large measure through parents' disciplinary practices and through their role as models of self-control, restraint, as well as caring, loving human beings.

During the early childhood years, the child acquires two complimentary sets of internalized standards which, as they grow older, become increasingly absorbed as part of the child's own moral conscience. Through internalizing his parents' prohibitions and restraints, the child begins to develop conscience—an inner voice that continues to guide him even when his parents are absent. As he internalizes the positive goals, values, and aspirations of his parents as well as their caring attitudes, the child also begins to develop an ego-ideal which becomes the root of a child's sense of pride, his desire to perfect certain moral principles and motivation to learn and to achieve. Whereas the emergence of conscience subjects the child to the necessary feelings of guilt when he has behaved in ways he should not, the development of an ego-ideal allows him to feel self-worth when he does what he should and shame when he has fallen short of these internalized goals. Clearly, the development of healthy self-control and social knowledge involves the internalizations of the parents' demands and restraints as well as their loving, caring attitude towards the child.[17]

Important identifications also take place during the early childhood years. Quite early in life, children begin to imitate routine aspects of their parents' speech, gestures, likes, and dislikes. These initial imitative behaviors are normally fostered by encouragement from parents who take pleasure in having their children act like them and who often reward these behaviors. Later, these specific imitations gradually develop into an identification process, a largely unconscious progression whereby the child is not aware that he is evolving from merely doing what his parents do, to being like them. During these years, the child correctly perceives his parents as being more powerful and competent than he is and as having certain privileges, skills, and opportunities that he does not have, yet would like to possess. In striving to be like the parents, the child absorbs, as it were, their moral standards which provide the rudiments of his own functioning moral conscience. The critical point, here, is that for the child to strongly identify with his parents he must feel that his love for them is returned by their love and acceptance of him. Thus, it is the strong emotional tie—established in infancy and reinforced during the early childhood years—that is the fundamental prerequisite of this identification process.

Werner, in her review of cross-cultural child development, points out that an estimated fifty percent of children under the age of six in many developing parts of the world are malnourished.[18] In addition to constituting the most critical health problem in the world today, the lack of food and a society's struggle towards economic self-sufficiency also has far-reaching consequences for children's social development. As Levine suggests, in populations with high rates of infant mortality and childhood malnutrition, and where there is a relative scarcity of resources for subsistence, parents will have as their primary concern their children's physical and economic survival, and childrearing customs will reflect this priority.[19] In such cases, parents exert strong pressure on their children to be obedient and compliant to adult authority because the family's livelihood may depend upon them assuming a number of work-related responsibilities at an early age; boys often help in the fields and girls often assume a number of household and child-care duties. Again, adult care-givers must be able to depend upon their children fulfilling these duties even when they are not present to supervise them. In addition to producing more compliant behavior, these socialization patterns also result in the sharper sex role differences found in many of these societies.

Levine believes that it is only in economic environments of stable affluence that parents are able to adopt childrearing philosophies that are less tightly linked to the dangers of child mortality and economic failure. In such societies, for example, family care-givers may often be more permissive in the socialization of their children. Through the use of both praise and criticism, parents may encourage attention-seeking in their children, which is believed to result in more autonomous, independent, and self-assertive types of behavior. Not only are these qualities expected and valued in certain societies, but they may even be necessary for the child's future competency within a highly technical, more diversified economy.

A child's emerging sense of self may also be influenced differently by the settlement pattern of his particular society. In hunting and gathering or fishing communities, for example, family care-givers often adapt relatively permissive social systems and socialization patterns. These patterns tend to encourage the kinds of autonomous, self-reliant attitudes and behavior necessary in living a somewhat isolated, mobile existence. As a result, children develop what is referred to as a "field independent" cognitive style, in which psychological as well as perceptual development is shaped along more independent lines and the child comes to rely more strongly on himself when seeking answers and solving problems. Children in urban and industrialized parts of the world generally develop this cognitive style as well.

In contrast, in traditional agricultural and pastoral societies, which are more sedentary, parents often demand greater adherence to social rules and reinforce through childrearing training the need for sharing, communication, and interdependence, rather than independence among community members. These patterns of socialization encourage comparatively more conformity in children's behavior and a higher degree of mutual cooperation and group affinity, all of which are needed in growing, harvesting, and storing basic crops. In such societies, children tend to develop a "field dependent" cognitive style in which there is a greater psychological dependency between the self and others.[20]

Ideological beliefs which are embedded within and reinforced through the language of a given culture also becomes a significant factor in children's perceptions and views of reality. While all languages appear to be similar with respect to basic syntactic structure and grammatical rules that dictate relationships, they differ significantly in the ways in which they code experiences, as well as in the ways individual units of meaning are linked together. Cultures also differ from one another in the transmission of language which ranges from oral to written, and if written, from pictorial to abstract-symbol representation.

Even more significant from the vantage point of children's thinking processes is that each language tends to embody and perpetuate a particular world view. Concepts, for instance, like "freedom," "ambition," and "responsibility" take on social and religious connotations which largely acquire their meanings through their being embedded within the explanatory verbal network of a given language. How a child comes to understand these and other concepts and his relationship to them, is connected to the cultural specificness of the language, and it is this, then, that plays an important role in shaping the child's perceptions and understandings of himself, others, and the environment.[21] Language must thus be seen as a prime culture-bearer, paving the way for the child's acculturation into a given society.

MIDDLE CHILDHOOD: CONTINUING SECURITY

Around the age of six, children in all societies start to learn the rudiments of skills required by their particular culture. In most societies, children begin attending schools where teachers initiate the process of teaching reading and writing. In other societies, either family care-givers turned teachers or other adults in the community assist children in learing the appropriate aspects of technology, farming, fishing, homemaking, etc. In this manner, adults outside the family become important role models and sources of knowledge for children. Peers, too, begin to assume a position of importance during these years. Peers are needed for self-esteem and serve as criteria for measuring the boy's or the girl's own successes and failures, and among them, children find another source of extra-familial identifications.

Piaget has found that sometime around the age of seven there is a shift in children's intellectual development from what is referred to as "preoperational" to "concrete operational" thinking.[22] Previously, there is an intuitive stage in which children begin to give reasons for what they do, yet tend to do so in a way that is dominated by what they are able to see before them rather than by a set of principles. At the beginning of operational thinking, however, children start to learn and to use more general rules in solving problems. They are not as easily misled by immediate impressions because they have firmer concepts of what should be, clearer skills in appreciating serial relationships, in holding mental images, and in realizing invariances of weight, volume, and number despite external changes in appearance.

Thus, with increasing age, the advent of higher mental functioning, more accomplished socialization, and somewhat greater independence, children's drive towards mastery evolves into a more complex orientation which includes both the motivation to achieve or to attain some long-range goals, as well as a sense of competency with respect to their acquired skills and ability to interact in new social situations. This complexity of attitudes and feelings has led Erikson to describe the middle childhood years as a time when the child's sense of "industry" can be enhanced or his sense of "inferiority" reinforced.[23] To the extent that a child's abilities, motivation, and attainments are reinforced by a reasonable amount of success, he will develop a positive sense of personal competency, a feeling that he can succeed in his efforts to learn, complete tasks, and make friends. If, on the other hand, the child is not afforded age and culturally appropriate learning opportunities, or if he meets with too many failures along the way, his feelings of inferiority are reinforced, which can undermine his self-esteem and sense of competency.

While this period of life is marked by a widening circle of social relationships and cultural learning, the family continues to serve and protect children in strong and important ways. Stable parental relationships afford the child the needed emotional security and encouragement necessary in forming new relationships with adults and friends outside the home. The adult-child relationship remains the root of the child's self-knowledge and respect for the social order and is critical in fostering healthy, independent, intellectual, social, and moral development. There is, for instance, a direct link between the child's identifications with his family care-givers' attitudes about work and community and his own efforts and attitudes towards learning, social involvement, and responsibility.

During this period of life, children also develop a sense of pride, or conversely, a sense of shame for their families. When a child grows up in a family which is basically warm and harmonious, he tends to develop feelings of pride for his family, feelings which necessarily reflect back on himself since he is also a member of the family. When, however, the child's family life is marked by discord and disharmony, he tends to grow up feeling ashamed or embarrassed about his home as well as about himself. Again, Erikson stresses that many of the individual's later attitudes toward work habits can be traced to the degree of self-esteem and industry which is fostered by the family and which the child expresses in the wider community.[24]

In infancy and the early childhood years, societal and cultural influences are largely transmitted to the child through the family itself. Now, the neighborhood, the wider community, and societal institutions are more directly involved in shaping the child's social and cognitive development. In their comparative study of *Children in Six Cultures,* the Whitings found that the complexity of the socioeconomic system of a given culture has profound effects on children's social behavior.[25] They suggest that in societies with relatively simple socioeconomic structures, children are often expected to help their parents by doing economic chores and caring for younger siblings. As a result, children in these societies often maintain closer psychological ties to the family and develop more responsible, nurturant, and helping types of social behaviors. In contrast, in more complex socioeconomic systems children spend a greater portion of their day in school, with peers outside the home, and parents tend to stress independent academic achievements. The Whitings noted that in these societies children often develop comparatively more "egoistic" or self-oriented types of behavior marked by the tendencies to seek help and attention rather than to offer support and assistance. In their final analysis, the Whitings suggest that the decision to keep a child at home or to send them to school may have a more profound effect on the child's social development than do actual differences in child rearing patterns.

Exposure to urban ways and formal education also has important implications for children's cognitive development. Piaget's conclusions on the development of concrete operational thinking, for instance, were derived largely from observations of European children. Since then, studies of children in other cultures suggest that, while the underlying structures or operations which allow for the development of concrete operational thought appear to be universal, whether and at what rate they become functional seems to depend to a large extent on other factors. In this regard, Werner cites several studies of Asian children which indicate that some of them develop "conversation" earlier than do Western children.[26] This somewhat earlier development is believed to be linked to the Asian children's instruction in an ideographic language and to the emphasis on patience and formality in their socialization. Yet, as Werner reviews, several studies noted that in more isolated cultures, in which there has been little exposure to Western ways and no formal education, some children may not perform concrete operations even by the ages of twelve to eighteen.

In general, then, it is believed that urbanization and exposure to formal schooling engender an approach to concept formation that focuses on particular properties of an object taken out of context. While abstract reasoning is necessary for developing mathematical and scientific skills which are needed in more complex, technologically sophisticated societies, it is far less important, and may even be antithetical to adaptation, in communities distinguished by extended families and subsistent economies. Here, the stress upon mutual cooperation and interdependence among community members fosters more descriptive, context-oriented, and relational thinking processes.

ADOLESCENCE: GRADUAL INDEPENDENCE

Adolescence, as the last stage of childhood, can be divided into three independent yet related subgroups. First, early adolescence, from roughly twelve to fourteen years, is marked by a rapid period of physical growth and sexual maturation. These major changes in body appearance can often result in increased feelings of insecurity, emotional turmoil, and self-consciousness. Secondly, during middle adolescence, fifteen- and sixteen-year-olds are primarily concerned with achieving greater psychological independence from their families, developing closer relationships with peers, and furthering the integration and mastery of work-related skills. As the adolescent presses towards greater independence and assumes more responsibility for his own actions, it is not uncommon for both the youth and his parents to experience feelings of ambivalence over this transition which can lead to problems and inconsistencies on the part of both. And finally, later adolescence is often dominated by increasing liberation from the family and increased efforts to forge an occupational identity. Later adolescence, which begins around the age of seventeen or eighteen, continues until the young person has managed to consolidate a reasonably clear, consistent personal identity and has committed himself to some relatively well-defined social roles, value systems, and life goals.

Indeed, Erikson suggests that the major development task of later adolescence is the consolidation of a "personal identity."[27] This sense of identity brings together into a meaningful whole much of what the youth has learned about himself during the long childhood years as a son, student, friend, and member of a particular community, ethnic, and religious group. The integration of these various roles, attitudes, skills, and beliefs allows the adolescent to achieve a sense of personal continuity with his past and helps prepare him for the future. "Ego identity," writes Erikson, "is the accrued confidence that the inner sameness and continuity prepared in the past are matched by sameness and continuity of one's meaning for others, as evidenced in the tangible promise of a career."[28]

The danger of this age is that some adolescents may be unable to integrate past experiences into a coherent sense of self and be so confused in their search for satisfying roles that they experience "identity diffusion," a sense of not knowing who one is, where or with whom one belongs, or what one is to do in the future. To varying degrees, these are common struggles of adolescents especially in rapidly changing, industrialized societies. Yet, too much identity diffusion can result in unusually withdrawn and apathetic behavior which is dominated by persistent self-doubt, loss of ambition, and an inability to make the kinds of decisions, commitments, and career choices that are necessary for identity consolidation. Moreover, some adolescents may attempt to alleviate this continuing confusion by adopting ethical and moral attitudes or social behaviors which are in opposition to family and societal expectations. A "negative identity," as Erikson suggests, may be preferable to remaining a nonentity, and reflects " . . . a desperate attempt at regaining some mastery in a situation in which available positive identity elements cancel each other out."[29]

While in most societies adolescence is generally perceived as a psychosocial transition from childhood into adulthood, it is normally a gradual process, during which time the family continues to serve important functions. While peer relationships are necessary and take on increasing importance during these years, studies have found that the adolescent's parents generally remain the most trusted and admired people in his life.[30] Whereas the adolescent does possess the intellectual capacity to reflect upon and make choices independent of his family's influence, he continues to rely on his parents for advice, security, and material support as he enters the many personal, social, and vocational choices that mark this period of life. This would appear to be even more so in subsistent economies in which a youth of fifteen or sixteen may already be permanently working and married with a child of his own. In such societies, the extended family continues to be a primary source of support, often playing a direct role in helping raise the newborn and in providing material assistance.[31]

Cultures may differ significantly in the degree of conformity versus independence they demand of their adolescence.[32] In many cultures, parents remain the primary decision makers in an adolescent's life, determining, for instance, what career will be pursued and whom the marriage partner will be. In these cultures interaction among male and female adolescents is also closely controlled, and if dating does occur, it is often with the expectation that it is a prelude to marriage. In other cultures, on the other hand, parents encourage greater autonomy in their adolescents and expect them to make more independent career and marriage choices. Often, dating is seen as a necessary aspect of personal growth and adolescents may have a number of different heterosexual relationships during these years. Moreover, the view of adolescence as a "moratorium," a time of playful experimentation between childhood and adulthood, is more prevalent in affluent societies than in subsistent ones. In many parts of the

world, adolescence is not so much a time for delayed career choice as it is a continuation of what was already begun in childhood. Within pluralistic societies, too, children who are members of ethnic, racial, or religious minority groups may, as adolescents, experience more acutely the problems and constraints which can result from
segregation, discrimination, and prejudice. The denial of equal social and economic
opportunity not only limit an adolescent's career choices but may also have a significant influence on how he views himself and others.

While adolescent development is indeed affected by these and other cultural and
socioeconomic factors, the developmental task, nonetheless, remains moving into
society as an interdependent member. Along with continuity of family attachments,
an adolescent's efforts to bridge the psychological gap between the past and the future
is greatly aided by continuity of cultural ties. Here, it is important to realize that while
issues regarding personal identity tend to culminate during later adolescence, their
roots are in earlier family and cultural experiences. At the same time, for example, the
child is progressing from the symbiotic attachment to family care-givers in infancy, to
the more autonomous ego functioning of adolescence, he is also forming complex psychological relationships with his community, culture, religion, and native homeland.
The sights, smells, tastes, and kinesthetic experiences the child is exposed to, the shape
of his home and the architectural design of his village, community, or city structures
and buildings, and the particularness of the topography, climate, and ecological setting, are all internalized and become part of the child's emerging identity. Moreover,
as we have seen, cultures and families differ from one another in how they raise their
children, in their expectations and emphasis of certain social behavior, and in how
they organize and understand various life experiences in terms of religious and other
ideological beliefs which are often embedded within the culture's language. In this
regard, Erikson has described adolescence as a "natural" period of *uprootedness* and
suggests that as the young person gradually leaves the heavy dependency of the family
and attempts to achieve full status as a member of adult society, he looks to his culture's values, religion, and ideology as a trusted source of support.[33]

12

Children in Emergencies

THE FAMILY: FIRST RING OF SECURITY

As we have seen, children come to rely on parents and families in many conscious as well as unconscious ways. In the course of everyday life parents are the child's primary source of emotional security and serve as essential representations of their own efforts and ability to cope with and successfully master new, difficult, and sometimes troubling experiences. It is not uncommon, moreover, for infants and young children to experience anxiety when their parents are preparing to leave or are temporarily unavailable when the child wants or feels he needs them. When approached too suddenly by a strange person or when placed in an unfamiliar setting, these youngsters may also become frightened and upset, retreating to their parents' arms for comfort and protection. Nor is it uncommon for older children, adolescents, or adults, for that matter, to seek out the security and suppport of the other family members when sick, under stress, or frightened.

During emergencies, anything that signals the threat of separation is likely to increase separation anxiety in children of all ages. Several studies on children's responses to natural disasters, for example, have observed that not only do children tend to withdraw from sudden, unexpected events which endanger their physical safety, but that they make every effort possible to find and remain close to their parents.[1] These studies point out that during times of unusual stress, children have an increased need for intense physical contact with their parents and that their predominant fear is separation from parents and other loved ones. As Morris Fraser states, in his review of this subject:

> There is a certain universality about a child's response to disaster. The varying realities of the event may well add details to nightmare and fantasy, but the child's fear is always, in essence, that of loss of the factors that make for physical and emotional security. He dreads the prospect of separation from his parents, as much, if not more, than he does bodily harm to himself . . . an aspect of preventive psychiatry often forgotten in the rush to evacuate children from disaster areas.[2]

As a general rule natural disasters do not disrupt family life; on the contrary, members tend to come together for mutual security and protection. It appears, too, that of all the potential determinants of children's reactions to tragedies, those in the category of the parent-child relationship are most significant.[3] Younger children, for instance, because they do not fully comprehend the inherent danger of a disaster, often exhibit only minor symptoms of anxiety when they are able to remain physically close to their parents and when their parents are able to remain relatively calm themselves. Middle-aged children and adolescents are obviously more aware of the potential dangers and consequently often show more marked signs of anxiety. Yet they, too, tend to key their own emotional responses and efforts to cope off those of trusted family members and other familiar adults.

This is not to suggest that children who remain with their families during a disaster will not be affected by stressful events and circumstances. Anxiety reactions are relatively common in children during and after situations which make unusual demands on their physical and emotional resources. Norman Farborow and Norma Gorden, in their *Manual for Child Health Workers in Major Disasters,* have summarized many of the ways in which children's anxiety may result in changed behavior, such as trembling, shaking, weeping, nightmares, general restlessness and mood swings, or in regressive behaviors such as thumb-sucking, loss of bladder or bowel control, and reversion to childish, sometimes aggressive behavior.[4] It is important, however, to realize that these are normal reactions to stressful events and that they rarely last long when children remain with their families. Parents, in turn, help their children by remaining calm, affording them reassurance as well as accurate factual information (in the case of older children and adolescents), and by responding to their children's increased need for physical closeness, constant reassurance, and their sometimes clinging, demanding, or regressive behavior with patience and understanding.

There are, of course, certain circumstances and events which may place children and their families at greater risk during and after major emergencies. In some cases, the family's equilibrium may be temporarily upset by stressful events and thus for a time is unable to fulfill its child-care functions in its usual manner. This disequilibrium may be longer lived if a family member has been lost through death or separation. When a parent dies, for instance, not only will the child have his own grief reactions, but his reactions will be influenced by those of the surviving parent. Several studies on disasters have found that community-based social support and counseling services for widows are important preventative mental health measures which eventually benefit the child as well.[5] Moreover, when the death of a parent or the emergency itself leads to a significant reduction in a family's means of economic survival, children are often placed at greater risk physically as well as psychologically. The danger of this occurring is especially great in developing parts of the world where families may already be struggling to provide children with the basic necessities of food, shelter, clothing, medical care, and education, and where rates of undernutrition and malnutrition may already be high. In this regard a series of reports by The International Union for Child Welfare on post-disaster programs in East Bengel, Bangladesh, Nigeria, and Sicily, point out that an integral aspect of caring for and protecting children in emergencies is providing them with assistance through their families.[6] The timely provision of food, material, or cash assistance can assist vulnerable families in their own efforts to maintain a regular income and to continue supporting themselves and their children. Such actions can prevent separations from occurring in the first place, ensure better care for children in families, and promote family reunifications when separations have already happened.

During war and refugee situations where the stressfulness of events and the kinds of deprivations individuals are exposed to are often chronic rather than acute, it has been found that family bonds also tend to take on increased importance. In Germany in the aftermath of World War II, for instance, a study on refugee family relationships noted that members of such a family often felt themselves to be united by a common destiny, and tended to cling to their mutual affection as their last remaining possession.[7] Direct observations of children who passed through the experiences of emigration, expulsion from homelands and even concentration camps revealed that those who had remained with their parents showed much less psychological disturbance than did children who had been separated from their families while enduring these same events.[8] As Maria Pfister, who made a detailed study during the war of refugee children received in Switzerland, writes:

> Children under school age are generally not deeply affected by war or flight if they have not been separated from the mother or some other person they love, and if the mother or this other person has been able to compensate for the shock they have received by extra care and affection. If they have been separated from the mother or foster mother, however, and above all, if the mother has died or has been taken away in their presence, serious psychological disturbances, particularly in the form of enuresis, stubborn silence and excessive shyness are likely to arise. The same is true of children of school age and adolescence.[9]

The importance of family cohesiveness during wartime is well supported by comparative studies of evacuated and nonevacuated children. The vast majority of these studies offer convincing evidence that the purposeful separation of children from their parents was more traumatic than the actual exposure to bombings and the witnessing of destruction, injury, and death from air raids. One study that focused on children in and from major cities in England found neurotic symptoms in fifty percent of the evacuated children as compared to only twenty percent of those children who remained with their families during the bombing.[10] In Bristol, it was noted that while one-half of the children assessed showed nervous symptoms immediately following a major air strike, these symptoms persisted in only ten percent of these children, all of whom were under the age of five, and that children's reactions to subsequent bombings were comparatively slight.[11] Another study of English children under the age of five reported more severe and long-lasting disturbances among evacuated than among nonevacuated children.[12] It appears, as Anna Freud and Dorothy Burlingham summarize in their classic volume, *War and Children,* that "war acquires comparatively little significance for children so long as it only threatens their lives, disturbs their material comfort, or cuts their food rations. It becomes enormously significant the moment it breaks up family life and uproots the first emotional attachments of the child within the family group. London children, therefore, were on the whole much less upset by bombing than by evacuation to the country as protection against it."[13]

Reports are also rather conclusive on the fact that when evacuations did occur, continuity of family attachments was essential to children's psychological well-being. During World War II, it was found that exposure to war and uprooting did not produce as deep or as long-lasting effects in children who arrived in Switzerland as did separation from parents and the absence of family care following these losses.[14] In France, a report on evacuated children noted increased emotional fragility in children sent from cities in peer groups as compared to those sent with their mothers or sisters and brothers.[15] Referring to the difficulties experienced in repatriating Finnish children received in Sweden between 1941 and 1944, most of whom remained there until

1945 or 1946, the report of the Swedish Save the Children Association stated that it was better to assist children without separating them from their mothers.[16] In support of this statement, they cite the fact that out of 68,000 Finnish children transferred to Sweden and subsequently repatriated, 400 children experienced serious readjustment difficulties and therefore created disturbances in approximately the same number of families. Finally, in comparing several reports and studies on unaccompanied children who were part of the larger exodus from Cuba to the United States between 1959 and 1962, it appears that those children cared for by relatives or Cuban family friends already in America showed less severe signs of initial separation anxiety and made easier transitions back to their own families several years later than did those children who were cared for by unfamiliar adults and in non-Cuban communities.[17]

With evacuation, then, it can be predicted that separation will be traumatic for the child. Whether this trauma will have a more harmful effect on the child than remaining with the family in the emergency will depend on both the likelihood of physical danger in the original locale and the conditions in the evacuation destination. For German Jewish children on the eve of World War II, for example, the balance tipped decidedly in favor of evacuation.

The same may be true of Nigerian children evacuted during the 1967–1970 Civil War.[18] In Biafra food was scarce and many children were in need of *kwashiorkor* treatment. Moreover, a large number of the 5,000 children who were sent abroad had already been separated from their families and were being cared for in understaffed and ill-equipped sick-bays, orphanages, and refugee camps.

As in all evacuation programs, there are a number of unanswered or unanswerable questions concerning whether the Nigerian children could have been fed without evacuation as well as how they were selected for evacuation, why some were chosen over others, and, once arrangements had been made to send them, whether or not they could have been accompanied by available family members. Nevertheless, in this instance those children who were evacuated were probably more helped by the emergency care they received while living abroad than they were harmed by the evacuation experience itself. Surveys, for example, undertaken by Nigerian social workers after the children were repatriated found that ninety percent of them returned home in "good" condition, indicating that nutritional and medical rehabilitation efforts had largely been successful.[19] After repatriation, the children's parents, foster parents, and guardians were also asked to make comparisons about conduct, impulse control, and general social behavior between evacuated and nonevacuated children. Most respondents rated the repatriated children as being equally competent or slightly better than children who remained in-state in all three areas. Another survey within the same report found no significant differences in the children's school performances. Based on these and other findings, Dr. D. S. Obikeze, the principal author of this study, concluded that "the influence of camp experience (after evacuation) had been overwhelmingly more positive than negative" and that "we did not find evidence of the 'usual' psychological and behavioral problems often associated with an early childhood separation and institutional care, particularly in times of war. On the contrary, our data seemed to show that camp experiences had exerted a favorable influence on the behavior of the children concerned."[20]

In sum, from World War II through more recent crises, the nuclear as well as the extended family has consistently emerged as the principal source of support for children during wars, natural disasters, and refugee movements. The majority of available evidence clearly suggests that children are better off remaining with their families in emergencies even though they may witness bombings, destruction, or have their food rations cut. Accordingly, most organized evacuation programs which have intentionally separated children from their families in order to protect them from potential psychological or physical harm have been judged to be historical mistakes (see Part I and Chapter 14). The exception to this finding has involved emergencies in which the purpose of evacuation was to save physical lives. In such situations, evacuation, as suggested by Obikeze's study, has had certain advantages, at least for the relatively few children involved.

THE COMMUNITY: THE SECOND RING OF SECURITY

Normal physical and psychological growth does not occur without causing a child a certain degree of inevitable inner turmoil. This inner instability needs to be offset by the security afforded by a stable external environment. Along with family, the larger social network of the community itself can provide the child with the opportunity to form stable object relations, firm identifications, and enduring values and goals necessary for smooth psychological, social and moral development.

During emergencies, the community often continues to be a support system of enormous importance to children. How a child reacts to the inherent stress of an emergency will depend to a large extent not only on the support and mental predictability he derives from his own parents, but from that afforded him by relatives, teachers, and other familiar adults. Familiar adults provide the child with a representation of his own ability to exert inner control in the midst of chaotic and changing circumstances. Community, cultural values and practices, and familiar geographic locality also offer the child a measure of security and predictability that enables him to better cope with upheaval.

Several studies on community responses during natural disasters point out that rather than becoming paralyzed, disoriented, or reverting to self-survival types of behavior, members of an affected community may often become more altruistic, going to considerable lengths to help family members, friends, and neighbors.[21] During World War II it was noted that children who remained in major cities during air raids not only benefited from family relationships, but (to a lesser degree) from the presence of other familiar adults and peers, and from the generally resilient morale of the community.[22] Several studies on the effects of repeated shelling of border Kibbutzim settlement in Israel also yield data which indicates that the relatively low incidence of acute anxiety symptoms in children was in part connected to the presence of a trusted adult population and the common bonds that evolved while facing these adversities together.[23]

As noted earlier, studies clearly indicate that the presence of familiar adults is essential in protecting separated children from further psychological harm. There is some evidence, too, that suggests sibling and peer relationships, as well as cultural factors, take on increased importance for unaccompanied children. World War II accounts describe how siblings were often inseparable, refusing placement arrangements which did not include other brothers and sisters.[24] These reports found that in the absence of adequate adult care, unaccompanied children tended to band together on the basis of past common experiences, former communities, and to a lesser extent, nationality. In a camp for Vietnamese refugees in Southeast Asia, Markowitz noted the same tendency among unaccompanied adolescents who were not provided adult guardians or organization relief assistance; their peer alliances were often based on ethnic and linguistic similarities.[25]

Finally, a longitudinal study of unaccompanied Cambodian children currently in progress indicates that most children benefited from a one year stay in the Khao I Dang holding center in Thailand.[26] In addition to the food, medical services, and shelter provided in Khao I Dang, the resilient strivings of the camp's adult population to restore fragments of religious and cultural traditions shattered during the Khmer Rouge years also provided the unaccompanied children with an important measure of stability. The ability to remain within a familiar Cambodian community, attend Khmer schools, participate in ceremonies for dead ancestors and other rituals at the Wat or Buddhist Center all assisted the unaccompanied children in their own efforts to recover psychologically from previous experiences of upheaval and violence.

13

Separation, Trauma, and Intervention

Having identified the family and the community as the two principal sources of emotional security and mental predictability for childen in emergencies, this chapter focuses specifically on the psychological effects of family separation during war and refugee situations. From the standpoint of policy and intervention, it is useful to view the psychological struggles of unaccompanied children in emergencies from an "interactional" perspective. Simply put, the psychological vulnerability that results from family separation or, conversely, the ability to cope with this inherent trauma, results from the interaction of structural features within the children themselves and situational factors that surround and follow the initial separation.

To begin with, children of various ages differ significantly in terms of their developmental needs, abilities, and limitations (see Chapter 11). It is not surprising, then, that age at the time of separation will affect the extent of initial trauma the child will experience as well as the ways in which this trauma manifests itself in terms of specific distress reactions. In general, younger children seem to suffer the most adverse effects, while older children (especially those who had a previous history of family warmth and affection) often possess internal resources which help them better cope with the stress of family separation.

Family separation or loss, however, never occurs in isolation of other factors which can increase or decrease the psychological vulnerability of unaccompanied children. On the protective end of the scale is the presence of familiar family members, other adults, peers, and cultural practices (see Chapter 12). On the risk end of the scale are other kinds of trauma and deprivation—exposure to violence, persecution, hunger, uprooting from native sociocultural settings—which can compound the inherent stress of family separation or loss and lead to far greater psychological suffering. Thus, the second component of this interaction perspective of unaccompanied children in emergencies involves situational factors that occur after the original separation or loss and before the child reaches the "safety" of the relief community.

Of major importance for children unaccompanied during war and refugee situations is the assistance and care provided by other adults. While families within affected communities usually take care of many of their unaccompanied children, outside intervention, as discussed in Part I, has often been required for others. Historically, when such efforts have been present they have resulted in a number of different placement options. The presence or absence of special assistance for unaccompanied children as well as the kind and quality of these alternative placements will affect the subsequent psychological fate of each child and is the third component of this interactional perspective.

With this basic framework in mind, this chapter begins with an overview of the psychological effects of family separation. It then turns to a discussion of how the initial effects of separation can be exacerbated by other war and refugee trauma. In both sections, the ways in which outside assistance and alternative placements have affected children's outcomes are examined.

THE EFFECTS OF SEPARATION

Clinical studies, questionaire-based surveys, and descriptive reports clearly indicate that children separated from their families during times of war are placed at increased psychological risk. They further suggest a number of distress reactions that are common to separated children who are placed with unfamiliar adult caretakers. These reactions, it must be noted, are only the visible manifestations of deeper emotional and mental struggles. The most frequently noted sign of distress among separated children is enuresis, which, in England for example, is estimated to have risen fifty percent over prewar World War II figures.[1] This, however, may be due to the fact that bedwetting is more easily recognizable than other symptoms of distress and thus more often reported by the children's caretakers. Observations of separated children of differing ages have also noted the following common distress reactions and potential developmental dangers.

Infants and Toddlers

Initial separation is often followed by:

- Periods of intense crying
- Initial reluctance to accept the substitute caretakers
- Food refusals
- Digestive upsets
- Sleeping problems

Developmentally, the greatest danger of separation at this age is that it can weaken the child's attachment behavior. With repeated disruptions, a child's subsequent attachments become progressively shallow and indiscriminate. Such a child tends to develop socially withdrawn or socially diffuse behavior and finds it difficult to respond with positive emotion towards others. Studies have found that the speedy reestablishment of the broken attachment or the creation of a new attachment with at least one adult is essential in protecting infants and young children from further developmental harm.[2]

Children Under the Age of Five

In addition to the above reactions, childen between about two and five years of age often suffer regressions in previous developmental attainments. Examples include:

- Thumbsucking
- Bedwetting
- Poorer impulse control
- Temporary regression in verbal skills

Moreover, four- and five-year-olds have exhibited marked increases in:

- Nightmares
- Night terrors
- Fear of actual objects (loud noises, animals, etc.)
- Fear of imaginary objects (ghosts, witches, etc.)

It is believed that early physical, mental, and social attainments evolve through the everyday intimate exchanges between the child and his parents. Separation tends to undermine this development, and regressive behavior occurs most often during the time parental attachments have been broken and before new ones with other adults are formed.[3]

School-Aged Children

In addition to the above, school-aged children can become:

- Withdrawn from substitute caretakers
- Depressed
- Irritable
- Restless
- Unable to concentrate
- Disruptive at school
- Withdrawn from play and peer groups in new settings

It is believed that a child's own attitudes and efforts towards learning and social responsibilities are linked to identifications with their own parents' prohibitions, demands, and values about work and community. Several studies of evacuated children point out that these problems tend to increase when children have been moved from placement to placement and are unable to form a secure relationship with at least one adult.[4]

Adolescents

Adolescents have reacted to family separations by becoming:

- Depressed
- Moody

- Withdrawn
- More aggressive
- Developing more frequent headaches, stomachaches, or other psychosomatic problems.[5]

One study found that thirteen- and fourteen-year-olds were more vulnerable to separation than were fifteen- and sixteen-year-olds.[6] The author suggested that the younger adolescent is still approaching puberty and lacks confidence, while the older one has, developmentally, already begun to detach himself from his family.

These distress reactions were usually transitory, however, and did not necessarily evolve into more serious psychological or behavioral problems when: (1) children came from stable families and, (2) during separations, they were afforded the opportunity to form new attachments with other adults and continue age-appropriate pursuits.[7] Thus, while one study in England did find behavioral disturbances in sixty-one percent of children assessed after initial separation, seven months later, these problems were noted in only eleven percent of these same children.[8] Again, it appears that distress reactions were most prevalent during what Freud and Burlingham have referred to as "the no-man's land of affections": that is, during the time old attachments have been given up and before new ones were formed.[9]

In providing attachment opportunities essential for ameliorating postseparation distress, foster families proved to be effective. Social scientists in England were suprised by the unusually low percentage of children—an estimated six to seven percent—who did not successfully adapt to foster family placements.[10] It was noted that most of the failed placements involved children with unusually withdrawn or aggressive personality characteristics; the former children often coming from somewhat anxious, overprotective families and the latter ones from families marked by earlier family disruptions or parental discord. Yet, as Susan Issacs found, in her *Cambridge Survey*, even a significant number of these problematic children were able to make adequate adjustments when placed in families who could accommodate their special needs.[11] Shy, withdrawn children tended to do better when placed in more structured family settings, while aggressive or more active children often adapted better when placed in flexible, less-demanding families. Issacs also found a sharp increase in maladjustments in evacuated adolescents, rising from thirteen percent at age thirteen to twenty-four percent for those fourteen years down. The behavior of many older adolescents improved when they were placed in group settings, especially when accompanied by other siblings, friends, or familiar teachers.

In comparison, Katheryn Close, in a report prepared for The U.S. Committee for the Care of European Children found that all but 85 of the 861 English children evacuated to the United States adjusted to family placements.[12] During the first year, however, Close stated that a number of family replacements had to be made, particularly among those children whose foster parents had taken them out of a sense of obligation to a British friend, and as time passed, grew tired of their "international" responsibility. In other cases, family breakdowns were linked to difficult behavior within the children themselves, or because the child and the family were ill matched.

Unfortunately, there are no figures available on how many children went through more than one foster family placement during their first year in America. Yet, in general, it is reported that after an initial period of distress, school-aged children made relatively positive adaptations to family life. Again, however, adolescents had a more difficult time, tending to resist the foster parents' efforts to come close or to assume a parental role with them. Many adolescents openly expressed resentment over having been sent away in the first place; appeared to feel guilty about their own relative advantages, comforts, and safety over those of their families and friends in England; and became especially despondent when letters from home were overdue. Nearly all expressed the desire to reach the age of seventeen when they would be allowed to return and become part of the war effort. Some fifteen- and sixteen-year-old adolescents never adjusted well to foster family placements. Their attitude and behavior improved when they were sent to boarding schools and were provided with a "sponsor" family which they visited during holidays or weekends. Most of the older adolescents in this latter group had come from relatively affluent English families in which the tradition of sending children to private schools at an earlier age was common.

Studies and accounts from other countries yield comparable findings on the adjustments of school-aged children cared for in foster families. Nic Wall reports on the favorable outcomes of Norwegian children who escaped to Sweden during World War II.[13] These children were placed in families, became fully integrated into community life, and according to Wall, showed no signs of psychological disturbance. Similar findings are offered by Eila Rasanen, in a study of Finnish children sent to Sweden.[14] Rasanen found that the receiving families "fairly quickly managed to form good and strong ties with the children ... (and) the children formed sound human contacts with their new parents."[15] Finally, while unaccompanied Cuban children sent to the United States between 1959 and 1962 were less distressed when cared for by relatives or other Cuban families, it appears that most adolescents placed in American families eventually made adequate adjustments. "We expected all kinds of problems and they just did not happen," states one social worker, "these children are not like the deprived children we see in our regular foster care program."[16]

SEPARATION COMPOUNDED BY WAR AND REFUGEE ADVERSITIES

The previous discussion focused on children whose separation experiences were usually not complicated by other trauma, deprivation, or abuse. While some of the communities receiving these children were indeed directly affected by war, often by air raids, most remained relatively stable. Equally important, these children were able to weather the bombardments with their substitute adult caretakers. In contrast, a great many children unaccompanied during war and refugee situations have not found stability following family separations and losses. Rather, these children have witnessed death and murder; suffered from cold, hunger, persecution, and emotional deprivation; and have sought refuge in one and then another country.

Infants and Young Children

During World War II, several child care workers used the term "totally uprooted child" in reference to the youngest of these victims. As a label, this term blinds the observer to the recuperative potential of such children. Yet as a descriptive term, it does portray the inner confusion and psychological disturbance that so often results when family separation and loss is compounded by multiple adversities. A director of a Children's Village in Trogen Switzerland offered the following account of the upheaval endured by a group of Polish children before their arrival:

> When the war began most of these children were in Silesia with their families or in Polish orphanages. They became German on annexation of Polish Silesia by Germany. Their education was given in German, and many of the older ones were made members of the Hitler Youth.
>
> At the time of the Russian advance in 1945, they were to have been evacuated, but ran into Allied troops, and finally arrived in Austria via Czechoslovakia. There, some of them were received in refugee camps; others became vagabonds. On arrival of the American and Polish armies, they were taken in charge of by the latter. During the years preceding their departure for the Pestalozzi Village, they were housed in military camps in Austria and later in Italy.
>
> In the course of a few years these children had thus left their home surroundings, and twice changed their language, environment, culture, religion, and in some cases, nationality. They had undergone the traumatic experience of war. The promiscuity of life with adult refugees in the camps had left few mysteries of which they were unaware. What is perhaps most striking, they had no memory of any past on which to build. They were, in other words, utterly uprooted, and this fact is at least partly responsible for the tremendous difficulty they have in learning and in adjusting themselves to new circumstances.[17]

Renate Sprengel, a psychotherapist working at the Bad Aibling Village in West Germany also distinguished two groups of children who she described as "totally uprooted."[18] One group consisted of children between the ages of eight and eleven who had lost both parents during their preschool years. These early losses were followed by several changes in caretakers and complicated further by human abuse and environmental deprivation. By the time they came to Bad Aibling, they were extremely distrustful of adults, and very active, often aggressively so, in testing and retesting the limits of their new setting. Many children did not remember their exact names, birth dates, nationality, or religion. In Sprengel's words, they were "completely without roots . . . they lack spontaneity and have no confidence in others or in themselves. Their sense of insecurity is increased by having no mother tongue, and not knowing any language well. Their care calls for the utmost patience and tact."

The second group Sprengel identified as especially vulnerable were children born after 1945 and abandoned as unwanted or illegitimate infants. None of these children had an opportunity to form secure early or selective attachments, and did not possess what Sprengel called "nest warmth." These children were observed to be either "apathetic and taciturn" or "markedly aggressive and unruly."

One of the most insightful accounts of the readjustment of young children who suffered early deprivation is Anna Freud's and Sophie Dann's study, "An Experiment in Group Upbringing."[19] Their study focuses on six children of German-Jewish background who were orphaned shortly after birth by Hitler's gas chambers and sent to the Tereszin concentration camp in Czechoslovakia. In Tereszin, these infants were placed in the ward for motherless children and raised by the collective efforts of interned women. While these infants received food and medical treatment on a par with other childen, there was little opportunity for sustained adult care; Tereszin was a transit camp and deportations of interned individuals were frequent. There were no toys, and indeed, these infants had to depend largely upon one another for companionship.

After liberation, and following a month of medical and physical rehabilitation in Czechoslovakia, these children (now about three years old) were sent to England. In light of their past experiences, the decision was made to keep them together for a year so "they could adapt themselves gradually to a new country, a new language and the altered circumstance of their lives." They came under the care of Sister Dann at Bulldogs Bank, a specially structured and staffed group home. It was during this year that Anna Freud and Sisten Dann made the following observations:

Initially, the children were described as being "wild," "restless," and "uncontrollably noisy." Within days, they destroyed all the toys and damaged much of the furniture. Compulsive behavior also marked each of these children: one scratched herself until she bled and them smeared the blood over her body; most were continuous thumbsuckers; all masturbated; and every one had excessive cravings for foods previously unavailable in Tereszin. Unless the children had an immediate, pressing need, they remained indifferent towards their new adult caretakers. When angered, however, the children would turn on their caretakers, sometimes hitting, biting, or cursing at them in German:

> In a good mood, they called the staff members indiscriminately *Tante* ("auntie"), as they had done in Tereszin; in bad moods this changed to *blode Tante* ("silly, stupid auntie"). Their favorite swear word was *blodder Ochs* (the equivalent of "stupid fool"), a German term which they retained longer than any other word.

At the same time, however, the children showed strong and positive feelings towards each other. Initially, their attachments were narrowly focused and did not extend to adults or children outside this immediate group:

> The children's positive feelings were centered exclusively in their own group. It was evident that they cared greatly for each other and not at all for anybody or anything else. They had no wish than to be together and became upset when they were separated from each other, even for short moments. . . . This insistence on being inseparable made it impossible in the beginning to treat the children as individuals or to vary their lives according to their special needs. Ruth, for instance, did not like going for walks, while the others greatly preferred walks to indoor play. But it was very difficult to induce the others to go out and let Ruth stay at home. One day, they actually left without her, but kept asking for her until, after approximately twenty minutes, John could bear it no longer and turned back to fetch her. The others joined him, they all returned home, greeted Ruth as if they had been separated for a long time and then took her for a walk, paying a great deal of special attention to her.

Here, then, is evidence of rather precocious peer relationships. The distress exhibited by John and the others over Ruth's absence is similar to the anxiety more often reserved for separations from mothers or fathers. It seems, too, that while in Tereszin, these attachments had kept the children psychologically intact, as Freud and Dann pointed out, they were not "deficient, deliquent, nor psychotic." Yet, the limits of these exclusive peer attachments are also clear. The disturbance in social behavior and in ego attitudes resulted from the children's lack of sustained adult care and adverse environmental experiences. The purpose of resocialization efforts at Bulldogs Bank was to gradually wean these children away from their exclusive peer attachments and towards increasing dependency on adults. Only through fostering adult attachments would the children be able to reorganize various aspects of mental, emotional, social, and moral growth in a more coherent and adaptive manner.

In summary, Freud and Dann report that by the end of their first year at Bulldogs Bank, the children had made substantial progress. All were able to speak English, bearing witness to their basically unharmed capacity for cognitive learning. To an impressive degree, too, they learned to control their aggressive behavior towards outsiders. All showed at least the beginning of more normal relationships with adults, including increased dependency, possessiveness, and signs of separation distress which initialy were clearly lacking. After leaving Bulldogs Bank, continuous nurturance and ongoing psychotherapy was provided for each child throughout his or her childhood and adolescence. All were able to live reasonably normal lives as adults.

The behavior of these children clearly differs from that of separated children of similar ages. Their initial disturbance and ongoing struggles resulted more from the abuse and deprivation experienced in Tereszin than from parental loss per se. In clinical studies, Albert J. Solnit found that the seriously deprived child's recovery efforts are often associated first with aggressive behavior and feelings, and only later with affection, pleasure, and attachment.[20] He sees the initial aggressive behavior as the child's demands for limits or boundaries and the more positive achievements as emblematic of the child's increasing capacity for object constancy and trust of others. Solnit's observations, as well as Freud and Dann's account, have important implications for the care of deprived children. When the child's aggressive or testing behavior is perceived by the child's caretakers as recovery, "an effort to reach out and hold onto the new love object," in Solnit's words, they may be better able to tolerate what can be the constructive use of aggression.

When family loss is coupled with sustained emotional and environmental deprivation, child-care workers are confronted with the often difficult task of caring for disturbed children. In other cases, however, relief workers within emergencies are in a position to prevent more serious harm following family separation or loss. In this regard, Boothby reports on the different outcomes of young Khmer infants abandoned in the border camp of Nong Chan, where periodic warfare and continuing population movements made living conditions unstable.[21] These observations are limited to eight of a larger number of infants abandoned in Nong Chan in 1981. All eight infants were between three and six months old when identified as unaccompanied. At that time, all were marasmic and may have been abandoned more than once before initial identification. Four infants placed with Khmer women in Nong Chan were subsequently reabandoned—three infants once and one infant four times after initial placements. All four babies developed failure-to-thrive symptoms marked by depressed appetites, dehydration, and continued weight loss, withdrawal from others, and serious delays in mental alertness and in gross motor skills. The one baby died three days after the fourth placement.

In contrast, four infants were sent to Sakaeo II, a holding center inside Thailand, where conditions, though far from ideal, were comparatively stable. In Sakaeo, child-care workers had already screened potential foster parents and as a result the infants were immediately placed in Khmer families. Adequate nutritional, medical, and social support services were also available. A six-month follow-up assessment indicated that all four infants—then between nine and twelve months olds—were at or near age-appropriate weight levels, appeared to be cognitively unharmed, showed only slight delays in gross and fine motor skills, and were demonstrating positive attachment behavior. The author pointed out that it was the advance preparation by the Sakaeo staff, and most importantly, the initial decision to send these children to the better available placement that spared these babies from more serious harm.

As we have just seen, there is indeed a distinction between developmental dangers specific to family loss and those that result from events following that loss. This distinction is further illustrated by studies and observations of infants and young children placed in large age-group institutions and orphanages during times of major crisis.

In 1951, John Bowlby completed the first comprehensive review of maternal deprivation in a monograph prepared for the World Health Organization.[22] To a large extent, Bowlby's findings were based on studies of institutions and orphanges for homeless European children during and after World War II. Bowlby reported that within several of these institutions infants became profoundly depressed, withdrawn from life, and sometimes lost their capacity for survival. He further noted that the physical, emotional, social, and intellectual development of all children under about the age of seven were adversely affected by institutional life, a finding he largely attributed to "the complete lack of maternal care."

As subsequent reviews on the subject have attempted to clarify, factors other than maternal deprivation were probably responsible for the poor prognosis of these children.[23] It appears, for example, that physical maturation and intellectual capacity (as measured by standardized I.Q. tests) is more closely associated with the presence or absence of sufficient social care and intellectual stimulation than to a direct one-to-one maternal relationship. More recent studies of institutions with reduced child-staff ratios and where adequate care, safety, and stimulation is provided, have not found the severe developmental delays reported by Bowlby and others. However, there is sufficient evidence to suggest that infants and young children who grow up in these institutions still do not develop as selective attachments as do children reared in families.[24] Moreover, not only do poor selective attachments impair the child's social capacities, but they can adversely affect intellectual attainment as well.

Illustrative of these findings is the work of Dorothy Tizard and associates.[25] Tizard has studied children reared in better institutions from infancy to the age of eight. Within the institutions, considerable effort was made to keep the child-staff ratios low and to provide for the child's physical and intellectual needs. Tizard found that by age two, the institutionally raised children were more clinging and diffuse in their attachments than children brought up in families. By four, these same children were still more clinging and overly friendly with strangers. Of those followed to the age of eight, less than one-half of the children who grew up in institutions were found to have close attachments with their caretakers and they continued seeking more attention than family raised children. Tizard reports that their behavior in school was equally striking. The institutionally raised children, all of at least average intelligence, were more restless, disobedient, unpopular, and attention-seeking than the control group.

Tizard's study indicates that the more generalized provision of adult care in institutions is not conducive to the formation of secure early attachments; clearly individualized and sustained adult care does make an important difference. It appears, however, that intellectual *potential* as measured by I.Q. tests is not necessarily affected by the lack of a one-to-one adult relationship. Yet, as Tizard's findings point out, the impaired social capacities which do result from the absence of adult attachments can indeed lead to poor school performance and thus adversely affect intellectual *attainment.*[26]

That Bowlby may have overemphasized the role maternal deprivation played in the developmental deterioration of children he and others studied does not undermine what they observed. In a number of World War II institutions, the quality of care was extremely poor, infant mortality rates high, and the physical, emotional, social, and intellectual development of children was jeopardized. Moreover, while systematic studies and observations are not available, there is sufficient descriptive information to suggest that the quality of institutional care in more recent crises has been of equally poor quality. The following accounts are offered as illustrative examples:

Korea: 1950–1960

South Korean social workers involved in the institutional care of children during and after the Korean War stated that there was often only five adults per 100 children, infant mortality rates were exceedingly high, and children suffered from a lack of stimulation. Food for children initially consisted of "leftovers" from the army.[27]

South Vietnam: 1970s

A report by Judith Coburn, cited by Edward Zigler, director of the Bush Center at Yale University, offers the following description of one of South Vietnam's many orphanages:

... The 273 children in the orphanage slept in four rooms so crowded that the children near the walls could not get out without crawling over the other children. In such crowded facilities, many babies lay in their urine and feces for hours while the overworked caretakers rushed from crib to crib attempting to change them. In such quarters, a single child's illness could quickly generate an epidemic. The senior nun reported to Coburn that the biggest problem in the orphanage was the expiration of young babies in their cribs. This nun felt that many of these dying infants had no visible disease and may have died "from simple lack of love or stimulation."[28]

Kampuchea: 1979 and after

T. Berry Brazelton, the noted pediatrician, offers the following account of orphanages in Phnom Penh in 1980:

Older orphanaged girls are assinged to babies. Each girl will have a room of four or five babies to care for. But these girls are survivors themselves. They have little energy to give to others. At best they carry one or two favorites around all day on their hips. I could stand at the door of a baby room and pick out the infant who was a favorite. That baby would look back at me if I called out across the room. He alone would smile if I smiled and cooed to him. He would reach for a preferred toy. And if I held him in my lap, he would gradually learn to transfer a toy from one hand to another as I encouraged him constantly. . . . And, if one kept playing with him, encouraging and rewarding him after each new achievement, his face might finally soften. He might lose his worried "old man" expression and look back at you softly when you said, "That's nice; you just made a lovely noise."

. . . The other babies in these large rooms were too depressed, too inexperienced in finding the internal reward of completing a task or of having an adult say, "That was great; now do it again." Most infants in these overcrowded orphanages lie around all day, flat on their backs, their heads flattened and their hair worn off the back of their heads.

The orphans are being physically salvaged in Phnom Penh, but the quality of their future lives is far from assured . . .[29]

The troubling fact is, that in the above crises, children who reached these insitutions were often considered to be "fortunate." In times of major crises, when there are large numbers of parentless children and the affected population is struggling for survival, orphanages may be a necessary emergency measure. Yet, clearly, children in these institutions are at extreme risk. While it is not fully known to what extent developmental damage suffered in infancy is reversible when children are placed in a better environment, it is safe to say that the longer they are subjected to these poor conditions, the greater the probability of the damage being permanent.

Relief agencies need to give special attention to unaccompanied children in large age-group institutions and orphanages during and after emergencies. The provision of child-care expertise, medical and material assistance, and, equally important, the recruitment of sufficient numbers of adults to care for infants and young children is essential. Moreover, efforts need to be undertaken as soon as possible to transfer these children to more viable placements. Langmeier and Matejcek, for example, report that a significant percentage of Europe's institutionalized children could have been cared for by grandparents or aunts and uncles had child care workers obtained accurate family histories and worked to create such opportunities.[30] In Korea, Foster Parents Plan International took the lead in returning institutionalized children to families by promoting community development projects and by attaching financial aid to the child rather than to the institution itself.[31] In this way, communities were prepared to receive these children and the child's financial support continued regardless of where he or she resided.

Older Children and Adolescents

For older unaccompanied children and adolescents, the quality of previous family experiences can play a significant role in their ability to endure the stress of separation and loss as well as other kinds of adversity. While Bowlby, for instance, did find that separation during infancy and early childhood has serious, far-reaching effects on a child's psychosocial development, he also noted that a child's capacity to bear separation between the ages of five and eight varies in accordance with the amount of maternal care received in infancy.[32] At a Conference of Directors of Children's Villages held in Switzerland after World War II, Rey offered the following comments:

A child's past history plays an important part in his adjustment. Certain fundamental aspects of character and personality are formed in infancy. A child who has had ideal parents and a long history of family affection, will suffer a great deal for a certain length of time from the disorganization caused by war, but nevertheless possesses values and habits that in the long run will facilitate his adjustment, especially if his early upbringing has taught him to rely on himself . . .[33]

Rey found that older children who had once enjoyed family warmth and whose parents had encouraged independence, possessed qualities which facilitated their psychological recoveries. Along these same lines, Robert Collins described a seven-year-old girl found in the Belsen Camp after World War II.[34] Collins (who was in charge of children liberated from Belsen) suggested that as long as Eva's mother was alive she was able to withstand, to some degree, the abuse and brutality of the camp. Several days after liberation, however, when the mother died, Eva became profoundly disturbed:

Eva looked at the stiffening form. Her mother was gone. She was in a world she could not face alone. She sank into a huddled heap on the floor. Her face sank into her arms. She lay practically senseless, unable to face the awful thing that had befallen her. . . . In what more desperate position could one imagine a child at the most formative time in her life, than to discover her lying, holding her dead mother's skirt, surrounded by ten thousand corpses?

Without denying the continuing pain of this loss and past camp experience, Collins, who later adopted Eva, goes on to suggest that it was the strength of the child's relationship with the mother that allowed Eva to bear her misfortune and to make substantial progress during the next three years:

. . . I am definitely of the opinion that the mother had imparted something of her own strength to her daughter during the years when they had been very close, which no external event, even the death of the mother herself, could destroy.

A study of unaccompanied Cambodian children begun in the Khao I Dang Holding Center in Thailand also found age at the time of separation or loss to be an important factor in children's outcomes.[35] Those children who lost both parents before the age of five, and were unable to secure another sustained adult relationship during the Khmer Rouge years, have tended towards greater psychological disturbances. In Khao I Dang, most of these younger children were not able to recall positive experiences spent with parents and have repressed painful memories of family deaths and other traumas. Impaired object relations, aggressive and oppositional behavior, and poor school performance have been common struggles for this group of unaccompanied children in both their camp and resettlement placements. In many instances, the stresses of resettlement have exacerbated these problems, especially for unaccompanied children placed in non-Cambodian families and communities in the United States.

This study suggests, however, that only a minority of the unaccompanied children in Khao I Dang exhibited these kinds of psychological disturbances. It attributes this observation in part to the finding that approximately seventy-five percent of the children in the study had been able to remain with one of their parents during the Khmer Rouge years until after the age of seven. In contrast to younger victims, these older boys and girls have been able to recall better moments spent with family and friends before the 1975 revolution and have usually not blocked out memories of parental loss and separation. While many of these older children continued to show signs of depression for over three years, all demonstrated at least outward signs of resiliency, including positive attachment behavior with adult caretakers, the maintenance of at least one close peer friendship over the course of several years, and satisfactory or better progress in school.

It appears that the ability to recall earlier, more positive experiences has been important for these older children. In assessments of more resilient children and adolescents, direct links were usually found between adaptive behavior in Khao I Dang as well as in resettlement placements and values previously acquired from their parents and community. Most of these children, for example, have consciously connected their efforts to learn and to achieve in the present to expectations and values bequeathed to them by their parents in the past. While this "connection" has been expressed in a number of differing ways, the main theme has involved the notion that "this is what my parents would expect of me." Many have talked and written of wanting to become teachers, nurses, doctors, mechanics, etc., because their parents had once held these positions. Others have stated that their parents would be pleased if they achieved one of these various statuses.

Not uncommon of more resilient children is the case study presented on a twelve-year-old girl referred to as "Mom." In a passage from an autobiographical account entitled "My Past," Mom described her parents (both of whom were killed by the Khmer Rouge before Mom's ninth birthday) as follows:

> My mother was beautiful with long hair and a kind heart. When she went to the market she took me. She let me pick what fruit to buy. She made me bargain for a good price. . . . We lived in a wooden house next to a pond. It was good because the water helped cool the house. . . . My father was kind. And smart too. He made my brother and sisters work hard in school. He knew the value of a good education and raised us to grow in mind and spirit so we would become good citizens.

The idealized quality of this recollection is apparent and, at times, has been conflictual for Mom, especially in her early efforts to establish new attachments with foster parents in the resettlement placement. Yet, the author believes that these kind of memories have served largely adaptive purposes. It was noted that Mom, unlike her brother and sisters mentioned in the above passage, was too young to have received any formal education in Cambodia. After a year and a half in Khao I Dang's school, however, Mom had indeed learned a great deal. The Khmer refugee who translated the passage, a former teacher in Phnom Pehn, found its grammar and word choice to be near age-appropriate level. Teachers in the United States have also commented on Mom's strong motivation to achieve, which, the author believes, is linked to Mom's memory of her mother as a determined woman and of her father as a bearer of educational, social, and moral traditions. It was further noted that while in Khao I Dang, Mom spent every afternoon practicing Khmer traditional dance. In a later interview, when Mom was asked why this was so, she recalled that her mother had once taught traditional dance in Cambodia. This connection with the past has continued to help shape Mom's future as well; some three years later, she was still performing traditional dances with a Cambodian troupe in the United States.

In sum, this study suggests that while recollections of family separation and loss have been the primary source of these older children's persisting grief and underlying depression, they have also been at the heart of their continuing personal stamina and resiliency. Through the internal processes of idealization and identification, memories of better times, along with previously acquired parental and cultural values, have continued to guide these older children, informing their efforts in school, helping them determine right from wrong, and encouraging them towards more hopeful futures. Unaccompanied children who lost both parents before the age of five, in contrast, did not have an opportunity to solidify primary attachments nor to develop the rudiments of independent thought and moral reasoning. Consequently, they have not been able to summon forth or to identify with earlier, more positive, family and community experiences.

While solid family backgrounds may indeed be linked to active coping efforts, it is clear that older children and adolescents are not immune to the psychological effects of family loss. And with older children, as with younger ones, much will depend upon circumstances which surround and follow these losses. In what follows, an attempt will be made to highlight some of the psychological disturbances which appear to be more specific to acute emotional trauma and those which are more likely to result from sustained trauma and deprivation.

During war and refugee situations, women, adolescents, and even young girls are especially vulnerable to rape, sexual abuse, and exploitation. Among many other emergency situations, rapes were frequent during and after World War II, in camps along the Thai-Kampuchean border, in holding centers inside Thailand, during boat flights from Vietnam, and in the United States' camps for 1980s Cuban and Haitian exiles.[36] Joost Meerloo, in a World War II report prepared by an Inter-Allied Psychology Study Group, pointed out that these victims require special psychological attention.

> ... in the case of people who have been unwillingly seduced, or raped, the main problem which arises is not the relative superficial and transient emotional disturbance—the assault on social status—but the deeper and usually unrecognized problem of guilt and shame in the individual. In other words, however unreasonable it sounds, the problem is to ... help the victims of such assaults to *forgive themselves* in relation to the very real but unreasonable sense of guilt which they possess over the incidents concerned ... [37]

Unaccompanied girls and adolescents have been among these victims. Alice Markowitz, for example, found the most vulnerable of unaccompanied minors in camps in Southeast Asia were female rape victims.[38] She contrasts the "neurotic" depression and emotional instability found among most unaccompanied children to the more "profound" depression and episodic thought disorders found among unaccompanied girls and adolescents raped during sea flights from Vietnam. Many of these rape victims struggled with suicidal ideations; a few made actual suicide attempts; all suffered from the trauma of the rapes as well as from a sense of shame over having dishonored their families. These girls tended not to talk about, and indeed, often attempted to hide the fact that they had been raped. Most were extremely isolated from the larger camp's population. Markowitz found that crisis counseling services established by expatriate Vietnamese social workers—services which were available in only a few of these camps—were of critical importance to the well-being of these young victims.

With Cambodian children, it was found that those who witnessed the death of a parent, especially when violence was involved, were prone to recurrent night terrors and nightmares.[39] Sometimes children awoke from night terrors with somatic complications: headaches, stiff necks, or respiratory complications unconnected to organic impairment. One child awoke from a night terror temporarily unable to see or hear.

Night terrors were noted among children between eight and thirteen, yet did not occur among older adolescents. It was found that when these children were provided individualized psychotherapy, the night terrors tended to subside and were not necessarily associated with thought disorders, behavioral problems, or social isolation.

In contrast, nightmares were relatively common among adolescents as well as middle-aged children. Unlike night terrors, nightmares were not linked to somatic problems, perhaps because the children were often able to recall in precise detail the troubling dream sequence. Again, preventive measures were deemed important. Effective treatment included active empathetic listening and the opportunity to reexperience, in the safety of drawing and play, the previously troubling event. Equally important, and at times, of greater assistance, were religious ceremonies performed by Buddhist monks and other traditional healers in honor of deceased loved ones. The study found that "troubled spirits," both within the child and in the perceived afterlife, were sometimes put to rest by ceremonies and rituals rooted in shared beliefs.

In some cases older children and adolescents later identified as "unaccompanied minors" have been without family support for several years. Some of these children sought shelter or were placed in large institutions; others have existed for varying periods of time "on the streets." These past deprivation experiences have often been associated with increased vulnerability and adjustment problems.

For example, special problems were noted among unaccompanied Cuban adolescents sent to the United States in 1980. In contrast to the 1960s evacuees who were from intact families, reports indicate that between sixty to eighty-five percent of these latter arrivals had been institutionalized for varying periods of time in Cuba. David Eddie, director of a children's program in Wisconsin, stated that one of the children's "greatest initial problems was overcoming the fears, mistrust, and aggressive behavior . . . which most workers felt were due to their previous institutionalization."[40] Joseph Szapocnik and co-workers also identified a number of problems among unaccompanied Cuban adolescents in three camps shortly after their arrivals in the United States.[41] Szapocnik found, among other problems, that approximately forty percent were suffering from chronic medical diseases (diabetes, epilepsy); thirteen percent were suspected of being mentally retarded; and sixty percent exhibited varying degrees of psychopathology at the time of the interview. The author expressed the opinion that over one-half of these adolescents would need some form of psychological treatment after leaving the camps and that their *differing needs* could not be met through one mode of placement. Placement recommendations included family units with support for thirty-three percent, group homes for eleven percent, transitional experience for another eleven percent, and institutionalization for five percent.

Little is actually known about the plight of children forced to live on the streets during and after major upheavals. After World War II, one of the primary objects of children's villages was to serve adolescents without any family or community support.[42] However, there are no available follow-up reports on their subsequent adjustments. In Korea, a social worker reported that because of overcrowding many institutions refused to admit children over the age of fourteen, and when younger children reached the age of sixteen they were told to leave.[43] Yet, aside from the generalized observation that many returned to the streets and tended toward antisocial and illegal activities, nothing has been written about the children's actual circumstances.

Kirk Felsman's work on street children in Colombia, although not specific to emergencies, is germane to this discussion. He identified three groups of street children.[44] The first, and by far the largest, are children who retain some contact with at least one parent. During the day these children survive by collecting and reselling rags, bottles, and other usable articles, and by begging in markets or cafes. At night, many return to a parent who, poor and vulnerable themselves, had set up lodging in one of the safer streets or alleyways.

The second group consists of children who have been totally abandoned by their parents. Although these children constitute the smallest percentage of street children, Felsman found they were the most vulnerable:

> The truly abandoned child may frequently be the one the family is least able to care for, perhaps because he suffers from some physical impairment or neurological disorder. Such children are least equipped to face life in the streets, but there are few care facilities that can offer an alternative. There is no doubt that many simply perish.[45]

The third group of children have not been abandoned; rather, they have intentionally left families behind. Often, the decision to leave was promoted by family poverty or by the presence of a new, sometimes abusive stepfather. In interviews with mothers of such children, the author found they were often described as the one who "never listened" or who "always asked too many questions." "In short," writes Felsman, "he is the troublemaker, a child with a difficult temperament in an intolerable environment. Possibly, he is a family scapegoat, who is constantly being given the message to move on."

Felsman believes that relief and rehabilitation programs for street children need to take into account their differing situations. He notes that the first group of children should not be encouraged, through assistance, to completely sever existing ties with parents. Rather, the children need to be provided protection, medical treatment, food, and clothing through the family unit. Most importantly, assistance needs to aim at rehabilitating the family's ability to secure an adequate livelihood which, in turn, can ensure better care for the children.

Programs for the second two groups of children must take into account how long they have lived on the streets. Some children who have only recently left home or been abandoned may be able to make immediate use of alternative family or group placements. Others, however, will need transitional arrangements such as halfway houses. These placements offer the child an opportunity to visit the facilities, receive a good meal, come to know the staff, and return to his companions on the street as needed. In some cases this may be as far as the child is able to move. Yet, for many, this graduated entry approach has lead to firmer commitments and to permanent residencies within these programs. In all cases, peer attachments need to be recognized and respected.

Felsman is especially critical of programs which perceive street children solely as deviants or delinquents and, as a result, establish restrictive and dictatorial programs for them. The author found that the runaway rate from such programs is extremely high, and fuels the myth that street children are uneducable and beyond sustained help. Indeed, one of the most successful programs described is a "family-like" home established by an adult and seventeen children after a larger institution was closed down. Before and after school the children helped with household chores and contributed to the family's income by selling pencils or shining shoes. Later, a family stationery store became the primary source of income. All these children remained in the home, have graduated from high school, and currently are either in college or working full-time.

Among the most troubled unaccompanied adolescents are those who have been deprived of family life while being subjected to persecution and other forms of human abuse. This is clearly indicated in reports and studies of adolescents who survived Nazi concentration camps. As noted in Freud and Dann's study, the youngest of these survivors suffered physical and particularly psychological effects on a global scale. In contrast, disorders among adolescents were sometimes less obvious at first. Some managed to adapt relatively quickly in their new settings and their behavior stabilized. Langmieir and Matejcek, for example, report that within a relatively short period of time, approximately sixty percent of older children and adolescents liberated from Tereszin showed no symptoms of frank mental disorders.[46] Yet, as these authors write, "almost all of them . . . were scarred in certain ways by their great suffering; this was reflected mainly in their increased vulnerability and mental instability and in the fact that minor changes in their life situations tended to produce breakdowns."

For many adolescent survivors, character structure had been deeply affected. The most marked symptom of disturbance was the adolescent's distrust of other people; they tended to suspect the motives and intentions of others, and acted as if they expected to be betrayed. As a result, they were seen as being "hypercritical," "envious," and "incapable of gratitude."[47]

Most of these adolescents had better relationships with peers than adults, especially companions who shared a similar fate. Often, it was possible to begin resocialization efforts at this point; these adolescents appeared to benefit most from intensive language, and educational and vocational programs which were group-oriented. UNESCO, in its post-World War II review of placement programs, reports that while individual differences were noted, group care was more successful than foster care in meeting the special needs of these adolescents:

> Foster home care may possibly be the riskier procedure, as most if not all refugee adolescents who have been deeply affected by their experiences have forgotten the value of family life, and are in some respects so mature as to be unable to tolerate parental authority. The proposed foster parents, moreover, may not possess the necessary educational and psychological insight to be in a position to help them. The attempt may result in bitter disappointment for both parties. . . . Placement in groups, under the leadership of specially trained adults, seems to offer better opportunities for re-education where the emphasis should be on . . . the fostering of a sense of security by giving opportunities for taking root and becoming useful members of society.[48]

REFUGEE AND DISPLACED PERSONS CAMPS

Unaccompanied children in refugee and displaced persons camps are placed at grave risk when they are not assured adequate care and protection. After World War II, relief officials in a number of camps in Europe were slow in recognizing the special needs of unaccompanied children. As a result, these children were largely left on their own and were especially vulnerable to the influences of Allied armies and various nationalistic groups.[49] A number of children were "adopted" by soldiers as troop "mascots," provided special favors for a time, moved from place to place, only to be reabandoned in the end. Unaccompanied adolescents were also recruited by resistance organizations intent on overthrowing newly established regimes in Eastern European countries.

In Nong Chan, a border camp run by a Khmer resistance group, children over the age of thirteen were recruited as soldiers.[50] Often the offer of food, better clothes, and a place to live was sufficient enticement. Sometimes the pain of past family deaths was exploited, and adolescents encouraged to become resistance fighters to "get even" with either the Khmer Rouge or Vietnamese who were responsible for these deaths.

Inadequate protection in camps is not limtied to the distant past nor to current wartime situations. David Eddie, for example, offers this 1980 account of Fort McCoy, a compound used for Cuban exiles in the United States:

> Many of the Cuban youth were subject to extreme psychological stress and some to physical and sexual abuse during their stay at Fort McCoy . . . Many observers and participants reported of a camp frequently dominated by a group of Cuban adults who intimidated and abused others, including minors.
>
> The September, 1980 report of the Governor's Fact-Finding Commission puts it bluntly: "Internal security for the protection of Cubans in the compound is for practical purposes nonexistent. Cubans are inside the compound with no protection afforded them by the U.S. Government." The report also found "substantial evidence of sexual and psychological abuse of both male and female juveniles housed in Fort McCoy."
>
> This atmosphere compounded the anxiety already experienced by many of the minors, who were coping with separation from family, confinement, and a foreign culture. For many of the minors, this anxiety was exhibited by suicidal gestures, such as cutting arms and wrists or swallowing dangerous liquids and materials. Most of these gestures were seen as manipulative in nature by psychiatric staff as a way of physically demanding a change in their environment.[51]

Psychological problems arise when camp practices and resettlement policies are not oriented toward preserving children's ties to extended or foster families. This was clearly documented by psychologists and psychiatrists working with Vietnamese refugees in the United States.[52] A number of children fled Vietnam with relatives and family friends. Once in camp, these families became anxious over obtaining resettlement sponsors. Sensing this anxiety and believing themselves to be unwanted burdens, many children left their "unofficial" families. Rather than attempting to negotiate their return, relief personnel placed them in a separate compound for unaccompanied children. Vietnamese psychologists deemed these measures to be harmful:

> These children became an acute mental health problem, for in camp they became separated often from those care-taking persons, either because the children chose to leave them or because they were abandoned, usually because of strict immigration laws. The psychiatrists recommended that the children be allowed to stay with families to which they had attached themselves. Instead, they were placed in a district group facility.
>
> (Within weeks) there were depressive themes in conversations of the children, feelings of hopelessness, regressing to bed, insomnia, somatic complaints; there were strong feelings of isolation from other children and Vietnamese people in camp

(partly due to the physical separation of their building) . . . partly due to the stigma
of not having a visible family in a family-oriented culture.[53]

Even when adequate protection is afforded, the length of time spent in refugee
and displaced persons' camps affects psychological adjustments. At the end of World
War II, H. B. M. Murphy found that the establishment of camps following a major
exodus was initially a psychologically sound practice.[54] The early use of camps provided a kind of moratorium between the past and future, and refugees seemed to benefit from group support. Murphy found these camps to be undesirable thereafter as
refugees were treated as a political mass rather than as individuals with common misfortunes. He cited the main problems within displaced persons' camps as being a lack
of privacy, overcrowding, segregation from the host population and the fact that the
entire course of the refugee's life is restricted to a circumscribed area. Under these
conditions, refugees are given a clear signal that they are unwanted and being controlled, which reinforces feelings of dependency and frustration.

Bakis has referred to this reinforced sense of helplessness among displaced persons as "D.P. apathy." He noted how after people spent a year or more in camps,
absenteeism from work, alcoholism, and crime increased, while participation in community and cultural activities declined.[55] Similar findings emerged from reports on
Ugandan exiles in England, Vietnamese refugees and Cuban entrants in the United
States, and Cambodians in Thai holding centers.[56]

For unaccompanied children, the possibility of successful camp adjustment can
be associated with the children's knowledge of their parents' fate and their limited
capacity for retaining hope of rejoining them.[57] In 1980, standardized interviews were
administered to seventy Cambodian unaccompanied minors in Khao I Dang, a camp
in Thailand. At that time, fifty-one of them indicated they preferred remaining in
camp, while nineteen stated they were writing to Western embassies requesting resettlement opportunities. All fifty-one who did not show a preference for resettlement at
this time had also stated they did not know whether or not one or both parents were
still alive. In contrast, the nineteen who wanted to leave Thailand stated they either
had witnessed their parents' death or were told by other Cambodians that their parents had died.

One year later, however, a striking reversal was noted among the first group of
minors. In 1981, the same interview indicated that forty-three of the fifty-one minors
were now seeking resettlement abroad. Analysis of this reversal indicated that most
of these minors had: (1) lost hope of rejoining parents; (2) felt they had to move on
with their lives; and (3) believed it was not safe to return to Kampuchea. Corresponding with this change in future desire were increases in depression and depressive
symptoms and decreases in school performance and participation in recreational
activities. While these children appeared to be preoccupied with thoughts of locating
lost family members throughout the first year in camp, by 1981, the constraints of a
life limited to seventy square acres of land, surrounded by a barbed-wire fence and
Thai guards, began causing its own kinds of psychological and social problems. Khao
I Dang, once a haven, was a place these children were now anxious to leave.

In a similar way, the length of time unaccompanied children spent in Khao I
Dang's group centers also affected their psychological adjustments. In 1979 and 1980,
the decision was made to place unaccompanied children over the age of five in group
facilities.[58] At that time, several advantages of group care were claimed by the

UNHCR and other relief agencies: children's centers offered a safe and efficient way of caring for large numbers of children, these children had been subjected to brutalities during the Khmer Rouge years and would be unable to adapt to family life, and most Cambodians were unfamiliar with the concept of foster care.

Early reports indicated that most children's behavior did stabilize under the improved care received in Khao I Dang. In 1980, Jan Williamson undertook the first cursory survey of unaccompanied children in this center.[59] At that time only ten children with severe psychological disorders were identified. Four months later, a similar survey noted a threefold increase in severe disturbance which was believed to result from delayed grief reactions. More in-depth assessments administered by camp social workers indicated that individual cases of acute emotional disorders and more serious psychopathology were most often connected to past trauma and not necessarily to group placement or camp life.

Yet, over time, the children's centers became less desirable and began creating their own particular kind of psychosocial problems. Unaccompanied children placed in group facilities by mid-1980 had, by the summer of 1981, at least six and sometimes as many as ten different adult caretakers.[60] Moreover, the children's teachers in school had an even greater turnover rate. In all cases, these important attachment figures had accepted resettlement offers abroad. While incidences of severe psychopathology were still rare, depression, emotional fragility, and socially withdrawn behavior was now common among children under about thirteen years of age. In 1981, one psychologist noted that younger and middle-aged children were "hungry for adult affection," and suggested that the continual loss of potential attachment figures was "perpetuating the children's recent life experiences."[61]

These findings contrast with those of unaccompanied children who were cared for in Cambodian families. As noted in Part I, children in group facilities were only a portion of the larger number of unaccompanied children in Thai camps. At the end of 1981, for example, while there were approximately 200 unaccompanied minors in Khao I Dang's children's centers, there were at least 400 other parentless children living in camp with nonrelated Cambodian families.[62] Most of this fostering had occurred "spontaneously" in Kampuchea or within Khao I Dang; it was not part of the official relief effort for unaccompanied minors.

In 1981, Thai social workers employed by Redd Barna (Norway's Save the Children) found that incidents of neglect and abuse in these families were the exception rather than the rule.[63] The majority of children had been with the same foster parents for at least one year, and in many cases, long-term commitments had been made. Most of these families, as well as other families in Khao I Dang, preferred resettlement over repatriation. More importantly, however, when commitments had been made, both the child and the adults expressed the desire to remain together regardless of future direction. In several instances children had already refused resettlement offers which did not include their new family members.

SUMMARY

When children are separated from their families during wars and refugee movements they are placed at increased psychological risk. Separated children tend to exhibit distress reactions and behavioral problems which are closely related to age and corresponding developmental stages. Yet, when separation is immediately followed by adequate alternative care—the opportunity to form attachments with other adults and the continuation of age-appropriate educational and social activities—these painful and potentially harmful distress reactions do not necessarily evolve into overt, long-term psychological disorders. Continuity of existing family relationships, language, and cultural ties are also critical.

When the initial trauma of family separation or loss is followed by other adversities, however, the likelihood of more serious psychological disorders increases dramatically. Some of these adversities—exposure to violence, death, abuse, hunger—have occurred during the children's wartime experiences. However, others—the absence of adequate physical care and protection, repeated changes in adult caretakers, multiple moves of sociocultural environment—have occurred after the children reached the "safety" of organized relief assistance. Much of this harm could have been prevented.

Finally, for unaccompanied children in displaced persons and refugee camps, two additional factors are likely to affect their subsequent psychological adjustments. First, groups of unaccompanied children, as well as individuals within the same group, differ with respect to what motivates their flights from their native homeland. During acute refugee flows, for example, many unaccompanied children seeking care and protection will enter first-asylum countries not knowing what has happened to one or both parents. Initially, at least, these children are motivated towards family reunion. With these children, camp placements can help stabilize behavior and provide the opportunity to locate and reunite with lost family members. In contrast, other unaccompanied children caught up in mass population movements may not be seeking parents, but leaving parents behind in native homelands. While "willful" separations may occur for a number of reasons (see Part I), these unaccompanied children usually arrive in asylum countries with clearer plans to accept settlement or resettlement offers. Even brief stays within camps may be harmful for these unaccompanied minors, as was found with Vietnamese youth in the United States. In all cases, the physical and psychological well-being of unaccompanied children is greatly endangered by the lack of adequate protection in asylum countries and by lengthy confinement in refugee and displaced persons camps.

14

Family Reunification:
A Child's Sense of Time

As discussed in Chapter 11, children normally derive enormous benefits from the continuous care of their families. For unaccompanied children, however, these relationships have been disrupted. History shows that family reunions are often possible. From the child's perspective, it is always best when this reunion occurs quickly. Serious psychological problems arise when children have been long separated from natural parents or long in the company of "substitute" parents. This chapter, then, looks at family separation and reunion from what Goldstein, Freud, and Solnit have referred to as a "child's sense of time."[1] It further discusses how language and cultural differences, changes in family membership and poverty, can adversely affect family reunions and place reunited children at increased risk.

As Donnald Winnicott noted during World War II, separated children have a limited capacity to sustain inner ties to absent parents:

> For days or even weeks all is well, and then the child finds he cannot feel that his mother is real, or else he keeps on having the idea that father or brothers or sisters are coming to harm in some way. This is the idea in his mind. He has dreams with all sorts of frightening struggles which point to the very intense conflicts in his mind. Worse than that, after a while he may find that he has no strong feelings at all. All his life he has had live feelings, and he has come to rely on them, taking them for granted, being buoyed up by them. Suddenly, in a strange land, he finds himself without the support of any live feelings at all, and he is terrified by this.[2]

In general, the younger the child, the more quickly separation from parents will be experienced as permanent loss and be accompanied by profound feelings of helplessness and despair. Freud and Burlingham, for example, found that most children under the age of four could not retain emotional links to their parents beyond a one- to two-month period.[3] Developmentally, this is when the child is first forming selective attachments and just beginning to be able to maintain relationships during brief separations. These authors observed that after an intense and harmful initial period of crying, distress, and withdrawn behavior, young children in their wartime nursery began accepting the care offered them by other adults. Shortly thereafter, memories of their own parents grew dim or appeared to vanish completely. Freud and Burlingham provide several case descriptions of young children who did not recognize their own mothers when they reappeared after several weeks or a month's absence.

For school-aged children, on the other hand, observations suggest that most were able to retain memories of natural parents over a longer period of time.[4] In part, this is because an older child possesses cognitive skills which allow him to understand that in this case separation did not necessarily mean abandonment or permanent loss of the relationship. In part, too, the older child is able to evoke in memory and fantasy emotionally charged images of parents which provide psychological continuity when they are absent. Yet, it does appear that within a year's time, intellectual memory and affectively charged images of parents were becoming vague and shadowy.

Susan Issacs, for example, asked older English children to write on the topic: "What I Like and What I Miss in Cambridge."[5] While most children mentioned their parents as being missed, the mentioning, as Katheryn Wolf has pointed out, is done in a conventional, unemotional tone, and that for their ages, the children's responses were vague and confused.[6] One seven-year-old girl, for instance, wrote: "I miss my doll Pram and my mummy and daddy and gramma and bathing her at night, and putting her to bed and putting cums down her throat, and bits of fish and dressing her in the morning and put in her plates. And cleaning her hair and nursing her." Even allowing for age appropriate grammatical errors and discontinuity of logic, it is quite apparent that a certain amount of displacement has taken place here, as it is difficult to discern whether the young girl is referring to her doll or her grandmother, while clearly, her parents have been placed in the background. Another example is provided by a thirteen-year-old boy, described by his teacher as "quite intelligent." All that is missed is summed up in one sentence: "Most of all I miss my parents." Along with being a rather stock answer, we learn from other information provided by Issacs that his mother had died when he was small, and that since then, he has lived alone with his father. Finally, in Close's report on English children sent to the United States, we note the struggle of a school-aged boy to reconcile within himself conflicted feelings of love and loyalty.[7] Rather ingeniously, he attempts to resolve the conflict of having an "English mummy and daddy" and an "American mummy and daddy" by stating he planned to get a job on a trans-Atlantic plane when he grows up.

In contrast, most adolescents appear to be able to maintain inner ties to absent parents to about the same extent as adults.[8] Again, this has to do with their more fully evolved capacity for retaining memories, anticipating future events, and sustaining themselves emotionally during separations from parents. Thus, while certain details about family experiences do change or become submerged within the adolescent's mind, he is nonetheless able to preserve the essential core of parental bonds longer than school-aged children.

These observations, however, pertain to children and adolescents who knew their parents were alive, and whose parents, for the most part, wanted them back. As already noted with respect to unaccompanied children in Khao I Dang, a similar period of time may have very different implications for a school-aged child or an adolescent who does not know what has happened to his natural parents. Different issues can also arise when children have been or believe themselves to have been abandoned. An abandoned child may not be able to call forth memories of family warmth and affection, and indeed, may be quite ambivalent towards family reunification attempts or reject them outright.

The case of Lim Yi, a ten-year-old Cambodian boy, is suggestive of some of these issues.[9] In early 1980, Lim Yi, along with his mother and two brothers, fled Kampuchea in favor of Thailand. Lim Yi's left leg, weakened by polio, prevented him from keeping up with his family during the flight, and he was left behind with other refugees. While his mother and brothers eventually reached and remained in the border camp of Nong Samet, several days later Lim Yi arrived at a different camp and from there was taken to Khao I Dang as an "unaccompanied minor."

Over the next six months, Lim Yi received physical therapy and regained considerable use of his leg. Eventually, the mother was located by the International Rescue Committee's tracing representative in Nong Samet. The mother explained that during the flight she had become concerned for Lim Yi's physical well-being and arranged to have another refugee family bring him to the border in their pushcart. She now wanted her son back.

Lim Yi, however, experienced these events quite differently. He was angry over being left behind and over the preference the mother had shown for his brothers, and recalled how the two of them had ridiculed him about his handicap. He further pointed out that although he could now "play almost as well as other children," he would not be able to run for shelter or keep up with his family should Nong Samet be attacked by Vietnamese soldiers, which in fact, had occurred in the past. The more Lim Yi was pressed to return to his family, the more upset he became, and the more his psychological state and behavior deteriorated.

Along with the length of time children are *away* from natural families, the time spent *with* foster parents also plays a significant role in children's reactions to reunification attempts. For example, the majority of "evacuated" children who eventually rejoined natural parents were separated from them for between two and five years. During this time, not only did children's memories of their parents grow dim, or, in the case of the youngest evacuees, vanish completely, but most had formed very real psychological attachments with foster parents.[10] Parting from these "substitute" families was often accompanied by the same emotional disturbance and behavioral deterioration that resulted from original separations from natural families.

Within England it appears that children's reactions to family reunion during World War II varied.[11] Where the separation was less than one year, the child's foster parents and natural parents had maintained a positive relationship, collaborated on the child's upbringing, and the child had remained in contact with his own family through visits, letters, and exchanges of photographs, reentry was fairly smooth. Yet, in many cases, especially when separations were longer than one year, the children's responses to family reunion followed a somewhat typical pattern of rejection, anger, and behavioral deterioration lasting for weeks to several months. Parents discovered that their early attempts to reestablish closeness and show affection were often met by icy indifference or outright rejection. Parents appeared to be especially hurt when, in a moment of frankness or anger, their children would tell them that "coming home was not as good as I had expected," or complain that the "other mother" was a better cook or that she had loved them more.

During the years of separation from natural families the vast majority of children had also become fully integrated within their host communities, absorbing, so to speak, the attitudes, habits, behaviors, and dialect or language of their particular environment. This would appear to be true even of English children who were evacuated to the countryside; they, too, had taken on many of the "regional" habits of their foster parents and rural communities.[12] Yet, for those children who were sent to or escaped to second countries, the significance of this inevitable acculturation process was even greater, and reentry into their natural families and native communities was often a more disturbing and complicated process.

Close's account, for example, of English children in America suggests that clashes over being "American" or "British" were sometimes stirred into the already stressful process of mutual readjustments:

> Children had difficulty with parents who were hurt at their un-British frankness and horrified by their American speech. Some . . . youngsters could not resist using all the slang at their command in order to tease. Sixteen- and seventeen-year-olds had trouble with mothers and fathers who kept telling them to "wash that paint off your face" and to "be in by nine o'clock."

> Parents had difficulty having to face children with healthy appetites and (what seemed to them) luxurious tastes, with the realities of British austerity. One mother received a shock when her twelve-year-old son, trying to be thoughtful on his first night home, told her that he didn't want much for breakfast, "just a banana and a glass of milk." Bananas, she had to remind him, had not been seen in England for six years, while fresh milk was available only for infants and tots.

> Some parents had to cope with restlessness among teenagers who had difficulty planning careers because it seemed to them the only opportunities lay in America. They sometimes found restlessness, too, because youngsters were having trouble in schools which required a background in some subjects neglected in America and were apt to be more stern in discipline . . . [13]

The case of the unaccompanied Cuban children sent to the United States in the early 1960s is quite different because upon arrival most spoke Spanish and not English and with few exceptions family reunions took place in America and not in Cuba. By the time parents arrived in the United States most Cuban children had already evolved through several of the well-known acculturation stages.[14] The majority of children had become relatively fluent in English while some of the youngest ones had forgotten their Spanish. Those children who had been cared for by relatives, Cuban families, or in the Miami area, where bilingual classes were established in the public school system, fared better in this regard. As a result, family reunions were often smoother.[15] Most children, too, had already begun acquiring at least the external manifestations of American "popular" culture as evidenced by changes in the children's eating habits, dress, hairstyle, music, and dance preferences. Again, children who remained in Cuban homes and communities were able to retain more of their original cultural practices which facilitated easier transition into their natural families.

Several studies on evacuated children point out that family reunions can be complicated by other stresses and deprivations which place children at increased risk. Ella Rasanen, for example, in a follow-up report currently in progress on the adult outcomes of Finnish children sent to Sweden during World War II, found that the "greatest problems of all those interviewed were connected with their return to Finland."[16] Again, the vast majority of these children had made solid attachments to foster parents and had become fully integrated into community life:

> Because of this, children were very often estranged from their biological parents, especially if the stay was prolonged over several years. Separation from the foster home, when the child returned to Finland, has for many children been a very trying experience.

The situation has been made even more difficult, after the return home, by the fact changes have taken place in the childhood home: either one of the parents has died, new brothers or sisters have been born, there is a new mother or father in the family, and the child does not understand the language spoken by the family members, because the child has forgotten his native language. Especially this last factor has caused many difficulties in human relationships; in addition, difficulties have arisen because the children have in Sweden become used to considerably better living standards than they were offered in their Finnish homes.[17]

While the Finnish children's reunions were complicated by time, language differences, and new membership within the family, reports on Nigerian children after family reunification also suggest the increased risks encountered by children returned to families and villages impoverished by war. A follow-up study provided by the International Union for Child Welfare points out that a number of the reunited children suffered from malnutrition, impoverished living conditions, and reabandonment after family reunification.[18] From the case examples cited in this report it appears that most of the difficulties involved children with physical handicaps or special medical problems, families weakened through death, significant declines in their standard of living, by malnutrition, and in many instances, by a combination of several of these factors. Again, follow-up services were essential in identifying and providing assistance to vulnerable children and families.

Finally, in the past children have been removed from long-standing family caregivers and returned not to natural parents but to unknown relatives or unknown adults from the child's original nation, religion, or culture. Van Grevel has discussed this issue with respect to Jewish children taken in by Dutch families and hidden from Nazi persecutors during World War II.[19] The case presented on "Benjamin" is indicative of the harm that results when essential psychological ties are not recognized and respected:

"Benjamin," born in 1942, was discovered in 1945 living with poor peasants who had four children of their own, aged 6 to 16. Members of the Resistance, they had managed to save the baby by snatching him from a German car, and the entire family had become deeply attached to the child. The mother was a remarkable woman and the harmonious atmosphere in this modest household may be ascribed to her influence. Because of Benjamin, the whole family had had to go into hiding, but they preferred the life of the Maquis rather than give up this particularly attractive child, who was obviously of good breed, intelligent and musical, and who got on well with the other children. When, in 1945, some members of the child's family were found, the Tracing Commission decided that it was in Benjamin's interest to return to his original milieu, a more cultivated one than that of the simple peasant family. He was given into the care of an aunt, but this first attempt at replanting was a failure and harmful to the child. He was filled with resentment both toward the aunt and the social worker who had established the contact with her. The changed environment and the different customs were too much of a shock for him, in addition to which the aunt had not succeeded in gaining the child's affections. If it had not been for the presence of one of his foster sisters who stayed with him at the aunt's, and the influence of his foster mother, the traumatic shock would have been still more serious. He was returned to his foster family. Later, another member of the family turned up who made a favorable impression on the Tracing Commission. This uncle was entrusted with the guardianship of the child but was advised to leave him in the care of his beloved foster family until he was eight.

However, the uncle preferred to take the child right away and soon after emigrated overseas . . . "[20]

In an effort to fulfill its mandate—returning children to biological families—the Tracing Commission deprived this child of the only real family he knew.

Unfortunately, there is very little follow-up research on the adult outcomes of unaccompanied children who returned to their parents. Maas' study, "The Young Adult Adjustment of Twenty Wartime Residential Nursery Children," is perhaps the most comprehensive report.[21] Although this study is based on a small number of adults who were looked at retrospectively through standardized interviews, it does afford some tentative insight into how these children fared as adults.

Maas identified a group of twenty individuals who, as children between the ages of one and four, had been separated from their own families for about one year and cared for in three different wartime nurseries. Through standardized interviews, Maas rated their adjustments as young adults in five areas: *life feeling* (apathy, inability to express feelings, depression, low self-esteem, etc.); *inner controls* (impulsiveness or overcontrol, lack of manifest anxiety, antisocial tendencies, etc.); *relationships with people* (lack of emotional attachments, social isolation, shallowness, short duration, etc.); *performance in key social roles* (as, for example, son, daughter, husband, wife, parent, employee, etc.); and *intellectual functioning* (low, unstable, etc.).

Maas concludes that those children separated at the age of one did show some evidence of psychological scars in their young adult years, and as a group fared worst of all. The two-year-olds appeared less harmed than the three-year-olds, while children separated at the age of four fared best of all.

Yet, as a total group, Maas suggests that:

> Although these twenty young adults may have been seriously damaged by their early childhood separation and residential nursery experiences, most of them give no evidence in young adulthood of any extreme aberrant reactions. . . . Where there is evidence in individual cases of aberrancy in the adjustment of these young adults, in almost every case the data on their families seem sufficient to explain it. Although our design called for the inclusion only of persons from intact families without gross pathology, as the families become better known, so did their disabilities.[22]

Rasanen's follow-up report on Finnish children is still in its initial stages of investigation.[23] To date, it has only reported findings based on generalized questionnaires and has not progressed to more in-depth case studies. With these limitations in mind, this pilot study suggests that most of these children have, as adults, been able to surmount many of their strenuous life experiences and difficulties. None of the interviewed persons had mental disturbances which demanded permanent care, and there appeared to be few overt problems with human relationships as evidenced by extremely low divorce rates. However, what may be of significance is the fact that two-thirds of the respondents have suffered from permanent disease or disability, including chronic back problems, endocrinological disturbances such as diabetes, deficient functioning of the thyroid gland, blood pressure, and troubles with the joints. As with the Maas study, it is impossible to connect these disabilities solely to the experience of parental separation. More likely, they are the result of the multiple separations from parents, foster parents, language, and culture this group had experienced.

In sum, there are several factors which affect family reunification efforts. First, from a child's perspective, the significance of family separation will depend upon its duration and the developmental stage during which it occurs. The younger the child the more quickly separation from parents will be experienced as a permanent loss and the more readily new and often more important attachments with alternative caretakers will be formed. Children who are removed from long-standing "substitute" parents and returned to estranged biological parents or other relatives are thus subjected to two painful and potentially harmful separation experiences.

Secondly, after reunions, differences in language and cultural practices can complicate the already difficult process of mutual adjustment between the child and his family. Children who are cared for within their own native milieu have a clear advantage over those who are sent to different countries.

Finally, simply returning unaccompanied children to absent parents does not guarantee relationships will evolve or that developmental needs will be met. Follow-up services aimed at identifying and providing social and material services to vulnerable children, families, and communities are essential.

15

Cross-Cultural Placements

JAN LINOWITZ AND NEIL BOOTHBY

During wars and refugee situations unaccompanied children have been placed in countries other than their own. In the past thirty years, cross-cultural placement of unaccompanied children has occurred most often through international adoption and organized resettlement programs. This final chapter in this section thus reviews studies on the adjustments of transculturally adopted children and recent literature on the resettlement of unaccompanied minors from Vietnam, Cambodia, and Laos.

TRANSCULTURAL ADOPTION

No matter how smooth the initial transition to a new home, and how fine the fit between the child and the adoptive family, adoption entails psychological risks. Adoptive parents assume full responsibility for a child they did not create and whom they must learn to love. Adopted children must accept new families to replace the ones they have lost and must forsake the histories of their biological families. This loss of origin causes some adoptive children and adolescents to suffer confusion about their place in the world.[1] However, most adopted children apparently cope well with the special difficulties they encounter; in a review of follow-up studies of adoption, Mech found that three out of four adoptions were judged successful after initial adjustment periods.[2]

Children who are adopted transculturally face unique risks to psychological development. Foreign children adopted by Western families must enter an unfamiliar world of unknown expectations and learn to define themselves as both distinct from and a part of the culture of their adoptive families. In all of this, transculturally adopted children must rely on the responsiveness of their adoptive parents and the flexibility of the larger community to help them know and love both their families and themselves.

Literature on transcultural adoption probes the extent to which these children find the acceptance and security they need in their new homes. Attention to the psychological well-being of the individual child is, however, frequently obscured by ideological battles over the ethics and politics of adoption practices. In all questions of child custody, write Goldstein, Freud and Solnit:

> The child's best interests are often balanced against and frequently made subordinate to adult interests and rights. Moreover . . . many decisions are "in-name-only" for the best interests of the specific child who is being placed. They are fashioned primarily to meet the needs and wishes of competing adult claimants or to protect the general policies of a child care or other administrative agency.[3]

Proponents and critics have more at stake in the discussion of transcultural adoption than the psychological needs of individual children; consequently, a thoughtful discussion of the literature on transcultural adoption must distinguish between serious attention to the psychological well-being of the children and ideological arguments for and against this practice. In what follows, an effort is made to maintain that distinction in order to gain an accurate sense of how much we know about the psychological well-being of transculturally adopted children and to consider the viability of these special placements for unaccompanied children.

IDEOLOGICAL ARGUMENTS

Supporters of transcultural adoption argue that the best interests of the individual child are protected by a placement outside the child's native land. "In many countries, individual human life is not particularly valued," suggests Joe. "The abandoned child is likely to die in the streets, or to grow up in an impersonal institution."[4] Transcultural adoption offers the possibility of "universal family insurance" for every child who loses his or her family:

> . . . as far as adoption is concerned, the entire world is a single system . . . The tendency is for social problems to be more alike than different in different countries despite real differences in cultural backgrounds, values and historical experience. This is not a consequence of cultural imperialism but rather the consequence of situational imperatives which generate similar problems and evoke experimentation with similar solutions.[5]

Homeless children require families in which to grow up; the needs of the individual child, argue advocates, rather than the political consequences of transcultural adoption, deserve primary consideration.

Critics of transcultural adoption believe that the practice offers, at best, a short-term solution to the problems of homeless children; it ultimately provides neither long-term solutions to the struggles of developing nations nor permanent security to individual children.[6] To the people of Third World countries, it reinforces the message that they are unable to care for their own children. For children adopted in Western countries, it creates a situation in which they are likely to suffer confusion about their origins, personal histories, and places in their adoptive families. According to Melone, "if . . . the aim . . . is to give the child concerned a family, it should also be to give the future adult a country. The adoptee is not a refugee."[7] By removing children from their native lands and raising them in families and communities where they will always be different, Western countries may condemn these children to the psychological status of refugees.

"The best interests" of the child thus receives different definitions from different spokespeople. Proponents of transcultural adoption stress the importance of secure family membership for psychological development. Ethnic identity is considered to be secondary. Critics argue that cultural transplantation deprives the child of fundamental knowledge about his or her place in the world. "What may be advantageous developmentally for the small child," states Berlin, "may rob him of his cultural heritage and be devastating to him in his later development."[8] In emphasizing the child's need for a family, advocates may overlook the lifelong need for affiliation with one's group and with the history of one's people.

REVIEW OF RESEARCH

Transcultural adoption has received less research attention than transracial adoption; this may be explained in several ways. Transcultural adoption has excited less controversy than transracial adoption in the United States and England, for example, and its practice has been less influenced by political rhetoric; the need for thorough research to inform policy has appeared less acute.[9] In addition, it may be more difficult for researchers to generalize in a meaningful way about the impact of transcultural adoption. Preadoptive experiences of children from different foreign countries may vary greatly. Ethnic heritage and the receptivity of the adoptive community to ethnic diversity strongly influence the child's adoptive experience. Some researchers have restricted study to populations of a specific national heritage.[10] Most studies, however, have included adopted children from a variety of national backgrounds.[11]

A number of these studies discuss the initial adjustment problems of transculturally adopted children based on parents' reports. Several problems in the first months of placement are consistently noted. Many children have voracious appetites; they are unable to leave food uneaten and may hoard food in their bedrooms.[12] Sleeping problems, including nightmares, night terrors, insomnia, and discomfort with Western beds, are common.[13] Frequent crying, clinginess and fear of separation, and fear and rejection of native foods, language, and visitors have all been noted.[14] Most authors, however, have emphasized the transience of these symptoms, and the capacity of the children to adapt to the linguistic and cultural demands of a new home and community.

There are several follow-up studies of transculturally adopted children. All stress the successful adaptations of most adoptees. Hochfeld reviewed social workers' reports on 1,399 children adopted in the United States between 1954 and 1959; by 1960 only twenty-five of the original number had been placed into a second adoptive family or an institutional setting.[15] He interprets this finding as an indication of the mutual satisfaction of adoptive children and their American families. Rathburn, McLaughlin, Bennet, and Garland interviewed the parents and teachers of thirty-three children, most of whom were of European background, who had been adopted by American families in 1958.[16] They found that sixty-four percent of the group was doing "adequately" or better in school and at home. Welter compared questionnaire responses of social workers on the adjustments of thirty-six foreign-born and thirty-six American-born adoptees who were matched on age at placement.[17] Although both groups were reported to be adjusting well, the foreign-born children appeared to be doing better than the American-born adoptees, both at home and in school. While these early studies of transcultural adoption provide encouraging evidence of resilience and successful adjustment, the range of ages, years in placement, ethnic backgrounds, and preadoptive experiences of children in the sample groups, and the reliance of researchers on parent, teacher, and social worker reports, limits interpretation of the findings.

Pruzan investigated the adjustment of sixty-eight Korean and ninety-two mixed-race German children adopted by Danish families.[18] The children were between the ages of eight and twelve at the time of the study. All had lived in Denmark for two years or longer. Pruzan interviewed parents and teachers and analyzed the results of classroom sociometric tests. The adopted children were compared to Danish-born, nonadopted peers. The adopted children were reported to be doing as well as the control group both at home and in school. Similarly successful adjustments of transculturally adopted children were reported by Hoksbergen in the Netherlands and by Gardell in Sweden.[19]

The Gardell study was distinguished by the inclusion of viewpoints expressed by the adopted children themselves in individual interviews. The children were between the ages of ten and nineteen at the time of the study and had lived with their adoptive families for at least five years. Over two-thirds of the group was of Korean ancestry. Gardell emphasized the desire of these children and adolescents to be considered as being the same as their peers:

> The children themselves maintain that they both are and feel like ordinary Swedish children. They cope well and they cannot see anything odd or distinctive about themselves. They dislike it when people regard them as different.[20]

Although they expressed the wish to be regarded as "ordinary," they did not deny the extraordinary course of their own childhoods. More than half of the children who were four years or older when they arrived in Sweden responded affirmatively to the statement, "I often or very often wonder about the conditions I lived in before coming to Sweden." Many of the children remained interested in and concerned about their pasts, while wishing to be acknowledged in the present as Swedes.

D. Kim reported a similar emphasis on the present among transculturally adopted adolescents from Korea living in America.[21] The young people "tended to lose Korean cultural patterns rapidly and to identify themselves as 'Americans' or . . . 'Korean-Americans.'" Kim reviewed questionnaire responses of parents and children, and the children's performances on the Tennessee Self-Concept Scale. He concluded that the 406 young people and their adoptive families were doing well, and that most adoptees had been able to make an "impressively healthy, normal development adjustment" over time.

While both Gardell and Kim reported a loss of interest in Korean cultural heritage among their subjects, it is important to note that both studies focused on adolescents. American and Western European adolescents of all cultural backgrounds often wear clothes, listen to music, and engage in the activities most favored by their peers, in efforts to be accepted. A wish to be "ordinary" may be a response to adolescent social pressure rather than a reflection of the rejection of cultural heritage. It is important to note, however, there are no follow-up reports on adults who were transculturally adopted as children; studies of these adults would be useful in achieving an understanding of the ways in which adopted young people ultimately come to terms with their pasts.

Several authors have compared the adjustments of children transculturally adopted in infancy and preschool years with those of children adopted at school age and later. Older children are more likely to have serious initial adjustment problems.[22] But follow-up studies report equally successful social and psychological adaptations over time. Gardell found that children adopted at an older age were, in fact, better accepted by their peers than children adopted when very young.[23] She suggests that older children may be better able to understand the adoption process; although they may be more active in initially rejecting new people and routines, they may be less profoundly overwhelmed by their move to a new land and less damaged by it.

While most studies of transcultural adoption stress the transience of adjustment problems and the resilience of adopted children, a few authors have focused on the risks of this kind of placement.[24] Most transculturally adopted children bring with them histories of deprivation, loss, and the psychological vulnerability that results from drastic cultural change. Arriving in their new homes, they face the task of adjusting to unfamiliar faces, voices, and expectations. Adoptive parents are suddenly responsible for welcoming, caring for, and comforting these new and bewildered members of their families. The expectation, writes Brazelton, that "a rather magical 'bonding'" will readily occur reflects a naive "myth of the adaptability of parents . . . and . . . the adaptability of a needy child to an entirely strange, new, and overwhelming set of stimuli."[25] The adjustment is often a difficult one: disappointment, anger, and a sense of failure are often experienced both by adopted children and by their new families.

In a recent article, two adoptive mothers describe the serious problems encountered in the five years since they adopted their Korean daughters.[26] The authors and their husbands had approached the adoptions with confidence and enthusiasm, but both mothers found themselves unable to comfort and communicate with their new daughters. Although the two girls had very different ways of expressing their confusion and unhappiness, the two mothers had similar feelings of inadequacy and reported similar tensions in their marriages and families as the "magical bonding" they hoped for failed to occur. As the girls grew older, their powerful initial distrust of others and their emotional volatility interfered with their relationships with siblings, peers, and teachers. S.P. Kim and Wolters described similar instances of the failure to establish trust and intimacy between parent and child early in the adoptive relationship, and subsequent feelings of guilt and anger in adoptive parents.[27]

The guarded conclusions of these case studies bring into question the optimistic results of most large-scale research on transcultural adoption. What could account for this clear difference in findings? It is possible that the children and parents selected for case study are unusual. The children may be particularly troubled, incompetent, or the match between parent and child unusually problematic. Wolters offers another possible explanation: in large studies, parents may be reluctant to report difficulties and disappointments:

> There can be a kind of *duty to be joyful and happy* about the adoption of their child prevalent in the parents of adopted children which makes this more negative side of adoption unacceptable. You must be happy; it is a wonderful thing to adopt a child from a far-away foreign country.[28]

Large-scale studies may reflect parents' needs to be positive about their decisions. It is interesting to note, however, that teachers and social workers generally evaluate transcultural adoptive placements in postive terms, as well, despite the fact that they are less invested in the success of the adoption than parents. The reason for the discrepancy between case studies of transcultural adoption and larger investigations thus remains unclear; however, the notes of caution sounded by Brazelton and others must be attended to by proponents of the practice. At the least, parents should be prepared to expect an extended period of mutual adjustment following the adoption.

DISCUSSION

The child who has lost his or her birth family "has already been deprived," write Goldstein, Freud, and Solnit, "of his 'best interests.'"[29] The task, thus, is not to find ideal placements for these children but to find placements that, in Winnicott's language, are *good enough* to allow children to recover as much as possible from the losses of their early lives and to provide enough love, guidance, comfort, and stimulation to enable children to grow into adequately functioning adults who have a sense of their place in the world.

Transcultural adoption is an imperfect placement option; it carries the risk of psychological and social confusion. Clearly, it is best when parentless children can be placed in families within native communities and homelands. When, however, such placements are not available the risks of transcultural adoption must often be compared with the most likely alternative—institutional care.[30]

As previously discussed, institutional care during time of major crisis is often of poor quality. For infants and young children, prolonged placements within institutions that lack adequate adult care and stimulation often result in impaired cognitive and social development. Seen from this perspective, then, research findings on transcultural adoption are encouraging; parents, teachers, and social workers largely agree that most of these children do well at home and school. It is not known, however, to what degree these children are, as adults, able to find fulfillment through commitment to other people, to their work, and to the pleasures of life; follow-up studies of adults who were transculturally adopted children have not been undertaken.

The controversy over transcultural adoption will and should continue. Critics and advocates will continue to look, and should continue to look, for better homes for these children. During emergencies, efforts need to be expanded to secure family placements for unaccompanied children within their own communities. Yet, when these families are neither present nor forthcoming policy makers must have the courage to act in the best interests of children and not according to ideological argument or political pressure. No infant or young child should be denied a good enough home while society searches for a better one.

RESETTLEMENT OF UNACCOMPANIED MINORS

Unaccompanied minors who cannot be reunited with their families or repatriated to their homelands are often resettled in other countries. The host country assumes legal responsibility for the minors it accepts and determines where and with whom each child will live. The placement decision has an important influence on the experience of the minor in the host country. Choice of placement affects the child's emotional and social adjustment, identity, and career plans. In this discussion of resettlement, different placement options will be explored, and the psychological and political consequences of their use will be discussed.

Since 1945 unaccompanied minors from Europe, Asia, Africa, and Latin America have been resettled in times of crisis. The circumstances leading to resettlement have ranged from natural disasters to wars. Some youngsters have been cared for in neighboring countries with familiar languages and cultures; others have been sent to distinctly different cultural and linguistic environments. It is difficult to generalize about the experiences of all modern unaccompanied minors. Instead, this discussion will focus on one recent group of unaccompanied minors in the hope that the lessons learned in their resettlement will be helpful in planning programs for other groups.

In the past decade, thousands of Southeast Asian youngsters have been resettled in Western countries. While most have been males over the age of fifteen, female adolescents and younger boys and girls have also been resettled. Although their personal histories vary, all have become separated from or lost their parents and have come under the care of the state in host countries. Several placement options have been employed with these minors from Vietnam, Cambodia, and Laos. Many live in foster care and participate in the daily life of a host family. Minors are usually placed with families of the host culture called majority families; however, an effort is sometimes made to place a minor in a family of his own cultural background. Other minors are placed in group or institutional programs. In group care, a number of unaccompanied minors live together under adult supervision and participate in the daily life of the host community. They attend school, hold jobs, and make friends outside of the group home. In contrast, in institutional care, all basic needs are met within the facility; the institution maintains primary responsibility for the academic, social, and occupational growth of the minors.

Foster, group, and institutional placements provide very different kinds of care. In what follows, the strengths and weaknesses of these placement options will be reviewed. There is little published research comparing resettlement programs for unaccompanied minors. Adjustment to a new culture is a complicated, personal process with no decisive end-point; differences in adaptation are difficult to quantify. This report will therefore review information gathered from published and unpublished studies, descriptive reports, as well as from visits to resettlement programs in Switzerland, France, the Netherlands, Great Britain, and the United States.

RECEPTION CARE

When unaccompanied minors arrive in a host country from Southeast Asia they may be received in one of several ways. Minors who arrive as part of a large group of refugees are usually sent to refugee camps or centers where they are received with their group and introduced to the host culture. Minors who are sent from transit camps in Asia to Western countries for resettlement are often received in programs specially designed for unaccompanied youngsters. Here, several models of reception care for these unaccompanied minors will be described and discussed.

Reception of Large Groups of Refugees

Unaccompanied minors often arrive in a host country as members of a large group of refugees which has been accepted en masse.[31] The group is temporarily placed in a reception center or camp, where youngsters who are not accompanied by parents or guardians are identified. Some of these youngsters may have been on their own since leaving home in Southeast Asia. Other minors may have formed close ties to refugee families during their journeys, but are identified as unaccompanied either because the relationships break down or because reception officials fail to recognize the special bonds between the minors and their informal "foster families." (See the previous discussion on Refugee and Displaced Persons Camps.) Youngsters labeled as unaccompanied are usually cared for by special staff, and housed as a group.[32]

While refugees are in reception care, information is gathered about their health, skills, and family ties, and plans for resettlement placement are made. In addition, the refugees are usually given some information about the host culture.[33] The length of reception programs varies among the resettlement countries. In Britain, for example, most refugees were in reception from three to six months; in Australia, the orientation program lasts one to two years.[34]

There has been little attention directed specifically to the experiences of unaccompanied minors in large reception programs. A study of the Dutch reception program reported that unaccompanied minors in the Netherlands tended to have longer stays in reception than did other refugees.[35] One investigation of the mental health of Vietnamese refugees found that the unaccompanied minors presented an "acute mental health problem" in one large reception camp.[36] Harding and Looney described these youngsters:

> Many of these children were significantly depressed. While somatic complaints were the most prevalent expression of distress, sleep disturbances, tantrums, violent antisocial behavior, and marked withdrawal were also apparent. . . . Being thrust together, they potentiated each other's sadness and hopelessness.[37]

Housed apart from other refugees, the minors felt isolated from the rich family life of the camp and became despondent over the prospect of being resettled alone in an unfamiliar land.

The transition from reception center to final placement also presents special problems for the unaccompanied minor. Refugee families are often able to move together into permanent placements; in contrast, unaccompanied minors are frequently separated from peers and familiar adults and placed in foster homes or group programs. The Australian government, which sponsors a one- to two-year orientation program for all refugees, has made a particular effort to ease the transition into resettlement care for the minors. At the end of their orientation stay, unaccompanied children are often placed in the surrounding community so that school attendance is not interrupted and friendships begun in reception can be maintained.[38]

Reception of Unaccompanied Minors

Many unaccompanied minors flee from their homelands to refugee transit camps in Asia. These youngsters are usually identified as unaccompanied in the camps before they are accepted for resettlement. When they arrive in host countries they usually enter programs designed for unaccompanied minors, rather than first being held in large refugee centers.

Sometimes these youngsters are placed in permanent care as soon as they arrive in the host country. In the United States, for example, many unaccompanied minors go straight from the airport to their foster families and do not enter a residential orientation program.[39] The primary advantages of direct resettlement placement are: first, that the number of transitions the minor makes before resettlement is reduced; and second, that the minor is immediately given a family and home to call his own. The major disadvantage is the difficult challenge that both youngster and foster family face in making the placement succeed. Local agencies rarely receive complete and accurate information about youngsters before they arrive, despite the fact that this information has often been gathered while the children were in camps. Carefully matching of minor and foster family is therefore impossible. Similarities or differences in expectation or temperament cannot be fully considered when making the placement decision. In addition, the minor is often separated from his peers and placed in a family where no one speaks his language. His understanding of foster arrangements and his knowledge of American customs are likely to be incomplete. Some agencies make a bilingual worker available to the youngster; however, since the worker does not live in the foster home and cannot be available to the minor twenty-four hours a day, confusion and misunderstanding in the first days of placement seem inevitable.

One way of minimizing the difficulties and of ensuring careful placement planning is to provide residential orientation care for arriving minors.[40] In a structured orientation program the minor is given a thorough explanation of his legal and financial status, and a brief introduction to the host culture by workers who speak his language. He shares initial reactions, disappointments and fears with peers who have traveled with him. The staff has time to assess the adjustment of the minor and to question him about his expectations of care before making a placement decision. If a foster family is selected for the youngster, he can have several visits with them and get to know them a bit before moving into their home. Proponents suggest that all of these measures make the adjustment to foster care easier for both youngster and family.

Supporters of orientation care stress the importance of regular contact with the host community. If the minors are placed in an isolated orientation program with only their Southeast Asian peers, they will gain little sense of the host country.[41] A study of a West German orientation program reported that minors placed in a children's hospital for an extended orientation program were unable to form a "realistic relationship" to German society, since they had little contact with German peers.[42]

France has developed an orientation program that differs from those described above; it consists of two residential reception placements.[42] When refugee minors arrive in France they join other refugees for a two- to three-week stay in a transit center. Youngsters who are unaccompanied are identified there, and efforts are made to find relatives who may be living in France. No French lessons, orientation classes, or structured activities are provided.

Unaccompanied minors leave the transit centers for placement in a Welcoming and Observation Center outside of Paris where they stay for two to six months. They live with forty to fifty Asian minors and attend classes for three hours a day with teachers who speak their native languages.[44] When they are not in classes, they are supervised by monitors, who observe their social and emotional development. Contact with French peers and with the large Asian community in Paris is very limited.[45] At the end of his stay, the minor chooses, with the help of program administrators, to enter foster or institutional care. Information gathered on the minor is recorded on a four-page form describing health, conduct, family background, social and emotional level, and scholastic performance, and passed on to the agency which makes the final placement determination. However, because minors have had little experience outside the reception center, they must choose between a familiar (institutional placement) and the unknown French foster care.

PLACEMENT OPTIONS

After reception the unaccompanied minor is usually placed in one of three forms of care: foster home, group home, or institution. Most minors remain in care until they reach adulthood. The placement decision is thus likely to have an important impact on the adjustment that the minor makes to the host country and the plans that he makes for his future. In what follows, the three forms of care will be described, and differences and similarities between programs will be examined. In addition, two forms of care which are used less frequently will be discussed: independent living and adoption.

Foster Care

Foster family care is the most widely employed placement for unaccompanied minors. The great majority of Southeast Asian unaccompanied minors in the United States live in foster homes.[46] Foster care is also used in Germany, France, the Netherlands, Australia, and elsewhere.[47] Youngsters usually live with families of non-Asian, host country heritage; these families are called *majority families.* When a youngster lives with a family of his own cultural background, his placement is referred to as an *ethnically similar family.* In both majority and ethnically similar foster care, the foster child is treated as a member of the family, rather than as a formal guest or servant. Foster care may extend to the minor's age of majority or longer, although placements may be terminated before the youngster becomes an adult if the foster family, the minor, or the agency is unhappy with the arrangement, or if the minor is reunited with his natural family.

When unaccompanied children are placed in majority foster homes, efforts are sometimes made to select several families within the same community or region so the children will be able to maintain relationships with other Southeast Asian youth. This practice is referred to as *clustering.* In the United States, for example, unaccompanied children placed in foster families are often clustered in an attempt to safeguard the minors' cultural traditions while promoting their integration into American society.[48]

In reality, however, the cluster concept takes on a variety of forms.[49] Sometimes, twenty to thirty unaccompanied children are resettled in foster homes in a tight-knit community; the children attend the same school, play on the same soccer team, help each other at night with their homework, and are able to maintain their native tongue. More often, however, geographic distance or weak links between foster families limit peer interaction to an occasional visit or an agency-scheduled monthly get-together.

Placement agencies use a variety of criteria in determining a family's suitability for fostering a youngster. Some placement workers emphasize family openness to new experiences and different cultures and flexibility in expectations.[50] Workers sometimes require that the parents have children of their own so that the foster child will not be isolated and will not receive all of the parents' attention.[51] Patience, interest in another culture, and humor are among other characteristics valued by placement workers.[52] Moreover, some agencies encourage foster families to accept more than one unaccompanied minor, and as will be shortly discussed, others give first priority to ethnically similar foster parents. Interviews, home visits, and even meetings with a psychiatrist are sometimes required for evaluation; however, even the most thorough investigation cannot eliminate the difficulty of estimating the ingenuity, enthusiasm, and patience of a prospective foster family.[53]

One measure of the difficulty of predicting successful matches between family and youngster is the replacement rate in care. The replacement rate indicates the proportion of unaccompanied minors who change foster care placements one or more times. This important statistic is not always available from placement agencies since it requires thorough information abut the course of care and the reasons for placement terminations.[54] A number of American foster care programs have gathered the relevant data and have released information about replacement rates. Most of these programs report that forty to fifty percent of placements break down within three to five years.[55] The rate is often lower for the first year of placement but increases as the years pass.[56]

What causes foster placement breakdowns? The explanations offered by agency and child care experts vary. Some causes—age, family background, and exposure to differing kinds of trauma and deprivation—have already been addressed; other causes—the haste of placement, cultural differences, cultural isolation, the lack of support services, the temporary nature of foster care, and individual circumstances—will be highlighted in this current discussion. It is interesting to note that research on domestic foster care in England and the United States yields comparable replacement rates; long-term foster care of American and British children is apparently as vulnerable to disruption as is the care of Southeast Asian children, although the causes of breakdown may be different.[57]

Experts suggest that the mutual adjustment of the minor and the foster family is a gradual process. During the first weeks or months of placement, the minor is usually compliant, agreeable, and easy to live with. As he settles into his new home, clashes in expectations and preferences are inevitable. Both the minor and the foster family may experience a range of emotions, including disappointment, anger, depression, and guilt.[58]

Boothby, for example, noted how unaccompanied children and foster parents initially come together with different sets of perceptions and expectations.[59] In Khao I Dang, unaccompanied Khmer children were interviewed and asked to draw pictures of their new homes just after being informed of resettlement plans. Understandably, most children did not have a realistic appreciation of what life in the West or in a Western family would be like. Rather, the children's drawings and comments clearly indicated that they envisioned new homes and families as being much like those they had enjoyed in Cambodia prior to the 1975 Khmer Rouge takeover. Equally important, the author notes, is that in Khao I Dang, most adolescents "were not so much preoccupied with thoughts of entering into relationships with new parents as they were with escaping what was for them a difficult life in camp."

In contrast, the study suggests that for foster parents in the United States a different set of expectations had often evolved prior to the arrival of the foster child. Without the aid of concrete information or photographs, many foster parents had already formed mental images of what the child would look like, what he or she would be like, and indeed, in dreams and reveries, had "rehearsed" what they in turn would say, think, feel, and do when the long-awaited union finally occurred. Most foster parents had high hopes of the children becoming full members of their families and had already begun thinking of them as their "sons" or "daughters." The fact that many unaccompanied children and foster parents have entered into new relationships with contrasting expectations has led to confusion, misunderstandings, and feelings of betrayal on the part of both parties.

Differences in perception and expectation may result in adjustment problems within foster care placements, especially when unaccompanied children remain in contact with their natural parents in Southeast Asia or other parts of the world. Jockenhovel-Schiecke, for example, noted that when foster parents attempt to assume a parental role with an adolescent, or expect him too quickly to become a "member of the family," the adolescent is likely to resent or misunderstand their intentions.[60] He may remain "emotionally distant" and focus on financial, academic, or vocational goals. Moreover, the minor's awareness of his natural family's continuing struggle in Southeast Asia can interfere with his enjoyment of foster family life. His feelings about his new home will be colored by his guilt and worry about the fates of other family members. Consequent confusion and disappointment may lead to the breakdown of the placement, and yet another disruption for the adolescent.

The impact of this conflict of loyalties is illustrated in a study of Vietnamese minors living in France.[61] Chi Lan Do-Lam compared ten unaccompanied minors living in French foster families with fifteen minors living in institutions, and thirteen minors who had come to France with their own families. She administered projective tests and interviewed the youngsters individually. She reported that the foster youngsters experienced more guilt and depression than either of the other two groups of children. They expressed concern about their families and felt undeserving of the care they were receiving. They had little nostalgia for life in Vietnam and spoke favorably about French culture and society. In contrast, the institutionalized youngsters expressed less guilt and depression, but were much more anxious about what the future held for them. They were nostalgic and idealistic about their lives in Vietnam and critical of France, although largely misinformed about French culture. The children living with their own families were most open about expressing both satisfaction and disappointment with French life, and were the least psychologically troubled of the three groups.

A number of placement workers have also noted the impact of differing styles of communication on minors' adjustments to care.[62] The problems cited by workers in the various resettlement countries were very similar; despite contrasts in cultural style, Westerners share an orientation towards family relationships that is different from that of most Southeast Asian youngsters. While Western culture places value on direct communication between parents and children, Southeast Asian youngsters are taught to defer to adults and to express dissatisfaction in an indirect manner. Misunderstandings are a frequent consequence of this difference in expectations. Western foster parents may express both affection and anger more openly than the minor. Physical affection and terms of endearment from Westerners whom the youngster barely knows may seem intrusive and ingenuine to the minor; a momentary explosion of temper may be taken far more seriously by the minor than the adult intended it. In contrast, the Westerner may be surprised to learn that after weeks or even months of apparent contentment and obedience, the youngster is unhappy with his situation and has not always been straightforward in discussions. The Western adult expects frankness in conversation; the minor expects cordial and smooth social interaction. These two styles are very different, and are only compatible if they are mutually understood and accepted. Obviously, language differences exacerbate the situation.

It is not known to what extent any one of these factors or the interaction of these and other factors contributes to placement breakdowns. Equally unknown is the extent of stress unaccompanied minors go through during this adjustment process, nor the ramifications of this stress on their long-term adjustments. However, several factors which appear to decrease the percentage of replacement rates have been identified.

A number of resettlement workers have stressed the importance of evaluating the minor's family ties when determining placements. Those who favor foster care suggest that foster parents must be carefully chosen and fully informed of the unusual circumstances of unaccompanied minors who remain in contact with natural parents. Rudnik and Molstad found that:

> The family which has been . . . able to understand that the minor will not be part
> of their family in the manner that natural children are, and that the youth may have
> a strong sense of allegiance to their own family of origin, will usually be the family
> that the minor finds it most comfortable to remain within.[63]

These authors argue that such a family can encourage the youngster to remain in close contact with family members and can help him accept the guilt and grief that result from being separated from them.

Even when unaccompanied minors and foster families are carefully matched, counseling and agency support play a crucial role in assisting the child and the foster parents through the adjustment process. Without this assistance placements may break down; a recent American study identified unaccompanied minors living in majority foster families as a "high-risk subgroup" of Southeast Asian refugees who are vulnerable to depression, family conflict, and replacement in care if they are not provided with adequate counseling and support.[64]

Two American programs for unaccompanied minors have reported a replacement rate that is substantially lower than that described by other agencies. Rudnik and Molstad describe one program in Minnesota in which only twenty-four percent of minors changed foster placements in three-and-one-half years in care.[65] The authors attributed this low rate of replacement to agency provision of intensive support and supervision to foster families. Each family received regular visits from a two-member team of a social worker and a bilingual Southeast Asian worker. The workers helped the American family and the Asian minor in their mutual adjustments. The authors stressed the critical role of the bilingual worker in the early stages of the placement, when the minor had little capacity to communicate in English. Secondly, Brown reported an equally low replacement rate in the Amherst, Massachusetts Program.[66] Brown attributed the relative success of this program to three factors: (1) the children are tightly clustered together and thus are able to draw support from one another; (2) the active involvement of the host community and its school system which acts as a support system of its own; and (3) the availability of professional counseling services for children and families who are experiencing difficult times which can help prevent foster care breakdowns.

Other authors have also emphasized the importance of regular contact between agency workers, minor, and foster family.[67] In a review of the Australian care system, Mathews attributed partial responsibility for a high replacement rate to the lack of professional support for families and foster children.[68] Mathews states that "the families ... complained about the lack of counseling and supervision; they had insufficient opportunity to discuss problems with qualified people and felt left alone." The French placement process has also been criticized for failure to provide regular follow-up and support to foster families.[69]

Ethnically Similar Foster Care

Foster placements in ethnically similar families have been implemented in several countries, including the Netherlands, Australia, and the United States. Often, the minor lives with a family which has only recently arrived in the host country itself. The advantages of ethically similar care are clear: the minor is placed in a family that shares his traditions and history. Critics suggest two disadvantages: first, that it may slow the minor's adjustment and participation in the new culture and, second, that it is difficult to find Asian families who wish to foster unrelated minors.[70] A variety of reasons are given for this difficulty in finding families; some resettlement personnel suggest that Southeast Asian families are unfamiliar with the foster care model; are too preoccupied with establishing themselves in the host country; or do not feel a commitment to unrelated refugee children. However, ethnically similar foster care has been successfully employed with families in some communities.

In Australia, for example, ethnically similar foster placements have been arranged for Vietnamese, ethnic Chinese, Laotian, and Cambodian children, and have been evaluated as "more positive than placements in an Australian family," although, "essentially more difficult for Australian social workers to assess."[71] Several resettlement agencies in the United States have also placed unaccompanied children in ethnically similar families. Some of these agencies use ethnically similar foster families primarily as back-up placements for children who are unable to adjust to majority families; others use ethnically similar foster care as the primary placement alternative. The Louise Wise Foundation in New York City, for instance, has placed all of the Cambodian minors that the agency has received in Cambodian families.[72] The agency workers report that the Khmer community feels a strong sense of responsibility and commitment to the parentless children.

A recent study by Nancy Schulz and Ann Sontz points out some of the advantages of similar foster care.[73] Using standardized questionnaires, the authors surveyed nearly 1,500 unaccompanied Southeast Asian minors resettled in the United States since 1975. A random sample of 420 of surveys formed the core of their analysis.

Not surprisingly, depression and depressive symptoms among the youth were relatively widespread: fifty-six percent of the youngsters in this study had suffered from incidents of depression or manifested depressive symptoms sometime during the resettlement process. When, however, the authors looked at the incidence of more severe depression, according to placement alternatives, a somewhat different picture emerged (see Table 15-1). Incidents of depression thus were lowest among children cared for in similar families. "The ethnic foster home," write Schulz and Sontz, "has been known to act as a sheltering and familiar cultural force following the trauma and difficulties surrounding escape and flight, camp life, and resettlement in a strange land. It can thus come to provide a strong and growing emotional resource during the course of an unaccompanied minor's adaptation to his country of emigration. This emotional and cultural framework may well be mirrored in the lower occurrence of depression among those placed in ethnic foster homes."[74]

It is important to note, too, that Schulz and Sontz found no significant differences between unaccompanied children in similar and majority foster families with respect to integration into American communities. Unaccompanied children appear to learn English, progress in school, obtain jobs, and make friends with American youth as quickly as unaccompanied children in majority families:

> Like many of their counterparts in American foster homes, children placed in ethnic foster families where close cultural ties and traditions are maintained and emphasized have evidently been able to initiate significant steps towards integration into local communities.[75]

Why do some agencies develop successful ethnically similar foster care programs while others reportedly have trouble doing so? Certainly, the concentration of Southeast Asian families in the placement community makes a critical difference. In this respect, the practice of resettling unaccompanied minors from camps in Southeast Asia to countries in which members of the child's cultural group do not already reside must be questioned. However, other factors within resettlement countries also influence the outcome of placement efforts. The experience of American social workers with the adoption of black children may be relevant here.

When public and private social service agencies began placing black children for adoption in the late 1950s, they had great difficulty finding black families who would adopt. Their recruitment campaigns in newspapers, radio, and television attracted

Table 15-1. Percentage of Children with Depression

Total Sample Group (420 children)	40%
Total in Foster Homes	36%
Children in Ethnic Foster Homes	22%
Children in Group Homes	57%
Children in Independent Living	52%

white rather than black respondents. The belief developed among the social work establishment that blacks were not interested in adoption.[76]

Critics of the adoption process in the United States have since argued that the policies and practices of the social agencies were responsible for the apparent lack of interest among black Americans.[77] The adoption agencies were largely staffed by white social workers who were unfamiliar with traditional black community networks and were thus unable to recruit black families. Those black parents who did express interest in adoption were uncomfortable with the intrusive approaches of adoption workers, the bureaucratic format of application procedures, and the rigid criteria for adoption, which often focused on economic position, and they were thus likely to drop out of the adoption process. In recent years, as blacks have become more prominent in the formulation of adoption policy, an agency practice has changed to meet the circumstances and needs of the black community, black adoptions have increased.[78]

The relevance of this experience to the use of ethnically similar families in foster care is clear. While many Southeast Asian families may be either not interested or not appropriate for foster care, there are indeed other families who can provide excellent care for minors. Western foster care practice may be unfamiliar to Southeast Asian families; this in itself is not a valid reason for dismissing the use of Asian families for the provision of care. Commitment to this approach, staffing, recruitment practices, worker expectations, and orientation procedures also have a profound influence on the availability of Southeast Asian foster families for unaccompanied minors.

Group Care

Unaccompanied minors in group care live together in a supervised home. Their daily lives include contact with the larger community. They go to school or work, go shopping, and meet friends in the city or town in which they live. Group placement is the dominant form of care in England; it has also been used in other host countries, including the Netherlands, Germany, and Australia.[79] The philosophy and structure of group care programs vary a great deal. The descriptions of two different group care programs illustrate the range of practices that fall within the definition of group care.

The Save the Children group home in Hampton Court, England, provides care for thirty-two Vietnamese minors ranging in age from five to eighteen. These youngsters have been together as a group since their arrival in England in the late 1970s. They now live on a large, lovely estate with several acres of gardens surrounding the house. A written description of the group program explains its purposes:

> It is considered of vital importance that these children should be kept together as a group in order both to preserve their cultural identity and to gain the support of others in similar circumstances. It is not the intention that these children should be fostered or adopted. The majority have parents or other relatives in Vietnam or elsewhere, and if they lost contact with their own language, culture and ways of life ... it would be difficult for them to settle back into their families should they be reunited.[80]

The effort is thus made to maintain Vietnamese tradition in the home. Most of the staff is Vietnamese, the meals are Vietnamese, and the children have the choice of sleeping on the floor, as Vietnamese youngsters do, or using Western beds. While they attend British schools and receive special tutoring in English, they speak both English and Vietnamese at home. They are encouraged to maintain contact with their families in Vietnam, and active efforts are made to reunite youngsters and families; to date, a number of these youngsters have left the home to rejoin their families in Britain or the United States.[81]

The Dutch group program in the city of Eindhoven exemplifies a different model of group care. The home is located in a large housing development which is a ten-minute bus ride from the center in the city. The house is in a row of modern brick houses and is furnished in a contemporary Western manner. From April 1982 through April 1983, two Dutch women lived with and supervised five Vietnamese adolescents on the weekdays; the mentors were not on hand on the weekends. The adolescents now live without supervision except for the scheduled visits of a Dutch social worker.

The minors who were selected for the one-year program were older adolescents who had not expressed an interest in foster care during their stays in reception centers. They attended local schools where they received special help with Dutch; some are continuing in school this year while others are now working. The goal of the group program was to introduce the minors to Dutch culture and to prepare them for independent living.[82] Emphasis was placed on acquiring experience with Dutch social, educational, and vocational situations.

The group home philosophy has undergone some changes in the Netherlands since the first minors were placed in group care several years ago. In the first group homes, an effort was made by the mentors to create a "family-like" atmosphere and to stress group activities. The minors, whose ethnic backgrounds, personal experiences, and aspirations differed greatly, were more interested in becoming self-sufficient in Dutch society than in forming close bonds with their housemates and mentors.[83] The group home program changed in response to the minors' expectations; the Eindhoven home thus stressed the development of independent living and decision-making skills. Mentors there did not serve primarily as parent figures or disciplinarians; instead, they used their authority to help minors anticipate the consequences of their own actions.[84] The mentor was the representative of Dutch culture, whose role it was to clarify and interpret Dutch customs to the minor.

The differences between the Eindhoven home and the Save the Children home in England are many. While the Dutch program provided short-term care, the British one offers long-term care. While the Dutch program emphasized the acquisition of an understanding of Dutch culture, the British program stresses the maintenance of Vietnamese customs. The Dutch mentors helped to prepare minors for an adult life in the Netherlands; the British counselors sought to reunify the youngsters with their Vietnamese families while encouraging gradual participation in British culture. Those features which are common to both homes distinguish them as group care programs: in both homes, the youngsters live with Asian peers, but participate daily in the life of the majority community.

There has been much less written about unaccompanied minors in group care than about those in foster care; the total number of Southeast Asian minors in group care is quite low compared to foster care. Supporters of group care argue that it affords the minor more independence and flexibility in the choices he makes than does foster care. It also provides continuity of both cultural and peer group ties. Such continuity may ease the shock that results from drastic change, and in the long run, is essential for the formation of a healthy identity. Follow-up research on unaccompanied children who spent time in group homes would be helpful in understanding the impact of this form of care on adult adjustment.

Foster Care or Group Care?

The controversy over what constitutes the best form of care for children who are separated from or who lose their parents did not originate with the recent migration of Southeast Asian unaccompanied minors. The debate over the merits of foster care versus other forms of care has been ongoing in Western countries for more than 100 years.[85] The principal arguments of this debate influence the course of placement policy for unaccompanied minors.

There is general agreement among parents, legal authorities, and child care experts in Western countries that almost all children are best off growing up in their natural families.[86] Disagreement arises only when critics try to decide what to do with the child who is not able to live with his own family. Until the middle of the nineteenth century, children who were separated from their families due to parental death, neglect, or abuse were raised in orphanages or almshouses.[87] In the past 150 years, foster family care has become an increasingly popular placement choice for dependent children who are not released for adoption, and is now the favored form of care in most Western countries.[88]

The popularity of foster care is often attributed to the impact of "family consciousness."[89] Foster care provides a substitute nuclear family for the dependent child; it thus replicates the accepted living arrangement for children in most cultures. In contrast, group care offers the child close interaction with a peer group but not with a family, and provides supervision by trained employees rather than by parents. This distinction can lead to the characterization of various forms of gruop care as "anti-family."[90] Proponents of group care are thus placed in the difficult position of defending an unpopular option; they do so either by criticizing the efficacy of the nuclear family for children with special needs or, more often, by promoting the "family-like" qualities of their programs.[91] This kind of promotion is reflected, for instance, in the terms directors use in describing their programs; children live in "cottages," rather than dormitories, and are supervised by "houseparents," rather than by counselors.

Critics of placement policy argue that the emphasis on "family consciousness" does not always serve the best interest of displaced youngsters. Meisel and Loeb suggest "that we do not accept without question the efficacy of family-like living simply because we are so impressed with the efficacy of the continuous natural family."[92] The problem, as they point out, is not so much one of comparing of our own family with other forms of care, but "whether the psychological criteria for the development of a healthy personality are met." These authors stress the importance of evaluating the developmental needs of the individual youngster when making a placement determination.

The confusion over foster care versus group care can also result from comparing good foster families with poor group homes, or good group care programs with poor foster care programs. "It is necessary," write Mayer and associates, "to compare good quality in either modality of care."[93] Observers agree that the quality of care provided in any placement will have a decisive influence on the adjustment of youngsters. In the debate between foster care and group care, the broad range in program quality is often overlooked.

Indeed, when better foster care and group care programs for unaccompanied minors are compared, several striking similarities emerge:

1. Both promote ongoing relationships with members of the child's native cultural group: foster care through the direct use of similar families, or by placing a number of unaccompanied children (often more than one child per placement) in majority families which are closely clustered together; group care through promoting peer relationships and the employment of ethnically similar caretakers and staff.

2. Both provide ongoing support services which include bilingual social workers, educational tutors, and professional counseling services when required.

3. Both utilize the support of the surrounding community; children in foster care and some group programs attend school, make friends, and hold jobs in the community.

Although there can be similarities between foster care and group care programs, a fundamental difference remains: the basic approach to the socialization of the child. Whereas foster care utilizes the nuclear family and adults within that family as the primary socializing agent, group care relies more heavily upon peer relationships. It is precisely this reliance on peer rather than adult relationships that can make various forms of group care inappropriate for infants and preschoolers and a less-desirable long-term alternative for most middle-aged children. Research suggests that when infants and children are separated from their natural parents, even for brief periods of time, the opportunity to secure attachments with alternative adults is crucial in preventing further developmental harm.[94] The emotional security derived from adult attachments eases the initial distress that results from separation or loss and allows the child to continue organizing and consolidating various aspects of his mental, social, and moral development in an adaptive manner. No placement alternative is able to guarantee that secure attachment will evolve between a child and an adult. In general, however, families are better able than group care programs to provide the kind of intimacy, security, and ongoing interaction between a child and an adult which are necessary to foster this complex psychological exchange.[95]

For infants and children, growing up without a family may present problems in social and emotional development, as well as in the formation of an adult identity. However, a number of child-care experts believe nonfamily placements can be appropriate for youngsters who grow up in families but are displaced in their adolescent years.[96] Although influenced by cultural factors, the main task of adolescence is generally believed to be the gradual separation from the nuclear family and the integration of a separate identity with physical, social, emotional, and moral self-reliance. The teenager who is struggling to separate from his own parents may find it difficult to become involved in a positive way with substitute parents in foster care. "Parental-like" attachments require that the child become emotionally and socially dependent upon the adult(s). This dependency, crucial for infants and children, may be counterproductive for adolescents who are already well on their way to establishing independent identities. Group care can allow adolescents more emotional distance from adults and more contact with peers, both of which may be helpful to them.

Group care may also hold advantages for unaccompanied adolescents who remain in contact with parents and other family members back home. Bennoun and Kelley suggest that such minors "do not yearn nor do they need a second family while living in exile. . . . The love they have is for their families and their country . . . and the heartbreak of losing these is not made less by the introduction of a mother and father figure."[97] Indeed, the presence of strong family ties and the ensuing conflict of loyalty leads other placement workers to reject the use of foster care for unaccompanied adolescents. Blacher, for example, found that:

> Foster care with Australian families may not be appropriate . . . The eagerness with which it was taken up by many Western countries of resettlement may in part be based on a culturally biased assumption that the nuclear family is the most legitimate form of care for these children.[98]

That group care can be a viable, if not, at times, a preferable placement alternative for unaccompanied adolescents seems clear. However, the preference of nuclear family care appears to be a bias which is not limited to the West. In this regard, Jockenhovel-Schiecke offers an interesting finding.[99] In Germany, group placements were being arranged for a number of unaccompanied adolescents who had left parents behind in Vietnam. When, through letters, the adolescents' parents were informed of this decision, many wrote back instructing their children not to accept group placement but to live with a German family. Apparently, these Vietnamese parents also preferred nuclear family care over group care.

Institutional Care

Institutional placement is the dominant form of care for unaccompanied Southeast Asian minors in France.[100] It has been used in other host countries as well, including Germany, Denmark, Switzerland, and the United States.[101] All of the minor's daily needs are met in an institution; food, housing, physical and emotional care, and educational and social opportunities are provided.[102]

Unaccompanied minors have been placed in various kinds of institutions. In Germany and Switzerland, some unaccompanied minors live in children's villages.[103] In a children's village each child belongs to a group or cottage within the village community. For example, in the Pestalozzi Children's Village in Switzerland, youngsters attend school and play with children from a number of different nations. Although the community is international, each youngster lives in a cottage with peers and houseparents from his own homeland and shares the traditions of his people with them. The institution thus provides a "family-like" setting.[104] In France unaccompanied minors have been placed in several different types of institutions that provide care for the mildly retarded, the predelinquent, or for young neglected children as well as for the minors.[105]

Advocates of institutional care for unaccompanied minors point out the broad range of services that an institution can provide. Since an institution serves more children than a foster or group home and potentially can employ staff members with many areas of expertise, it may be more able to meet the special needs of Asian minors.[106] Skilled educators can design programs that address the minors' academic strengths and weaknesses. Social workers and counselors can ensure that minors receive treatment that is sensitive to their cultural and personal backgrounds. Within the self-contained institution, minors can be gradually introduced to the expectations and traditions of the host country; the shock of radical culture change can thus be minimized. Proponents suggest that a total program can be implemented which is tailor-made to fulfill the academic, social, and cultural needs of the unaccompanied minor.

In reality, the quality and quantity of services offered in institutions, as in other forms of care, vary a great deal. The self-contained nature of an institution makes it difficult to evelute the provision of care from the outside; this may lead, suggest critics of the French placement program, to a "strange conspiracy of silence" which protects the programs that fail to meet the needs of the youngsters they serve.[107] Labe and Croissandeau describe one institution in which the poorly qualified staff and the lack of sufficient recreational apparatus and private space for the youngsters resulted in inadequate care.

In the United States institutional care is sometimes used for those minors who are unable to adjust to other forms of placement because of special social or emotional needs.[108] In such cases, minors are placed in treatment programs with American children with special needs.

One common criticism of institutional care is its lack of permanence. Employees and youngsters come and go in most residential programs, interrupting the development of emotional ties. The youngster's connection with the institution is frequently ambiguous; the institution is where he lives, but is it a home? The Council of Europe addresses this question in its 1971 report:

> Even if the [C]hildren's [H]ome is supposed to replace the parental home ... in no countries do the authorities ... seem to realize the full responsibility and full consequences of the upbringing supplied. In practice, a parental home gives support and refuge even to grown-up children ... Children are seldom forced to leave at the age of 16 or 18 years. Even when children return to visit the [H]ome in which they grew up, the staff will probably have changed and other children will have taken their places.[109]

The child who grows up in an institution may easily lose contact as an adult with the counselors and teachers who have cared for him.

There is a consistent body of clinical and empirical findings which indicate that for infants and young children, institutions do not provide the stimulation, intimacy, and *consistency* of adult care necessary for healthy development.[110] "There is ample evidence," writes Michael Rutter, "that even in good institutions the maternal care provided differs both in quantity and quality from that experienced in a family setting."[111] The emphasis on group participation, the frequent rotations and changes in staff, and the large number of children needing care limit the amount of sustained, individual attention that each child receives.

As with group care, however, a number of authors have suggested that institutional care may be appropriate for adolescents who have lost their families.[112] Yet, an additional question arises with respect to unaccompanied minors: to what extent are institutions able to prepare these minors for futures either in their native homelands, should they return, or in the host society, should they remain? On this essential question there is only scant information.

First, there are no studies which provide follow-up information on unaccompanied refugee children who, after spending time in age-group institutions or children's villages, returned to original homelands. This was pointed out in 1952 by the authors of the *UNESCO Report* with respect to displaced children at the Pestalozzi Village in Switzerland:

> Its children . . . despite the fact that they are grouped together according to nationality, will undoubtedly experience the need to adjust themselves afresh when they return to their own country, and it's too early yet to know for certain whether the international education they have received will have given them something which will be an asset to their country of origin when they return.[113]

Recognizing the broken attachments and disorientation that result from the uprooting of children from the village and rerooting them in original homelands, the Pestalozzi directors are currently in the process of shifting the program's orientation.

On the other hand, several reports on unaccompanied minors reared in institutional settings found that they are *not* well prepared to assume adult responsibilities when they remain in the host country.[114] Minors who are placed in this form of care sometimes recognize this drawback themselves.[115] A review of the West German program for unaccompanied minors noted that, initially, the minors seemed pleased to be living as an isolated group. As time went on, however, the minors "expressed their wish to have more contact with German people" and complained of a hunger for life."[116] And again on this subject the *UNESCO Report* concluded:

> At all events it is of the highest importance to ensure that children's villages remain open and in contact with the broader setting into which the child must one day integrate. Every effort must be made to prevent such communities from becoming islands endowed with a rich and harmonious internal life, but cut off from regular, normal contact with the world around them. Not only must the child learn the local language, but they also need regular intercourse with the local population, for their adjustments to be complete."[117]

Independent Living and Adoption

The needs of individual minors are sometimes best served in other forms of placement. Two kinds of care which are used in special cases are independent living and adoption.

Not all unaccompanied minors need or want the close supervision that foster families and group homes provide. Independent living programs allow mature adolescents to live alone or with peers. Such programs have been developed in several resettlement countries, including Australia, the Netherlands and the United States.[118]

Independent living arrangements have been structured in several ways. In some cases, minors live with host families as boarders rather than as foster children.[119] The boarding youngster is not considered a member of the host family, and the host parents do not have primary responsibility for his care and well-being as foster parents would. The minor and the host family reach agreement on meal arrangements and household commitments and maintain a friendly but restrained relationship. While the minor may seek informal advice from the host family, he receives regular supervision from an agency worker.

Australian authors have described the use of the community house, another form of independent living.[120] Groups of minors are placed together in homes without live-in mentors. They receive supervision from agency workers and may also look to neighbors for advice. The authors stress the importance of strong relationships between the minors who are placed together; they recommend that minors be placed together with siblings or with close friends since cooperation between house residents is essential to the success of the home.

In the Netherlands similar arrangements have been made for large sibling groups. The youngsters live in their own home without a mentor or foster parent. They are supervised regularly by a neighborhood mentor family, which receives payment from the agency for their involvement with and availability to the sibling group. Dutch authorities note that this arrangement works well for older adolescents who are in close contact with their parents in Southeast Asia; although they profit from the support and attention of the mentor family, they are not asked to become a part of that family and may remain in contact with their parents.[121]

Schulz describes an American program that provides independent living arrangements for certain adolescents.[122] All minors are first placed in foster care where they learn American customs and expectations. Those adolescents who wish to live more independently and are judged to be mature enough by agency workers are then allowed to move into rental apartments with one to four peers. The minors are supervised by agency social workers and one resident must make daily phone contact with the agency office. The author compared the backgrounds of those minors who choose to live in apartments to the backgrounds of those who stayed in foster care. The minors who seek independent living are slightly older than the other males and are more likely to have lost their fathers at an early age. In addition, they are more likely to list the desire for academic achievement as a primary reason for coming to America. The minors who are best prepared for independent living appear to be those who have strong motivation to learn and succeed in the host country and who have not had paternal supervision for a number of years.

The primary advantage of independent living is that it allows the older minor to take responsibility for his own care. Adoptive placement serves the needs of a very different group of minors who seek permanent ties with host families. The adoption of unaccompanied minors by host families is usually discouraged by agency personnel.[123] There are a number of reasons for this. First, most unaccompanied minors are never formally released for adoption by their parents or guardians. This fact introduces both psychological and legal complications into the adoption process.[124] Second, there are many unaccompanied minors who are not interested in being adopted. For example, minors who are in close contact through letters with their natural families may not wish to become permanent members of host families because their strongest emotional ties remain with their parents. Similarly, minors who have lost track of their families may maintain hope of being reunited with them in the future, and do not necessarily want to commit themselves to permanent relationships with adoptive parents. Moreover, certain adolescents who know that they are orphans may be more interested in developing relationships with their peers and pursuing their education in the host country than in becoming permanent members of new host families.

The serious psychological, legal, and ethical issues that can arise when unaccompanied children are adopted in haste are illustrated in the 1975 Western airlift of Vietnamese children (see Part I). Because of these complications and the growing awareness among relief agencies that many unaccompanied minors are not orphans, resettlement governments and international organizations have been quick to emphasize that subsequent groups of unaccompanied children are not available for adoption. This does not mean, however, that adoptive placement is never in the best interests of an unaccompanied minor. British and American research on domestic foster care and adoption stresses the importance of permanent ties with available adults for healthy emotional development.[125] Foster care is not a permanent arrangement; legal and temporal limitations may do the minor a disservice by restricting the emotional commitment between him and his foster family. Consequently, when tensions arise in the home, both the minor and his foster family may be more likely to end the placement than to work out their differences.[126]

Indeed, Fanshel found that the single best predictor of the number of moves a child makes in foster care was the length of time he spent in care; the longer the child spent in foster care, the more likely he was to experience changes in placements.[127] This finding, confirmed in other studies, suggests that foster care frequently fails to provide the permanency of relationships that the natural or adoptive family offers.[128]

Thus, the high replacement in foster care of unaccompanied minors may reflect this lack of mutual commitment. The result for the child can be a series of short-term placements in care and a failure to establish ties in the host country while awaiting reunion with a distant or silent natural family. The negative influence of this kind of drifting on the emotional and social development of children in domestic foster care is well documented; the consequences for the Southeast Asian minor has not been recorded by follow-up studies but are likely to be equally harmful.[129]

Host countries are thus faced with the difficult task of balancing the minor's need for permanence with respect for his ties to his homeland. Temporary care denies permanence; traditional adoption may deny continued commitment to the natural family and homeland. Several authors have suggested policy options which would help resolve this dilemma. For instance, in 1975, the Vietnamese Children's Resettlement Advisory Group in the United States recommended that unaccompanied minors be placed in prospective adoptive homes.[130] The group felt that permanent placements, preferably with ethnically similar families, are in the best interests of most youngsters. They suggested that the American government should initiate tracing procedures to find natural families as soon as minors arrive in the host country, and that the minors be returned to their families automatically if the families are located during the first eighteen months of placement. After this period the adoption could become permanent and changes in custody should be decided by the state courts in individual cases. Adolescent minors must be permitted to choose long-term foster care rather than adoption if they preferred it, but they must not be denied an adoptive arrangement because of their ages. The authors thus stressed the need for permanent placement in the host country but also encouraged the use of "individual, case-by-case" consideration "where special circumstances are involved."

The provision of permanent care is also discussed in a French government report on unaccompanied minors.[131] The authors recommend the use of a special form of adoption—"adoption simple"—for minors who are placed in families. Under French law, "adoption simple" differs from "adoption plénière" [full adoption], which is permanent, irreversible adoption. In adoption simple, the adoption remains potentially reversible. The adopted youngster is permitted to keep his birth name and his nationality. Natural family members and the court maintain involvement in the planning of academic, religious, and vocational training. The arrangement thus provides a permanent home for the youngster but allows him to retain connections with his past and his natural family. For an unaccompanied minor adoption simple allows the possibility of return to the natural family or the homeland while supporting a legal commitment between host family and minor. This form of care clearly places the stress of uncertainty on the host family; thus a family that cannot manage the strain is not considered appropriate for placement.

For the reasons discussed, many unaccompanied minors resettled in third countries are not appropriate candidates for adoption. However, for those minors and foster families who do form mutual attachments, the desire to make these connections both permanent and legal may be very powerful. A policy that forbids adoption, or limits adoption only to very young minors, can deny older children as well as adolescents an important source of security and identify.

HOST COUNTRY EXPECTATIONS, NATIONAL POLICY, AND CULTURAL IDENTITY

Refugee resettlement, writes Virginia Dominguez, is "a kind of mirror of . . . society"; the welcome the refugee receives reflects the host country's response to newcomers.[132] The resettlement nations differ in the plans they make for unaccompanied minors. These placement patterns illustrate the expectations which different cultures hold for the refugees from Southeast Asia.

Anthropologists have identified a continuum of cultural responses to newcomers.[133] At one extreme are policies which promote the assimilation of the refugee into majority culture. Assimilative policies assume that the newcomer wishes to become a member of the host culture. True assimilation requires a change in the personal identification of the newcomer; he stops thinking of himself as a citizen of his homeland and embraces a new identity as a participant in the host culture.[134] Assimilation also requires majority culture acceptance of the newcomer as a member; it may thus be more easily accomplished in heterogeneous societies in which the newcomer's physical or cultural attributes do not stand out. Assimilative philosophy is illustrated in the following quotation from a 1983 speech by Ronald Reagan, President of the United States:

> We can be proud and thankful that . . . [Vietnamese refugees are] joining us today
> in parades and ball games and backyard barbecues as young members of an old
> family.[135]

Most families in the United States were immigrants to the country; Reagan and many other Americans assume that the Vietnamese refugees are voluntary immigrants who wish to join the American "family."

At the opposite end of the continuum of cultural responses is the multicultural model. Refugees are viewed as displaced persons rather than as voluntary immigrants and they are encouraged to maintain the culture and traditions of their people rather than to adopt Western patterns. Such resettlement policies stress the conservation of

Asian identity and community within the Western country. Colin Hodgetts of the British Refugee Council explains the multicultural philosophy of resettlement:

> Refugees are torn from the land, and to them this is like the loss of a limb. They are no longer whole people . . . In any refugee programme, it is essential to try and preserve as much as possible of family and culture.[136]

Assimilation requires "the committing of partial suicide" according to this model, since it involves a partial or complete change of identity, and the acceptance of different traditions and expectations as one's own.[137]

Between the extremes of the assimilative and multicultural models is the acculturative perspective. Acculturation stresses both an internal continuity of identity and an external adaptation to the host culture. In other words, the refugee continues to think of himself as a Vietnamese or Cambodian, maintaining the traditions, religion, and community of his homeland. However, he is encouraged to familiarize himself with and to participate in the host culture. The refugee's sense of himself remains relatively constant, but his social skills, language, and everyday activities reflect his familiarity with Western society.

In practice, refugee policy only suggests and cannot legislate the adjustment of the individual refugee; every refugee finds his own balance of continuity and adaptation in resettlement. However, resettlement policy certainly influences the course of individual adjustment. Unaccompanied minors, who cannot rely on their parents to provide balance and to reinforce the value of their cultural background, are particularly vulnerable to the orientations of placement workers. Different kinds of placement reflect different expectations of refugee participation in the host culture and encourage different kinds of social and emotional adaptation.

Majority foster care, widely employed in many countries with unaccompanied minors, is on the assimilative end of the continuum.[138] The minor lives among people who do not speak his languge and who are not members of his culture. While many foster parents allow or even encourage introduction of Southeast Asian customs into their home, emphasis is on the minor's adaptation to the host culture. Unless regular contact is maintained with other refugees of similar background, the minor becomes isolated from those who share his history and traditions; he loses this important source of identity. Some minors in majority foster homes temporarily reject reminders of Southeast Asian culture and fully embrace the traditions of their foster families.[139] In contrast, minors in foster care who maintained contact with Asian peers and who engage in refugee community activities will maintain more of their Southeast Asian identities and may become acculturated rather than assimilated.

At the other end of the continuum are various forms of institutional care. The minor who lives, eats meals, and attends school with other Southeast Asian minors, and has little contact with host society, will maintain much of his cultural identity. He will not experience the same conflict and confusion that meet minors who are in more regular contact with the host culture. His life in this "kind of refugee reservation," as Dominguez has referred to them, will be predictable and protected.[140] It often fails, however, to prepare him for the life he will lead in the host culture once he leaves the institutional setting.[141]

Between the two extremes of isolation and assimilation lie community-based group care programs and ethnically similar foster care. In community-based group care, minors live with Asian peers but spend their days in the schools, stores, and offices of the host community. They may maintain some aspects of traditional life in the home and experience the stresses and successes of adaptation in the community. Often, minors speak the host language in school and work, but use their native language at home among peers. Some group programs encourage gradual adaptation to the host culture at home, while others support the preservation of Asian traditions; some thus lean more towards the assimilative end of the continuum while others reflect a multicultural model. All permit a more gradual introduction to the host culture than majority foster care usually does, and more regular contact with the community than institutional care usually provides.

The same kind of gradual introduction to the host society is possible in ethnically similar foster care. Southeast Asian youngsters are placed with families of their own cultural background; some of these families have recently arrived in the host country themselves; others have resided there for several years. In any event, the unaccompanied minor is offered the familiarity of Southeast Asian home life and the support of Southeast Asian adults in his entrance into the host society. Critics of ethnically similar foster care argue that refugee families may be unable to offer the guidance that the youngster needs as he interacts with Western peers and adults; to date, these claims have not been substantiated.[142] Moreover, the emotional support provided by a shared history and culture may be more important to the displaced youngster than specific information and advice about life in the host country.[143]

The kind of placement that a nation or an agency endorses reflects expectations about the future of the refugees. Some program directors focus on host country adjustment on the assumption that the minors will stay in the West. They say the goal in resettlement is to promote competent, independent participation in the host country; life with a majority family provides a fast and thorough introduction to the host culture. Faster, however, is not necessarily better, and in fact, it can entail more psychological risk. If unaccompanied minors are to remain in the West, placements in similar families and community-based group homes can ease the shock of cultural change, promote the continuity of native language and cultural ties, and still enable the minors to acquire new language, social, and educational skills necessary to become viable members of the host society.

Other program planners do not see host adaptation as the primary goal of resettlement. The focus of placement policy becomes the eventual reunification of the minors with their families, either in the original homeland or in the host country. Placement workers may try to prevent assimilation or acculturation by isolating the minors from the host country so they will be more comfortable with their parents' way of life when they rejoin them. In reality, however, workers can only guess the future course of the minor's life. "It is impossible to predict if a return will be possible in the foreseeable future," concludes Jockenhovel-Schiecke; it is equally impossible to predict when and if a minor will be reunited with his parents.[144] Indeed, if the past is an indicator of the future, the vast majority of minors from Southeast Asia are here to stay; since World War II only a small percentage of unaccompanied minors resettled in Western countries have returned to original homelands or to natural parents during their childhood or adolescent years (see Part I).

III

UNACCOMPANIED CHILDREN IN COMPARATIVE AND INTERNATIONAL LAW

There is law regulating and informing all major aspects of the problem of unaccompanied children in emergencies—wars, natural disasters, and refugee situations. Law affects the interaction of parents, family, children, and community, prescribing what rights and obligations each has with respect to the others. The law bears on all phases of the plight of unaccompanied children, from the prevention of family separation through protection, care, and assistance, to permanent placement.

For children living with their natural parents, law recognizes the fundamental importance of the parent-child relationship but also specifies certain conditions under which that connection can or should be severed. For children already separated from their natural parents, law sets the terms of their future relations with the parents, other family members, and those from society at large who would act upon the children. The law specifies the duty of the community members to protect and aid unaccompanied children, the standard to guide their actions, and the form of care and protection they must provide. It also bears upon questions of family reunion and adoption or other permanent placement, including the role of the child in such decisions. Furthermore, the law itself defines who has the right to act upon and for the child. Law, then, is central to all phases of the treatment of unaccompanied children.

The first and most important level of law which affects unaccompanied children is *national:* statutes or codes, regulations, and administrative practices of nations. Nations are sovereign within their borders, and the care and protection of unaccompanied children falls within their general domestic jurisdiction. In general, the legal situation of unaccompanied children is governed by the law of the country in which they are located.

National law may, of course, be customary law: unwritten but well-established practice. Before the creation of formal written law, parent-child relations and relations between children without parents and the larger society were regulated by custom. Customary law still exists in some countries, including several where more formal legal systems have also come into being. Although much of the discussion here concerns national law in the more formal sense, customary law is included within the term as well.

The second level of law bearing on unaccompanied children is *international.* Nations have agreed among themselves on certain principles of conduct relevant to both children in general and to children without families in particular. These agreements are found in conventions, protocols, declarations (and their draft versions), and the law and practice of States and international organizations, and scholarly writings. A convention is a treaty among States creating legal rights and obligations among them concerning a particular subject, such as the protection of victims of war under the Geneva Conventions of 1949. A protocol, which amends or supplements another treaty, has the same legal effect as a treaty among States party to it. A declaration is a statement of principles, endorsed and recommended by several States acting together, which are not generally considered to be legally binding though they may reformulate already existing international customary norms or contribute to creation of new customary norms.

Customary international norms are practices which are generally recognized as obligatory on States, even though they are not formally established by a convention. While scholars are not in total agreement, uniformity, consistency, and generality of the practice and acceptance of the practice as law are required for practices to become part of customary international law. State practice and that of international organizations, conventions, protocols, and declarations are among the evidentiary sources of customary international law. The question of when a practice has attained the status of customary international law is a matter of dispute among scholars, as both a general matter and in particular instances.

As noted above, a convention or protocol is legally binding among States which agree to be bound by it, usually by ratification or accession. Thus, the impact of a convention will depend on the extent to which it is ratified. Further, some conventions and some countries require the passage of national legislation before a convention has legal effect domestically.

Though the other sources of international law mentioned above, such as declarations and draft instruments, do not generally create legally binding rules, their principles constitute an important source of norms for the formulation of international recommendations about unaccompanied children. A short definition of the sources of international law and their legal effect is contained in Appendix B.

In brief, then, the law governing unaccompanied children in emergencies is generally the national law of the country in which they are located. That country may, however, subscribe to international conventions which it is legally *bound* to follow; the convention may even require that the State conform its national law to particular requirements. A State also may *choose* to follow nonbinding international recommendations. The basic level of law affecting unaccompanied children is thus clearly the law of the various sovereign nations, but international instruments may have an effect on that national law.

* * *

Within the national/international structure just outlined, the plight of unaccompanied children in wars, national disasters, and refugee situations involves three main substantive bodies of law:

1. *family and child welfare law:* the law concerning the relation of the child to his family and the community or State;

2. *the law of emergencies:* the law of natural disasters, armed conflict, and population movements;

3. *jurisdiction and choice of law:* the law concerning the authority to take action on and for the child, and which national law to apply.

Children become separated from their parents through death or other causes in all societies, and their situation is usually addressed by a society's family and child welfare law. What makes the legal situation of unaccompanied children in emergencies unique and especially problematic is the interaction of family and child welfare law issues with both (1) the law of disasters, armed conflict, and population movements, and (2) issues of jurisdiction and choice of law arising from the frequent transnational movement of children in emergencies. The "law of emergencies" and jurisdiction and choice of law can thus be seen as overlays on the basic family and child welfare law which applies to all children separated from their families. Unaccompanied children in war situations, for example, may be subject to the law of armed conflict in addition to the general family and child welfare law of their locale; displaced children may come under aliens or refugee legislation as well. For children found or moved outside of their native lands in emergencies, issues of jurisdiction and choice of law may be implicated in the basic family and child welfare law questions.

Part III will therefore discuss these three bodies of law as they pertain to unaccompanied children in emergencies. The approach is comparative and international; that is, the chapter attempts to combine an examination of national law and practice from a comparative perspective with a review of international instruments. In general, the discussion will not attempt to evaluate whether a principle or practice has attained the status of customary international law. It will, however, attempt to indicate broadly the degree of acceptance of various principles as they are discussed. Although family and child welfare law, the law of emergencies, and jurisdiction and choice of law are treated separately, they are brought together at various points in this Part and in Part IV of the book. In addition, because of the important role played by international and voluntary organizations with unaccompanied children in emergencies, the final chapter of this Part examines the major legal issues involved in their activities.

This Part is addressed primarily to lay men and women rather than to lawyers or jurists. Its organization reflects this approach by placing jurisdiction and choice of law—which in a legal text would probably come first, but which are of little interest to the lay reader—toward the end. Furthermore, the discussion avoids overly technical explication and does not attempt the kind of exhaustive or analytic treatment that might be found in a legal treatise. In the interest of both simplicity and brevity, propositions may be stated broadly without mention of certain existing exceptions. What follows, then, does not pretend to be the complete word on many of the subjects touched on here. The focus, instead, will be to explore the major legal doctrines which should inform any treatment of unaccompanied children in emergencies.

* * *

The main point of these chapters is that there is already a great deal of law bearing on the protection, care, and placement of unaccompanied children in emergencies, much of it consistent with what is now known about children and their physical, psychological, and social development. On the international level, numerous conventions and declarations, including several still in the draft stage, now concern themselves with children and their relations with parents, family, and the larger community. Children have been a primary focus of global aspirations for human rights and personal dignity in this century. At the national level, too, family and children's legislation has undergone rapid expansion and development in the past few decades.

The international instruments can be grouped into three general categories: (1) those concerning human rights, both of children in particular as well as broad human rights instruments containing provisions about children and family; (2) those concerning the guardianship, placement, and adoption of children; (3) those providing individual rights and State obligations in emergencies. In the twentieth century the development of the first group begins with the Declaration of Geneva of 1924 (also known as the Declaration of the Rights of the Child of 1924), which, though privately initiated, was later adopted by the assembly of the League of Nations. In the United Nations era, the Declaration of Human Rights of 1948 contains several provisions relevant to children, including recognition of the family as the natural and fundamental group unit of society, the right to freedom from arbitrary interference with family and home, and entitlement to special protection for family, motherhood, and childhood. These guarantees and others, such as education, achieved conventional recognition in the International Covenants on Civil and Political Rights (ICCPR) and Economic Social and Cultural Rights (ICESCR) of 1966. Regional human rights conventions and declarations contain similar provisions.

Meanwhile, in 1959, the General Assembly of the United Nations unanimously approved a Declaration of the Rights of the Child (DRC). This declaration attempts a comprehensive statement of society's obligations to children and their corresponding rights. An effort to embody its principles in a convention began in 1978, and since then the United Nations Commission on Human Rights has considered several drafts of a Draft Convention on the Rights of the Child (DCRC). A number of articles have been agreed upon by the Commission, but at the time of this writing, the draft of this convention had not been completed. Given the difficulty in reaching agreement on the contents of the draft, and the specificity of some of its articles, it is questionable how widely this convention will be subscribed to if and when it is finally completed.

A second group of international instruments bearing on unaccompanied children concerns legislation on guardianship, foster placement, and adoption. There have been two efforts of the Hague Conference on Private International Law to regulate guardianship of children outside their countries of nationality: the Convention on Guardianship of 1902 and the Convention on the Protection of Infants of 1961. With respect to adoption there is also a Hague Convention on Adoption (1965) as well as the European Convention on Adoption (1967) and other regional instruments. By and large there has not been a great deal of adherence to the conventions in these areas. The ponderously named Draft Declaration on Social and Legal Principles Relating to the Protection and Welfare of Children with Special Reference to Foster Placement and Adoption Nationally and Internationally (DDFPA) was completed in 1982 and has been considered by the United Nations General Assembly in several sessions but has not been adopted as of this writing.

Finally, among the international instruments relevant to unaccompanied children in emergencies are those concerned with protection of civilians in emergencies. Most notable are the provisions for the protection of civilians in armed conflict in the Fourth Geneva Convention of 1949 and its two Protocols of 1977 and the United Nations Convention Relating to the Status of Refugees of 1951 and its 1967 Protocol.

The result of this rather active growth of international legal thinking on issues that concern unaccompanied children in emergencies will be examined issue by issue in the following chapters. The legal significance of the instruments depends, of course, on the nature of each instrument, its status (with conventions and protocols; for example, whether or not they are in effect), and the number of States adhering to it. These matters are included in notes to the instruments as they are discussed further on.

Viewed broadly, the law about children has developed in two complementary directions: protection of the child's well-being and growth, and respect for the child's individuality and maturing personal freedom. Thus, unaccompanied children have the right to special protection and assistance, including aid in reunion with their natural families or substitute family care because of their immaturity and vulnerability. And, by virtue of their humanity, they also have human and civil rights to freedom, dignity, and individual treatment consistent with their age and maturity.

These two strands of law come together in the concept that adults acting on behalf of children without families must act "in the best interests of the child," a principle now contained in virtually all national and international law. The "best interests" standard dictates that adult authorities place the child's welfare above all other considerations; respect his individuality; and meet his physical, psychological, and social developmental needs. In this way, the child's rights to adult supervision and assistance and to individual dignity and freedom can be honored to the fullest possible extent.

Unfortunately, history amply demonstrates that the treatment of unaccompanied children in emergencies has frequently disregarded or violated fundamental and well-recognized legal principles. With unaccompanied children in emergencies, as with all people in many contexts, basic human rights and the dignity and worth of the human person have frequently been subordinated to narrower and more selfish individual, institutional, or governmental interests. Despite clear legal direction to the contrary, unaccompanied children have been subjected to such maltreatment as abduction, physical abuse and exploitation, neglect and abandonment, and enforced military service. In a few emergencies there has been no legal system governing their treatment; in others the law has been unclear. More often, governnments have ignored their own law pertaining to such children, particularly when the children were not their own nationals. Voluntary agencies and international organizations, too, have, on some occasions, ignored or failed to implement relevant legal principles in their treatment of unaccompanied children in emergencies, following instead political, social, or religious agendas.

In general, there is less need for the development of new legal theory or the creation of new legislation for unaccompanied children than there is for adherence to, and implementation of, existing legal standards. Much law relevant to these children is legally binding on those who would act upon them. In other cases, international legislation to protect children without families is already written but is not widely ratified or acceded to and therefore lacks the force of law. General accession to these instruments would solidify and clarify the legal situation of unaccompanied children in emergencies. Even where the existing legal principles are not formally binding on the actors, however, these principles can and should inform the decisions individuals or institutions will take. The main legal challenge is to make the protection, care, and individual treatment already demanded by many sources of law a reality for all children who are separated from their families in war, natural disaster, and refugee situations.

16

The Family and Child
Welfare Law Framework
for Unaccompanied Children

Many of the legal questions raised by unaccompanied children in emergencies involve basic family and child welfare law. Such critical issues as the right of parents to raise their own children, when and how this right may be lost, the duty of the community to protect and aid children without parents, the form of protection and temporary care, the reunion of parents and children, the role of the children themselves and of other relatives, and the conditions of adoption are not unique to emergencies, and are addressed in many national legal systems. This general body of family and child welfare would by its terms ordinarily apply to unaccompanied children in emergencies. In addition, a number of international instruments contain provisions relevant to the same issues.

This section, therefore, examines the family and child welfare law issues which have arisen most frequently with unaccompanied children and presents the salient doctrines governing their resolution. These principles are derived from a comparative examination of national law and practice,[1] from international instruments and practice, and from what is known about the psychological needs of unaccompanied children as described in Part II of this book. The doctrines discussed here comprise the most basic legal background against which actions concerning unaccompanied children take place.

In discussing these doctrines no attempt is made to specify which have acquired the status of customary international law. As Pappas has stated, "[T]he complex analysis required to address these questions . . . forms a most important topic for further study. Suffice it here to emphasize generally that internationally proclaimed standards . . . can and often do crystallize into generally legally binding international rules."[2] The collection of sources here can be seen, however, as part of the process of determining which rules are acquiring such obligatory force, and the reader who is interested in this question is referred to the discussion of the various sources of international law in Appendix B.

In general, while there are certainly differences in national laws, what has been said about family law in Europe may be applied to the entire international community: "In the last quarter of the twentieth century the laws of the majority of countries

215

... tended to converge towards a uniform model, and this evolution has been very rapid."[3] Undoubtedly, the growing collection of regional and international legal instruments concerning children, family, and community has contributed to this trend. At the most fundamental level there is a growing global consensus that legal institutions should protect the integrity of the family, ensure the best interests of children who have been separated from their parents, and respect the individual humanity of every child. This agreement also results, one may surmise, from increasing transnational contact and the attendant crosscultural influences. Reproduction and child-raising are universal human experiences, of course, so the similarity in legal forms is ultimately only a reflection—and perhaps a reassuring reminder—of our common humanity.

THE BASIC LEGAL RELATIONSHIP BETWEEN PARENT AND CHILD

The basic legal relationship between parents and child is this: by virtue of their biological parenthood, the parents automatically acquire the rights and obligations of guardianship of the child. They retain this "natural guardianship" unless and until (1) they voluntarily surrender it in whole or part, or (2) it is limited or terminated in order to protect the well-being of the child. The natural parents have, then, the initial and presumptive right to the guardianship of their own child.[4] Those who would remove the child from the parents or otherwise intervene in the relationship between parent and child must therefore show substantial and lawful justification.[5]

Guardianship Rights and Duties

The adult guardian(s) of a child have the following rights and duties with respect to the child:

> a. The *right* to control his residence (or custody); education, both secular and religious; and discipline. These rights are all generally subject to limitations imposed by the community, acting through the State.
> b. The *duty* to protect the child physically, mentally, morally, spiritually, and socially and to help him develop in a healthy and normal manner.[6] This includes the duty to provide food, shelter, care, education, discipline, and generally to protect the child's interests.

The Role of Natural Parents in Comparative and International Law

In the national law of virtually all States "parents have as they always had, the right to keep and bring up their child as *their* child."[7] In civil law countries, codes will often explicitly vest "parental authority" in the natural parents;[8] in the common law, it is implicit in the entire family law structure. In socialist law, though the parents are responsible to society for the upbringing of the child, the parents have the right to take personal care of their child and exercise parental rights and duties.[9] In all societies the parents' rights typically include the right to keep the child and control its upbringing, to determine its education and religion,[10] and to discipline the child, though often within bounds set by the State.

International legislation repeats the theme that the parents have the initial and presumptive right to raise their own children. The Declaration of the Rights of the Child (1959) states that the child "shall, whenever possible grow up in the care and under the responsibility of his parents," and that "a child of tender years shall not, save in exceptional circumstances, be separated from his mother."[11] The Draft Convention on the Rights of the Child (1985) provides that the "[p]arents or, as the case may be, guardians, have the primary responsibility for the upbringing and development of the child"[12] and that "the child should enjoy parental care and should have his place of residence determined by parent(s)" absent exceptional circumstances.[13] The Draft Declaration on Foster Placement and Adoption (1982) affirms that "the first priority for a child is to be cared for by the biological parents."[14]

No less than two conventions and four declarations, including the Universal Declaration of Human Rights (1948), have stated that "the family is the natural and fundamental group unit of society and state."[15] This is certainly meant to confirm the initial and presumptive right of the biological parents to keep and raise their own children, since there can be no other "natural" family group.

It is clear, then, that natural parents have a presumptive right to guardianship of their biological children in both national and international law.[16]

Community Support for the Natural Family

Community support for the natural parents' ability to care for their own child is a logical extension of parental guardianship. In recent years there has been widespread recognition of the need for, and wisdom of, support of the parents' capacity to meet their obligations adequately. This support has taken the form of child allowances, tax concessions for parents, social services, counseling, and education.[17] The International Covenant on Economic, Social and Cultural Rights (1966) recognizes that "[t]he widest possible protection and assistance should be accorded to the family . . . particularly while it is responsible for the care and education of dependent children."[18] Material support for needy parents is recommended in the Declaration of the Rights of the Child (1959): "Society and the public authorities shall have a duty to extend particular care to children . . . without adequate means of support. Payment of State and other assistance towards the maintenance of children of large families is desirable."[19] The Draft Convention on the Rights of the Child states "that the family . . . should be afforded the necessary protection and assistance so that it can fully assume its responsibilities within the community."[20] That Convention would therefore require States Parties to "render appropriate assistance to parents and guardians in the performance of the child-rearing responsibilities and . . . ensure the development of institutions for the care of children."[21]

PARENT-CHILD SEPARATIONS

The law recognizes the fact that, despite its general promotion of family integrity, some parent-child separations will inevitably take place. Because they can occur under a variety of circumstances, such separations have different legal consequences. This section defines in legal terms the kinds of separation of children from *both* of their parents; as noted in the Introduction, children separated from one parent are not considered here as unaccompanied children. It is most useful to group these categories as *voluntary* or *involuntary,* according to the intent of the parents at the time of separation. Although this overall grouping is based on the intent of parents at the time of separation, the child's intent is also relevant and is not always the same as the parents' intent, as the categories make clear.

It is also important to recognize that the definitions given here relate to the intention of the parents and child at *the time of their separation.* After a period of separation the parents' desire to be with the child may change, as may the child's situation and/ or his desire to be with the parents. In other words, over time a child can move from one category to another. The child's legal status at the time of a proposed reunion, then, may be different from what it was at the time of separation.

Children Involuntarily Separated from Parents

1. *Orphan:* a child whose parents are both dead. Obviously, the death of the parents involuntarily separates the child from them. Although the death of one parent may raise legal questions about whether the other parent automatically succeeds to guardianship,[22] for purposes of this discussion, children living with one parent are considered to be neither orphans nor unaccompanied children.

2. *Lost:* a child unintentionally separated from the parents. This category includes children accidentally separated, for example, during armed conflict or in flight. With a lost child there is no parental intent to separate or remain apart from the child, in contrast to *Abandoned* children, discussed below.

There is also another group of "lost" children, those whose parents are not present due to incarceration, hospitalization, or other involuntary causes. These are referred to as "dependent" children in some national legal systems. With these children, not only is the separation unintentional on the part of the parents, but there is no element of fault or failure to meet parental obligations.

3. *Abducted:* a child involuntarily and illegally taken from the parents. All removals of a child from its parents against their wishes are illegal, except when done by the community or State solely in order to protect the child's well-being (see *Removed* children below).[23] Kidnapping and child abduction are crimes in most national laws, and parents have a civil law right to return of their child from anyone who holds the child without a basis in law.[24] On the international level, the Hague Convention on the Civil Aspects of International Child Abduction (1980) aims "to protect children internationally from the harmful effects of their wrongful removal or retention and to establish procedures to ensure prompt return."[25] The Draft Convention on the Rights of the Child (1985) would obligate States "to take appropriate measures to combat the

illicit transfer and nonreturn of children abroad."[26] In the most extreme case, the forcible transfer of children of one group to another group constitutes the crime of genocide when done with intent to destroy a national, ethic, racial, or religious group.[27]

4. *Removed:* a child removed from the parents as a result of legal suspension or loss of parental rights. Almost all States provide that the parents' presumptive right to raise their own child will be lost, either temporarily or permanently, if they grossly fail to meet their responsibilities to the child. In that event the community, usually acting through the State, may intervene and remove the child from the parents' custody in order to protect the child's interests. The legal standards for such intervention are phrased in such terms as "in need of care and protection,"[28] "ill treatment or neglect of care or guidance,"[29] "gross dereliction in one's parental duties,"[30] and "neglect or abuse."[31] Whatever the terminology used, the aim is to protect children when parents are seriously derelict in the performance of their parental duties.

International legislation accords with this national law and practice. The Declaration of the Rights of the Child (1959) states: "The child shall be protected against all forms of neglect, cruelty and exploitation."[32] The Draft Declaration on Foster Placement and Adoption (1982) states: "It must be recognized that there are parents who cannot bring up their own children and that the children's right to security, affecting and continuing care should be of greatest importance."[33] The Draft Convention on the Rights of the Child (1985) would permit the involuntary removal of a child from his parents, in accordance with national law, where "such separation is necessary for the best interests of the child," such as cases of "abuse or neglect."[34]

The law, then, is clear that the presumptive right and capacity of parents to raise their children is not absolute; it must yield to State or community intervention when necessary to protect the child. This is usually done by temporarily removing the child from the parental home and placing him elsewhere, thus limiting the parents' right of control of the child's residence and upbringing. When the parents are unwilling or unable to resume proper parenting, most States also provide for complete termination of their rights.

It is important to note that parental right cannot be interfered with simply because the child will be "better off" in some other arrangement. As noted above, there is a strong presumption in favor of the natural parents. To be sure, this presumption may be based in part on the assumption that "the best interests of the child will be [parents'] basic concern,"[35] and no doubt this is factually most often the case. A family that remains intact from the birth of the child is recognized as the ideal arrangement to provide the continuity of security, affection and stimulation that children require.[36] Yet, the right of the parents and children to be together rests on something more than the assumption that parents will usually provide the best care: since the family is the fundamental group unit of society, parents have a basic right as human beings to raise their own children. It is only when parental behavior threatens the welfare of the child that parental authority will be called into question by the larger society.

5. *Runaway:* a child who intentionally left his parents without their consent. From the parents' point of view this is an involuntary separation, but it is a voluntary one on the child's part. Although the parents generally have the right to demand the child's return, the right to reunion may not be absolute. As discussed in greater detail

later, the child's best interests or other rights under the laws of war, refugees or immigration may stand in the way of his return to the parents.[37]

Child Voluntarily Separated from Parents

1. *Entrusted:* a child voluntarily placed in the care of another adult or institution by the parents who intend to reclaim the child. As part of their right to control the child's residence, parents can place the child in a different residential arrangement; for example, with a relative, in a boarding school, or in another child-care institution. They can, of course, also delegate control of the child's religious and secular education. Such "entrusting" placements are temporary, with the parents intending to have the child return to them, and may vary in the degree of control given up by the parents. They can, for example, give up custody (actual physical presence) while retaining other guardianship rights such as authority over major decisions for the child.

Evacuation of children in wartime or other emergencies is an example of this parental right to control the child's residence. Typically, such placement involves a transfer of the right to supervise the child's day-to-day activities to another person or organization. During the Spanish Civil War, for example, Basque parents living as refugees in France signed the following agreement regarding their children who were living elsewhere:[38]

> I, parent of x, now residing in y, exercise my parental rights by giving, freely and spontaneously, to the Basque Committee on Evacuation, the responsibility of education and Christian-Basque instruction in the way they judge best. I promise to receive my children when the cited committee judges this to be appropriate.

Where parents place the children with others they can retain the right to demand return of the children in the placement agreement, as was done in the evacuation of Spanish children to the Soviet Union during the Spanish Civil War.[39]

2. *Abandoned:* a child whose parents have deserted him with no intention of reunion. Abandonment is a voluntary, if informal and dangerous, method of permanently surrendering parental rights. There is no international definition of abandonment; national law governs the question of when a child is deemed abandoned. In most cases the child's circumstances when he is found and the parents' intent as inferred from their behavior are the salient factors in the determination. In Colombia, for example, infants who have been exposed (foundlings) and minors turned over to a social welfare establishment and not reclaimed within a period of three months are deemed to be abandoned and are therefore free for adoption.[40]

3. *Surrendered:* a child whose parents have permanently given up their parental rights. Parents can voluntarily surrender all parental guardianship rights in their child in a formal manner, usually through a written instrument which may be reviewed by a court. This consent to termination of parental rights is most often executed in order to free the child for adoption by others. In fact, in virtually all countries adoption cannot legally take place without the consent of the parents if they are available and are fit parents.[41] The European Convention on the Adoption of the Child (1967), for example, requires "the consent of the mother, and, where the child is legitimate, the father; or if there is neither father nor mother to consent, the consent of any person

or body who may be entitled in their place to exercise their parental rights in that respect."[42]

A valid consent to termination of parental rights must ordinarily be knowing, intelligent, and voluntary. Not surprisingly, in the unsettled conditions of many emergencies parents may consent to termination of their parental rights unknowingly or under duress. The issue of the voluntariness of parental surrender of their rights can thus be of great practical importance during and after emergencies.

The situation of the Vietnamese children brought to the United States in 1975 in the so-called Operation Babylift is a case in point. In subsequent litigation family members seeking custody of the children from American foster parents claimed that the parents had not consented to the children's adoption, had not fully understood or intended the consequences of such consents as they had signed, or had signed under the duress of the circumstances.[43] In addition, some consents to adoption were given by orphanage directors who apparently had no legal authority to do so.[44] In those few instances in which the parents or other family members managed to follow their children to the United States, they were generally successful in reclaiming them.[45] Courts found that the parents had never consented to termination of parental rights[46] or that the parents' consent in the circumstances of panic in 1974 and 1975 in Saigon was not binding.[47]

Recent draft international instruments attempt to ensure the voluntariness of parental surrender of rights and consent to adoption. The Draft Convention on the Rights of the Child (1985) would require the competent authorities to determine "that the adoption is permissible in view of the child's status concerning parents, relatives, and guardians and that, if required, the appropriate persons concerned have given their informed consent to the adoption on the basis of such counselling as may be necessary."[48] The Draft Declaration on Foster Placement and Adoption (1982) calls for all necessary consents to be in a form which is legally valid in both countries in intercountry adoptions.[49]

4. *Independent:* a child voluntarily living apart from his parents with their consent. Parents may permit their child, usually an older adolescent, to live on his own, and it is not uncommon for parents to permit children to live independently during and after emergencies. This can be done without legal process. Some countries recognize the legal emancipation of the child before he reaches the age of majority. "Emancipation has traditionally been seen as a release from the constraints of minority and a grant of the privileges and responsibilities of majority, by and with the consent of the minor's parents."[50] As a legal action, emancipation is usually possible only when the minor marries, joins the military, or is otherwise self-supporting. It is not generally a means for the parents to avoid their duty of support by shifting it to others in the community.

THE COMMUNITY DUTY OF CARE AND PROTECTION FOR CHILDREN WHOSE PARENTS ARE DEAD OR UNAVAILABLE

Though the family is universally recognized as the basic unit within which children should be raised, all societies face the problem of caring for and protecting children whose "natural guardians"—their parents—are dead or unavailable. The term "unavailable" as used here includes both intentional unavailablity, as for example with abandonment or gross neglect, and unintentional unavailability, as when parents are dead, imprisoned, hospitalized, displaced, or otherwise unable to locate or join the child.

Virtually all nations recognize a community obligation to care for and protect children separated from their parents, even where the resources to do so adequately are not available. This sense of obligation can be found in custom as well as formal law. Tribal societies in various parts of the world, for example, care for orphans out of a communal sense of belonging, with no need for a formal legal structure.[51] Societies governed by a more formal legal system have generally made care and protection of children without available parents an obligation and function of the State.

The State's responsibility is exercised most commonly through formal child guardianship court proceedings and by recourse to child welfare and protective agencies. In many countries a guardian *must* be appointed for a child whose parents are dead or unavailable; this has long been the law in Europe and the U.S.S.R.[52] In other countries, child welfare agencies are directed to take charge of children in "need of protection,"[53] or of those whose parents are prevented by reasons of a serious obstacle from exercising parental rights.[54] Though the statutory language may vary, almost every nation has a provision directing the competent authorities to intervene to assist such children. Historically, these provisions grew out of guardianship law (which was mainly designed for individual children with some known family members or property) and the law and practice of institutions for the care of foundlings and other homeless indigent children.

The obligation of the community to assist children whose parents are dead or unavailable stems, of course, from the fact that children are immature human beings who cannot completely care for themselves and who thus need adult protection and assistance. Internationally, the need for such special protection of children during their formative years has been recognized in a number of instruments: The Geneva Declaration of the Rights of the Child of 1924 ("Mankind owes to the child the best that it has to give" . . . "The child must be given the means requisite for its normal development, both materially and spiritually");[55] the Universal Declaration of Human Rights (1948) ("Motherhood and childhood are entitled to special protection");[56] the International Covenant on Economic, Social and Cultural Rights (1966) ("Special measures of protection and assistance should be taken on behalf of all children and young persons");[57] the Declaration of the Rights of the Child (1959) ("[t]he child, by reason of his physical and mental immaturity, needs special safeguards and care, including appropriate legal protection");[58] and the Draft Convention on the Rights of the Child (1985) ("[T]he child, due to the needs of his physical and mental development, requires particular care and assistance with regard to health, physical, mental, moral and social development, and requires legal protection in conditions of freedom, dignity and security").[59]

Awareness of the child's special needs and vulnerability leads inevitably to an explicit recognition of the necessity for special care and protection of children without parents. In international law this idea was first expressed in the Declaration of Geneva, which stated: "The orphan and the waif must be sheltered and succoured."[60] The Declaration of the Rights of the Child (1959) repeats the point in more contemporary language stating: "Society and the public authorities shall have the duty to extend particular care to children without a family. . . ."[61] This principle is carried forward in the Draft Convention on the Rights of the Child (1985): "A child permanently or temporarily deprived of his family environment for any reason shall be entitled to special protection and assistance provided by the State."[62] Similarly, the Draft Declaration on Foster Placement and Adoption (1982) states: "It must be recognized that there are parents who cannot bring up their own children and that the children's right to security, affecting and continuing care should be of greatest importance."[63]

International concern for the protection of children without parents also finds expression in the Hague Convention on the Protection of Infants (1961), which is designed to ensure guardianship protection for children living outside their native countries.[64] In addition, there are the specific provisions for protection of children without parents during war and other emergencies discussed in Chapter 17.

In sum, then, on both a national and international level, it is well recognized that (1) all children need adult protection and assistance; (2) in particular, children whose parents are dead or unavailable need special care and protection; and (3) the community, acting through the State, has a duty to ensure that care and protection are provided to all children.

UNACCOMPANIED CHILDREN: DEFINITIONS

The provision of special care for children who are separated from parents or other adult care-givers requires that the phrase "unaccompanied children" be defined. The object of such a definition is to specify those children whose especially vulnerable situation makes them *eligible* for special protection and assistance. Classifying a child as "unaccompanied" does not automatically require that a child receive any particular services; ordinarily, the assistance to be provided would depend on each child's individual circumstances and needs. There are two components to any definition of unaccompanied children: age and "unaccompaniedness"—the status of being apart from parents or other previous adult care-givers.

The child welfare laws of nation States provide their own definitions of children eligible for care and protection because of the death or unavailability of parents or other adult care-givers. Such definitions reflect the countries' ideas about which children require special protection in their particular social contexts, including such concepts as when adolescents are considered old enough to fend for themselves and the extent to which family members other than parents may be expected to care for minor relatives. National legislation will thus address both the age and the unaccompaniedness (or need for protection) aspects of the definition of children without families. The definitions in national laws will, of course, vary from country to country.

With respect to the question of who is considered a child as defined by age, although there are countries which still use a higher age (typically twenty-one), "[t]he lowering of the age of civil majority to eighteen represents a continuing trend in the legal definition of 'child'."[65] Ages for other purposes, such as marriage, work, and criminal responsibility are often lower, and the age used for child protection measures may also be below that of civil majority. Despite the growing convergence on the age of majority of the child, ages used for child protective services, however, vary from country to country. With respect to unaccompaniedness, national laws do not usually speak of "unaccompanied children" as such. Rather, these children would fall within the terms of the general legislation for children in need of community protection described in the preceding section.

There is no one accepted definition for unaccompanied children in the international instruments concerning children. In fact, international legislation has not even reached agreement on an age definition of a child. The Declaration of the Rights of the Child (1959) contains no specification of age; the same is true of the reference to children in other present United Nations instruments. The Convention on the Rights of the Child (1985) would define a child as "every human being to the age of eighteen years unless, under the law of his State, he has attained the age of majority earlier,"[66] but this Convention is still in the draft stage. Other international instruments bearing on the protection of children which specify age at all use a variety of ages from fifteen to twenty-one. The Geneva Conventions (1949) and Protocols (1977) use the age of fifteen for most protective measures, but disapprove military recruitment and prohibit execution of those under eighteen.[67] The Hague Convention on the Civil Aspects of Child Abduction (1980) ceases to apply when the child attains the age of sixteen.[68] The Hague Convention on Adoption (1965) applies to children eighteen years old and less,[69] and the Hague Conventions on the Law Applicable to Child Support Obligations (1956)[70] and the Recognition and Execution of Child Support Judgments (1958)[71] both apply to those under twenty-one years old.

On the issue of unaccompaniedness (or need for protection), however, international legislation consistently states or implies that it is children deprived of the care of both parents, at least, who are to be accorded special aid. The Fourth Geneva Convention (1949) thus refers to those "who are orphaned or are separated from their families."[72] The Declaration of the Rights of the Child (1959) refers to "children without a family,"[73] while the Draft Declaration on Foster Placement and Adoption (1982) concerns itself with children "who cannot remain in their biological family" or "who cannot be raised by their parents."[74] Under some of these instruments, then, a child who was with his "family" would not be considered unaccompanied. The term "family" is not defined, but it is fair to assume that it would include at least adult siblings, grandparents, and aunts and uncles.

The United Nations High Commissioner for Refugees makes a distinction in its definitions between an unaccompanied child and an unaccompanied minor. An unaccompanied *child* is defined as "a child under fifteen years of age who has been separated from *both* parents and for whose care no person can be found who by law or custom has primary responsibility."[75] An unaccompanied *minor* is one over the age of fifteen but below the age of legal majority of the country concerned, which maybe the country of origin, asylum or resettlement, who otherwise meets the criteria of an unaccompanied child.

The disparities in definitions among various countries and agencies present problems for unaccompanied children in emergencies displaced or resettled outside of their countries of origin. A child may be treated as an unaccompanied minor in one context but not in another; this may be particularly problematic when resettlement opportunities are created for specially defined groups of unaccompanied children.[76] As discussed in greater detail in Chapter 18, some countries will apply their own age definition to all children in their territory while others will apply the age of majority of the State of the child's nationality. Some of these latter States, however, hold that majority, once acquired, is not lost through subsequent change of domicile or nationality. The Hague Convention on the Protection of Infants defines "infant" as any person who has that status under both the laws of his nationality and that of his habitual residence.[77] This choice of law rule addresses but does not completely eliminate the differences in treatment experienced by unaccompanied children outside their native lands. Furthermore, measures for "unaccompanied children" may apply only to those meeting different special age definitions in different countries.

Also, past experience has demonstrated the practical difficulty of applying any age definition to an individual child. Children may not know their ages and young adults may falsely claim to be children in order to obtain perceived benefits. The obvious methods—asking the child, his siblings and acquaintances, physical observation, taking a personal history of the child—are all far from fool-proof. The difficulties are compounded in those cultures in which birthdates are often of no particular significance and therefore sometimes unknown. Age determinations for purposes of unaccompanied child identification will thus have to depend on the best method possible in the circumstances, recognizing that no method is perfect. Perhaps the best means of directing any special services to the desired population is by ensuring that unaccompanied children are not given such favored treatment that there is a powerful incentive for young adults to falsify their ages.

GUARDIANSHIP

Among the most important measures of protection and assistance for children without their parents is the assumption of guardianship rights and duties by another adult or adults. There are basically two kinds of guardianship: private and public. In both, the guardian is responsible for protection and care of the child; he administers his property and represents his interests.[78] The guardian has the right, and the duty, to control a child's residence, education, discipline and general upbringing.[79]

A guardian thus substitutes for the absent parents in exercising the "parental" rights and duties. The assumption or assignment of guardianship must be seen as an essential element of the community duty of care and protection which was discussed in the preceding section. Children who are unaccompanied outside their own countries have an especially pressing need for a legal guardian.[80] The Draft Declaration on Foster Placement and Adoption (1982) recommends: "The child should at all times have a name, nationality and legal guardian."[81]

Guardianship and physical custody of a child are conceptually different issues. Guardianship may thus rest with one person or agency while the child resides with another individual or institution. This derives from the fact that the guardian has the right to decide on the child's residence, but no obligation to have the child live with him personally. As will be seen below, this bifurcation of guardianship and residence is quite common where State or charitable officials or agencies act as guardians.

Private Guardianship

Private guardianship is guardianship by a private individual, often a relative of the child. It arose historically from the role of the extended family and clan. In present times a private guardian is not ordinarily responsible, solely by virtue of his guardianship, for the cost of protection and maintenance.[82] These expenses must be met by the minor's own resources, by those relatives who are legally liable for support, or by the State.

Private guardianship is the preferred mechanism in most societies. In many cultures, in fact, extended families spontaneously assume responsibility for minor relatives without any legal or other community involvement. In some legal systems, guardianship devolves on certain relatives by operation of law. Islamic law, for example, provides that in the father's absence the paternal grandfather becomes the child's guardian; if the grandfather is unavailable, a series of male relatives is prescribed.[83] In several countries, including France and Belgium, a family council may be made responsible for minor family members.[84] Even where relatives are not strictly mandated by law, they are often the first choice as guardians.[85] This widespread practice is reflected in the Draft Declaration on Foster Placement and Adoption: "Other family members should be the first alternative if the biological parents cannot care for the child."[86]

Many countries now require judicial confirmation of guardians, whether relatives or unrelated persons, in order to protect the welfare of the child. Thus "private" guardianship as used here does not necessarily mean the absence of State involvement or supervision. State supervision of private guardians is now, in fact, the majority practice. The degree to which such supervision actually affects these private arrangements is an open question, however.

Public Guardianship

A public guardianship is guardianship by a State official or agency or by a charitable agency or institution. Public guardianship has a long history, arising from the need to care for destitute, parentless children with no relatives or friends to support them.[87] In earlier times, charitable, often religious, institutions assumed this function; in the twentieth century it has usually been assumed by State welfare and child protective services.

Public guardianship may rest with one of the officers of the institution in which the child is placed, as in Italy,[88] with the institution itself, as in the Soviet Union,[89] or with a state social service agency, as in the Federal Republic of Germany and Colombia.[90] Many countries employ more than one form of public guardianship. Where the guardianship is held by a child welfare agency, the agency will typically place the child with a foster family, group home, or institution.

Guardianship of Recently Resettled Unaccompanied Children

Arrangements for guardianship of unaccompanied children settled or resettled outside of their countries of origin take a variety of forms, depending on the local legal system, the situation of the child, and the preferences of the authorities. A brief examination of the guardianship of unaccompanied children from Indochina who have been resettled in Australia, Europe, and North America in the past ten years gives some idea of the range of legal mechanisms.

1. *Australia:* Guardianship by the State Ministers for Community Welfare as delegated by the Federal Minister of Immigration.[91]

2. *Belgium:* Guardianship by local Social Welfare Center (Centres Publiques d'Aide Sociale) acting for the Department of Social Welfare (Administration de l'Aide Sociale), with the assistance of a Belgian and Indochinese family, or of a family council consisting of Belgian and Indochinese members.[92]

3. *Canada:* In Quebec, the majority of unaccompanied children have been placed under the guardianship of the Centre des Services Sociaux, while some were made the responsibility of voluntary organizations. Foster families there have been encouraged to become the legal guardians, but few have done so.[93] In Ontario, the authorities required that families taking unaccompanied minors into their care expeditiously assume guardianship.[94]

4. *France:* Several possibilities for guardianship exist: (1) delegation of parental authority to the Child Welfare Organization (Aide Sociale a l'Enfance); (2) organization of a guardianship with a family council or as a State guardianship, filled by a prefect who delegates it to the Departmental Office of Health and Social Affairs (Direction Departmentale des Affaires Sanitaires et Sociales) or by the head of the institution where the child is placed; or (3) placement under the protection of the Child Welfare Organization (A.S.E.) as "in temporary care" or "assisted" or "in custody." Foster parents may also apply for guardianship.[95]

5. *Germany (Federal Republic):* Generally, guardianship has been held by the Youth Office (Judendamt), sometimes in consultation with the sponsoring agency. There has been discussion of whether foster parents should assume guardianship of children settled with them.[96]

6. *Netherlands:* The De Opbouw Foundation is the guardian for all Southeast Asian minors under the supervision of a cantonal judge who is responsible for all placement decisions.[97]

7. *Norway:* A private citizen is appointed guardian by a probate judge together with the supervisor of guardians. A municipal employee is also appointed by the Child Protection Committee to advise refugee guardians.[98]

8. *Sweden:* A co-guardian or *god man* (trustee, tutor, "goodman") is appointed to take care of the minor's affairs. Municipal Social Welfare Committees will ensure care and accommodation for those needing it.[99]

9. *United Kingdom:* Voluntary organizations act as guardians for children they have sponsored; other children have been taken into care by local authorities under section 1 of the Children's Act.[100]

10. *United States:* Under the Refugee Act of 1980 the Director of the Office of Refugee Resettlement acts as guardian until protective legal responsibility is established according to the law of the state where the minor is resettled.[101] In most cases, the state courts appoint the state child welfare agency as guardian.[102]

THE PRINCIPLE OF DECISION FOR INTERVENORS AND GUARDIANS: "THE BEST INTERESTS OF THE CHILD"

In the words of one commentator, "[t]he unifying principle that runs through modern custody law is that custody of children is to promote the 'interests,' 'best interests,' or 'welfare' of the child."[103] This principle—referred to here as "the best interests of the child" principle—appears in the legislative or judicial family law of the majority of States, and in all recent international legislation concerning the care of children. It is used to guide decision in a variety of contexts, including custody of children due to divorce, guardianship, foster care, and adoption. The "best interests of the child" is thus the current international principle for decision by those who intervene on behalf of children whose parents are dead or unavailable. An indication of its widespread acceptance, at least in codes and judicial decisions, can be seen by the fact that all thirteen countries whose child welfare law is described in the United Nations Institute for Training and Research (UNITAR) Report on Law and the Status of the Child, which includes countries in North and South America, East and West Europe, the Middle East, Africa, and Asia, use a version of the best interests standard (e.g., "best interests," "most beneficial," "most favorable conditions for optimal development") for placement decisions.[104]

This consensus in national law has made itself felt in the post-World War II era in international legislation, starting with the Declaration of the Rights of the Child (1959): "In the enactment of laws for the special protection of the child, the best interests of the child shall be the paramount consideration"[105] and "The best interests of the child shall be the guiding principle of those responsible for his education and guidance...."[106] The Draft Convention on the Rights of the Child (1985) states:[107]

> In all actions concerning children, whether undertaken by public or private social welfare institutions, courts of law, or administrative authorities, the best interests of the child shall be a primary consideration.

The Draft Declaration on Foster Placement and Adoption (1982) can be seen in its entirety as an attempt to define the best interests of the child. Its basic approach, as noted above, is that "[i]t must be recognized that there are parents who cannot bring up their own children and that the children's right to security, affecting and continuing care should be of greatest importance."[108]

The Adoption and Child Abduction Conventions of the Hague Conference on Private International Law both incorporate the best interests standard. The Adoption Convention (1965) specifies that the authorities of a State Party "shall not grant an adoption unless it will be in the interest of the child," thus insisting on this standard in preference to any conflicting national rules.[109] The preamble of the Convention on Child Abduction (1980) recites that the States Party are "[f]irmly convinced that the interests of children are of paramount importance in matters relating to their custody." The Convention proceeds to impose a best interests test on the question of the return of wrongfully removed children.[110]

In the emergency context, the United Nations High Commissioner for Refugees has adopted a best interests standard for placement and family reunion of unaccompanied children within its concern. The best interests of the child, its Director of International Protection has stated, should be "the point of departure" for actions concerning unaccompanied children.[111] The implementation of this doctrine by UNHCR is described in the section on Reunion and in Chapter 19.

In light of this widespread international agreement on the best interests standard in matters of the custody of children already separated from their parents, it is worth examining some of its implications. Most importantly, the bests interests standard directs the court or other agency to put what is best for the child above all other considerations, including, of course, political, social, or religious considerations. The aim is thus to do the best possible in the circumstances to meet the needs of the child.

Second, this standard means that the child must be treated as an individual and that his particular circumstances and needs are the most important factors to be considered. Individual assessments and individualized placement decisions are therefore required; what is truly best for a given child cannot be determined by a general formula. Nor can it be decided without consulting the child, as is discussed at greater length below.[112]

Third, those seeking to provide for the best interests of the child must have as their goal the satisfaction of the child's developmental needs: in the words of the Declaration of the Rights of the Child (1959), "to enable him to develop physically, mentally, morally, spiritually, and socially in a healthy and normal manner."[113] Obviously, this includes a minimum of material assistance and security as well as protection against exploitation.[114]

In recent years, it has come to be recognized that a child's emotional and psychological needs must also be met by any action meant to be in his best interest. One of the moving forces behind this awareness has been *Beyond the Best Interests of the Child,* by Joseph Goldstein, Anna Freud, and Albert J. Solnit, first published in 1973.[115] The authors argue that the placement and the process of placement of children should safeguard the child's need for continuity of relationships, should reflect the child's sense of time, and should intervene in private child placement decisions only when absolutely necessary. The absorption of these ideas into current legal thought can be seen in a number of contexts.[116]

The "best interests" standard thus imposes several constraints on intervenors and guardians. They must (1) put the child's welfare ahead of all other considerations; (2) respect the individual humanity of the child; and (3) to the maximum extent possible, meet the child's developmental needs. Within these bounds, intervenors and guardians have wide latitude to assist the child according to the mores of the relevant culture.[117] To some degree, in fact, how a society chooses to best meet the child's needs will depend on how it perceives children in general within its cultural and legal framework. Different placement methods can thus be properly found among different countries under the same "best interests" rubric. It is also the case, however, that some countries nominally employing a best interests test in fact apply other bases for decision, such as preference for placement with family members or political, religious, or other agendas. The broad nature of the term "best interests of the child," has allowed it to be used in a standardless, highly discretionary manner. Inadequate resources may also cause decisions to be made which fail to meet the child's best interests. Nevertheless, if administered with the child's true interests as being paramount, and in light of the realities and developmental needs of each child, it can be applied in a way which furthers the child's well-being.

The next section explores the manifestation of this best interests principle in the law of interim placements: institutional care, group homes, and foster care. The sections which follow will discuss the implications of the principle with respect to the child himself, the parents from whom he was separated, his other relatives, and his State of nationality.

INTERIM PLACEMENT: INSTITUTIONAL, GROUP, AND FOSTER CARE

Institutional, group, and foster care are placement options available to guardians or others responsible for choosing the child's residence. Institutional care refers to institutions, such as orphanages, which house large numbers of children under the care of a staff of adults. In group care, a relatively small number of children live together under adult supervision and participate in the daily life of the community. Foster care (also called "foster family care") is the placement of a child in the home of an adult or adults other than his parents who are responsible for his day-to-day upbringing and care. Foster parents are not necessarily, or even usually, the child's guardians, though they can be in most legal systems. Legally, neither institutional, group, nor foster care is permanent; they can all be changed or terminated at any time by the child's legal guardian.

From a legal standpoint, an important question is whether the law itself specifies a preference for institutional, group, or foster care. The child welfare law may permit foster care only for certain categories of children, such as those who are likely to need care for a relatively long period of time. Israel, for example, limits the use of foster care to those children whose parents agree to their foster care, or for whom the court has withdrawn parental rights and appointed a guardian, or who have been found in need of protection.[118] Often, however, the law leaves the decision about the kind of placement to the guardians or judicial officers to whom responsibility for the child has been assigned.[119] These persons would then presumably use their discretion in placing the child in accordance with the best interests principle. In actuality, cultural preferences and the availability of space within the various placements will likely influence the choice as well.

Child welfare experts have generally come to prefer family care to the use of institutions for reasons relating to the child's need for attachment and continuity of care. National practice has been moving in that direction for many years, particularly where public funds are available to support the practice.[120] In certain countries, however, the idea of having an unrelated child in the home may conflict with cultural norms.

The only international instruments which mention institutional or foster care or explicitly take a position on the question of institutional care versus family care, however, are both still in draft form. (There is no mention of group care in any international instruments.) The Draft Convention on the Rights of the Child (1985) states in its preamble, "the child, for the full and harmonious development of his personality, should grow up in [a] family environment, in an atmosphere of happiness, love and understanding,"[121] but so far none of the articles adopted by the working group incorporates this viewpoint.[122] In fact, Art. 10(2) of the Draft Convention merely states:

> . . . a child who is parentless, or who is temporarily or permanently deprived of his family environment, or who in his best interests cannot be brought up or be allowed to remain in that environment, shall be provided with alternative family care which could include, *inter alia,* adoption, foster placement, or placement in suitable institutions for the care of children.

The Draft Declaration on Foster Placement and Adoption (1982), however, explicitly includes preference for family care in its text:[123]

> Every child has a right to a family. Children who cannot remain in their biological family should be placed in a foster family or adoption in preference to institutions, unless the child's particular needs can best be met in a specialized facility.
>
> Children for whom institutional care was formerly regarded as the only option should be placed with families, both foster and adoptive.

This Draft Declaration goes on to say: "Foster family care should be a planned, temporary service as a bridge to permanency for a child, which includes, but is not limited to, restoration to the biological family or adoption."[124]

In addition to these draft instruments, some support for a preference for foster care can be read into the statement that "the child . . . shall . . . grow up . . . in any case, in an atmosphere of affection and of moral and material security," contained in the Declaration of the Rights of the Child (1959).[125] A child is far more likely to find this "atmosphere of affection" within a family than in an institution. In light of what is now known about children and their developmental needs, it can well be argued that where foster families are available, or would be with reasonable effort by the competent authorities, placement of a child in an institution would be contrary to the best interests principle. On the other hand, where there are no other alternatives to aid homeless, needy children, then institutional care can be seen as the best available means to protect their welfare.

THE ROLE OF THE CHILD IN DECISIONS MADE IN HIS BEST INTERESTS

If the "best interests" principle is to guide the actions of guardians and other adults who act to protect and care for the child, the question then arises as to the child's role in the making of these decisions. With unaccompanied children in emergencies, these decisions concern mainly (1) reunion with parents and other relatives, (2) adoption, and (3) repatriation, settlement, and resettlement for displaced children. The major issues for each of these decision points are (1) when should the child's preferences be solicited and considered, and (2) in those instances when the child's preferences are considered, what weight should they be given. These issues apply to the role of the child vís-a-vís both legal guardians and other administrative and judicial authorities with power to make decisions about him.[126]

In considering this question, it must be understood that the child's choice will be made within the context of the available options, and that these options will be determined both by the nature of the situation and by actions taken by the competent authorities. A child may have a choice between placement with an uncle and resettlement, for example, only if steps have been taken to find the uncle and an opportunity for resettlement is provided by a receiving State. Otherwise, there may be no such "choice" facing the child. The development of options, then, is an integral part of the decision process, involving not the role of the child but the policy of the various entities acting upon the entire situation.

Furthermore, the child's competence to choose may not be the same for all kinds of decisions. The decision about whether or not to remain with present caretakers or rejoin parents or other relatives, for example, may involve different kinds of information, reasoning ability, and experience than the choice of whether or not to reset-tle.[127] In other words, children may be capable of making one kind of choice at one age and other kinds of choices at other ages.

This section will outline the role of the child in placement and adoption decisions in national and international law. Proposals regarding the role of the child in various kinds of decisions appear in the Recommendations in Part IV.

National Law

The issue of the role of the child arises most often in national law in connection with custody disputes between parents, guardianship proceedings, and adoption. In all three strands of law there has been a gradual evolution toward a greater role for the child,[128] but the change has been neither universal nor complete. At present, there is wide diversity in national law and practice in the role given the child's wishes in all three contexts.

At one extreme are the more recent family codes of Cuba and Israel which give substantial weight to the child's wishes. In Cuba, courts must seek and consider the views of children over the age of seven in guardianship and adoption proceedings.[129] In Israel, the courts applying the best interests principle in custody and guardianship proceedings will follow the wishes of a child over the age of ten or eleven unless his choice is clearly contrary to his own welfare.[130] At the other extreme are countries such as Greece and the People's Republic of the Congo, where the child's preferences are given little weight and custody decisions are made according to rules preferring one of the parents or another adult.[131] In the middle range, one might say, is China, which in applying the best interests principle in custody disputes will consult the child old enough to express a preference and follow his wishes "to the extent possible."[132] Similarly, Australian courts will consider the wishes of children under fourteen in custody matters and follow the wishes of those over fourteen unless the child's interest clearly requires the contrary.[133] In the United States, the preference of children over twelve or fourteen is often followed; below that age it may be considered but it is normally not controlling.[134]

There is more widespread agreement in national law and practice about the child's role in adoption proceedings than in custody and guardianship. A wide range of States require the consent of children over the age of fourteen to their own adoption.[135]

International Norms

The Declaration of the Rights of the Child (1959) does not specifically provide a role for the child in decisions concerning him, but the Draft Convention on the Rights of the Child (1985) has two such provisions:

> In all judicial or administrative proceedings affecting a child that [sic] is capable of forming his own views, an opportunity shall be provided for the views of the child to be heard, either directly or indirectly, through a representative, as a party to the proceedings, and those views shall be taken into consideration by the competent authorities followed in the State Party for the application of its legislation.[136]

The States Parties to the present Convention shall assure to the child who is capable of forming his own views the right to express his opinion freely in all matters, the wishes of the child being given weight in accordance with his age and maturity.[137]

To the same effect, the Hague Convention on the Civil Aspects of International Child Abduction (1980) permits (but does not require) courts to refuse to return a child "if the child objects to being returned and has attained an age and degree of maturity at which it is appropriate to take account of its views."[138]

In the sphere of international practice, the UNHCR Guidelines for the Promotion of Durable Solutions for Unaccompanied Minor Children in South-East Asia (1982) are the most specific about the role of the child in placement decisions:[139]

[P]ersons of 15 years of age or older are to be treated as adults. An unaccompanied minor child between the ages of 10 and 15 who is found to be of sufficient maturity to make an independent judgment shall also be treated as an adult with regard to the choice of a durable solution. It is presumed, in the absence of strong evidence to the contrary, that children below 10 years of age are not of sufficient maturity to make an independent judgment.

In addition to the specific provisions and practice just cited, it is important to note the participation rights included in the Universal Declaration of Human Rights (1948) and other international instruments. The Declaration states:

- All human beings are born free and equal in dignity and rights (Art. 1).
- Everyone has the right to recognition everywhere as a person before the law (Art. 6).[140]
- Everyone is entitled in full equality to a fair and public hearing by an independent and impartial tribunal, in the determination of his rights and obligations . . . (Art. 10).[141]
- Everyone has a right to freedom and movement within the borders of each State (Art. 13).

These general provisions, which refer to "all human beings" and "everyone," suggest that limitations on children's rights to make their own choices be only as extensive as necessary. Protection of the right to choose is, after all, an important aspect of respecting any individual.[142] Recent international legislation moves in this direction by weighting the child's opinion in accordance with his age and maturity.[143] The main question then becomes a factual one: at what age or ages are children capable of making various kinds of decisions?

While it is impossible to review this question fully here, studies have shown that children over the age of fourteen do not differ from adults in their capacity to make informed decisions on such questions as medical treatment when presented with information about the available options.[144] Children of the age of nine differ from adults in their capacity to understand and offer rational reasons for such decisions, but appear virtually equal to adults in their ability to make rational choices. These findings bear more on such questions as choice of repatriation or resettlement for unaccompanied children than on such emotionally-laden decisions as those concerning whether to rejoin a parent or other previous care-giver or stay with a foster family. Even with the former kind of choice, however, children must be provided information about the options available to them and be completely free from adult coercion.

Although they possess an increasing ability to make decisions as they mature, children are particularly susceptible to being misled by omission or misrepresentation of information and to being influenced by subtle or flagrant adult pressure.

With respect to choices between potential caretakers, children are capable of expressing themselves at earlier ages. But studies have shown that the trauma and guilt of choosing between the parents in divorce proceedings can be substantial.[145] While a choice between a natural family member and a substitute family may not carry quite the same degree of emotional involvement, it still needs to be handled with great sensitivity. At the least, skilled questioning of the child is likely to be required, and in some situations, it may not be advisable to put the burden of choice on the child.

REUNION: THE ROLE OF THE NATURAL PARENTS IN DECISIONS MADE IN THE CHILD'S BEST INTERESTS

Clearly, reunion of a child with his natural parents will often be in his best interests.[146] Attachment between the parents and child can remain even after substantial separation, at least for children over the age of five.[147] Biological connection and previous affection and care produce at the least a rebuttable presumption that the natural parents will be the best caretakers of their child even after a period of separation. In fact, in many countries the rights of a natural parent to reunite with a child after a period of separation is absolute, unless the parent is found to be likely to neglect or abuse the child. The precise boundaries of this right depend on national family and child welfare law, and intervenors must look first to the law of the State or States concerned with unaccompanied children whose parents desire reunion.

In recent years, however, there has been a trend on the international level toward recognition of the fact that the child's best interest may not always be served by reunion with parents after a substantial period of separation. A child's makeup is such that emotional ties to the natural parents can lapse and be replaced by new, equally valid, emotional ties to other care-giving adults. The longer the separation, the greater is the likelihood that this will naturally and normally occur. When this transition happens, it may be harmful to the child's well-being—physical, emotional, cognitive, and social—to remove him from his new substitute family and install him in a home that is no longer his own.[148] It might even be argued that such a break in his placement would violate his freedom from arbitrary interference with his privacy and home as guaranteed in the Universal Declaration of Human Rights.[149] Thus, when old family attachments have faded and new ones have formed, the best interests principle implies that the child remain with those new adult care-givers, even at the expense of the legitimate desires for reunion of his natural parents or other previous care-givers. If and when this happens will depend on the individual child, his age at the time of separation, and the extent and quality of care he is receiving from the new adult care-givers.

The interplay between the rights of natural parents (or others with legal guardianship rights) and the child's best interests in the legal context is well illustrated in the 1980 Hague Convention on the Civil Aspects of International Child Abduction.[150]

This convention concerns the transnational wrongful removal and retention of children in breach of legal custody rights.[151] Even where the adult keeping the child has violated legal custody rights of a natural parent or legal guardian who seeks the child's return, under this Convention the court of the State in which the child is found need not order the child's return if it is established that "there is a grave risk that his or her return would expose the child to physical or psychological harm or otherwise place the child in an intolerable situation" or if a child of sufficient maturity "objects to being returned."[152] In addition, the child need not be returned if "[t]he return of the child . . . would not be permitted by the fundamental principles of the requested State relating to the protection of human rights and fundamental freedoms.[153] Furthermore, a child who has been separated from the legal custodian for more than one year can remain where he is if "it is demonstrated that the child is now settled in its new environment."[154] Thus, the Hague Convention on Child Abduction gives priority to the child's ties to his present care arrangement even where that new arrangement resulted from his wrongful removal or retention by his present caretaker. Under this Convention, the claim to custody of the natural parent or legal guardian will be denied where the interests of the child so require.[155] The Convention, however, had only been ratified by four States as of December 1983.

The possible conflict between parental rights and the child's best interests in the emergency context is also addressed by UNHCR in its *Handbook for Emergencies* (1982):[156]

> Every effort must be made to keep the child with the same substitute parents until blood parents are found. The child will then need time to reattach to his or her blood parent(s). How long this takes depends on the child's age and the strength of the attachment to the substitute parent(s) which now has to be broken. Where years have elapsed, it has been found that the child's interest may even be better served by remaining with the substitute family.

While this directive recognizes the possibility of leaving the child with the substitute family, in contrast to the Hague Convention on the Civil Aspects of Child Abduction it contains neither standards for evaluation of when this would be appropriate, presumptive time periods, nor procedures for making the decision.

REUNION: THE ROLE OF OTHER RELATIVES

Placement, or "reunion," with family members when the parents are dead or unavailable, is treated in national law in several different ways. One approach is to make certain relatives responsible for the child's care by operation of law. In these legal systems, children whose parents are dead or unavailable would thus ordinarily be placed with the closest available relative according to a prescribed order. As noted previously, for example, Islamic law makes the paternal grandfather responsible for a child in the father's absence; a series of male relatives is mandated if the grandfather is unavailable. Other legal systems have similar provisions making family members responsible for minor relatives.

Under these systems, unaccompanied children would ordinarily join the specified relatives upon the latters' request. There would thus be little or no inquiry about the suitability of placement. In effect, these legal systems contain a general public policy decision that the children are best off with specified adult family members.

A second approach is to evaluate placement with a relative under the best interests standard, as in placement with an unrelated adult. In contrast to the first approach, this requires an inquiry into the desirability of the particular placement. Relatives, of course, will often be from an ethnic, social, cultural, and religious background similar to that of the child and have a serious commitment to the child's well-being. They may not, however, have had as much recent contact with the child as other adults, and therefore not be in a position to provide the continuity of relationship the child may require.

The possibility that a child may be better off with adults other than relatives is well illustrated by the experience of Jewish children in the Netherlands after World War II. Thousands of these children had lived with non-Jewish Dutch families—often for years—after their parents were taken to concentration camps or killed. At the end of the war some of the children were placed with relatives whom they hardly knew, even though the children had been well-integrated into their Dutch foster families.[157] It is clear that removing children from unrelated care-givers of longstanding relationships and "returning" them to relatives may be contrary to their best interests.

There is a third approach to the role of relatives which is something of a combination of the two already mentioned: preference for placement with relatives coupled with judicial or other examination of their suitability. Several countries which use family councils for placement of children without parents now oversee the resulting arrangements in order to protect the child's welfare.[158] The recommendation in the Draft Declaration on Foster Placement and Adoption (1982) that "[o]ther family members should be the first alternative if the biological parents cannot provide care for the child" is consistent with this approach, making other family members the "first alternative," but not the only one. In accordance with the best interests standard, placement of the child with these relatives would take place only if it was the least detrimental alternative for the child.

THE ROLE OF THE STATE OF THE CHILD'S NATIONALITY

Children have been found unaccompanied outside the country of which they are citizens numerous times in the twentieth century, and questions about the role of their national State have often arisen. Can countries "claim" children who are their nationals? If such "claims" conflict with those of the parents or relatives, whose wishes should be followed? Even in the absence of parents or other relatives, do national claims on the child outweigh the child's best interests or his or her own desires? Because these questions were a heated issue with displaced European children in the years after World War II, this section begins with an overview of the legal issues involved in placing unaccompanied children in that period. It then addresses these questions in the context of current international law.

Unaccompanied Children in Post–World War II Europe

A general history of unaccompanied children in post-World War II Europe appears in Part I. With respect to the legal issues, after World War II the victorious Allies agreed among themselves that the government of each country would have the guardianship of its own unaccompanied minor citizens displaced by the war outside their countries of origin.[159] Placement and reunion decisions were thus the ultimate responsibility of the authorities of the State of which the children were nationals. As discussed later in Chapter 19, though the United Nations Relief and Rehabilitation Administration (UNRRA) and the International Refugee Organization (IRO) were designated to identify and care for the children, neither had legal guardianship of unaccompanied displaced children or legal power over their placement.[160]

The first priority of the national authorities was to reunite the children with their parents (and sometimes other relatives) wherever these adults were then living. If this was impossible, the children ordinarily were repatriated to their country of nationality. This order of priorities was aptly summarized by a contemporary observer as: "children belong first to their families, and second to their countries."[161]

Thus, efforts were made to locate the parents or other family members for purposes of reunion. If these adults were located, children were sent to join them in the country of origin or wherever else they might be. Only if the parents were "unfit," or did not want the child, was reunion to be withheld.[162]

The placement of those for whom family reunion was impossible was decided by the authorities of the child's country of nationality. The vast majority of the children of known nationality were repatriated at the insistence of their national representatives. The national representatives did have authority to "release" children for resettlement, however, and Poland, Czechoslovakia, and Yugoslavia agreed to the resettlement of Jewish unaccompanied children who were their citizens. For the most part, though, national representatives were extremely reluctant to permit any placement but repatriation, mainly out of a desire to rebuild their populations.

Under this approach, it was quite important to establish the child's nationality, since responsibility for placement depended directly on citizenship. Children whose parents had been legally married were assigned the father's nationality; those born out of wedlock were given the mother's.[163] There were a number of children whose parentage, and hence their nationality, could not be determined. Some of these were given the nationality of their country of present residence, while others were declared stateless persons.

The operation of the general rule that family reunion took precedence over national claims but that national claims outweighed all other considerations is well illustrated by the experience in 1946 of two groups of school children. The first involved a number of children from Belgrade whose fathers were White Russian emigrés and whose mothers were Serbian. The children had been evacuated to Saxony by the Germans in 1944 and found themselves in the Soviet-occupied zone of Germany at the end of the war. Their parents, who were in the American-occupied zone, requested their return. Although the Soviet authorities had planned to repatriate the children on the basis of their citizenship before this request, the children were eventually reunited with their parents in the American zone.[164]

The second group of children were Yugoslavian citizens who had moved with their religious school to Austria during the war. At the end of the war, the government of Yugoslavia requested their repatriation as did some of their parents. The children, mostly adolescents, initially vehemently opposed repatriation on political grounds, but their objections were summarily rejected. A British officer in the prisoner-of-war and displaced person division involved in the case flatly stated the prevailing rule: "children are on the whole unable to raise objections for themselves against repatriation and so should go back to their countries."[165]

As the postwar period progressed, this rather absolute approach to placement of the unaccompanied children—family reunion or repatriation—was somewhat modified. In part, this was due to widespread disillusion with the practice of removing children from homes in Germany or Austria in which they had been settled for years. In 1948, the Economic and Social Council recommended that the IRO act:[166]

> to unite children with their parents wherever the latter may be, and in the case of orphans or unaccompanied children (*i.e.*, children who were abandoned or whose parents could not be reached) whose nationality has been established beyond doubt, to return them to their country, always providing that the best interest of the individual child shall be the determining factor.

It can be seen that this recommendation makes family reunion an absolute first priority, while repatriation, the second priority, can take place only when it is in the child's best interest. In 1948 also, the IRO modified its directives to permit children to remain in German, Austrian, or displaced persons foster families under certain conditions.[167] When responsibility for child placement in the United States zone in Germany was passed to the High Commissioner for Germany (HICOG) in 1950, his Law No. 11 specifically directed that the question of repatriation, resettlement, or settlement in Germany was to be determined by the child's best interests in the HICOG courts. In deciding the child's best interests, the court, pursuant to Article 14 of the law, was to be guided by the following factors:[168]

a) the existence or absence of a wholesome relationship between the child and its foster-parents or other persons,

b) the likelihood that the child would secure an adequate education,

c) the physical and moral welfare of the child including the probability of its obtaining adequate food, clothing, medical care and a desirable atmosphere,

d) the legal and economic protection of the child in the relation to rights of citizenship, rights to future public care and maintenance including medical and nursing care, opportunity to earn a livelihood, and the likelihood of discrimination or bias,

e) the wishes of the child if it had sufficient maturity and had formed its wishes without coercion,

f) the desires of a natural parent, foster-parent, or other near relative by consanguinity.

The original position of the Allies—repatriation as a matter of course of children whose parents could not be located—must be seen in the context of the times. The Allied authorities were faced with millions of displaced persons in the wake of a war which had devastated their economic, social, and administrative structures and decimated their populations. In order to sort out and place the unaccompanied children, they took the understandable measure of making national authorities responsible for their placement. Understandably, too, these authorities were anxious to rebuild their own national populations. Nevertheless, as time passed, such nationalistic considerations gradually yielded to considerations of the child's best interests. This trend has continued in the period since World War II.

The Role of the State of the Child's Nationality in Placement under Current International Law

Under current international legislation, the unaccompanied child's national State has no legal right to insist on his repatriation. The child's national State may, of course, *offer* repatriation, and it may also appoint a guardian under its own laws whose direction other countries may, in some instances, be obligated to follow. But even that guardian, whether an individual or governmental entity, must, under international legislation, act in the child's best interests. In short, international norms direct guardians to act in the best interests of the *child,* not in the best interests of the *nation;* when the two conflict, the interests of the child must certainly come first.

As discussed in greater detail later in Chapter 18 on Jurisdiction and Choice of Law, countries can, absent treaty obligations to the contrary, choose whether or not to recognize guardians appointed by the child's national State when the child is in their territory.[169] Some countries will recognize the acts of such foreign guardians, some will not; the Hague Convention on the Protection of Infants (1961) regulates this matter as between party States. Under that convention, measures "for the protection of the person or property" of that child taken by the authorities of the State of the child's nationality will generally be recognized in the State of his residence. However, the Convention also provides that decisions by the child's national authorities need not be enforced by the State where the child is located if they are "manifestly contrary to public policy" *(ordre public)* in that State.[170] Even in the absence of such treaty provisions, matters of child protection clearly come within the sphere of public policy or *ordre public.* As was stated by Judge Lauterpacht in the only International Court of Justice case concerning child protection:[171]

> Apart from criminal law, it is difficult to conceive of a more appropriate and more natural object of *ordre public,* as generally understood, than the protection by the State of infants, especially when they are helpless, ill, in actual or potential danger to themselves or to society, a legitimate object of its compassion and assistance, and an occasion for public resentment whenever the State fails to measure up to its responsibilities in this respect. . . . [T]here is a hard core within [the field of *ordre public*] which is not open to reasonable challenge. The protection of children, in the sense indicated above, is an obvious particle of that hard core.

Given the widespread adherence to the "best interests of the child" as the guide for decision in national law and the universal acceptance of this standard in all international legislation concerning children, measures taken by national authorities that are not in conformity with this standard should be regarded as contrary to public policy by States in which enforcement is sought. In other words, measures taken by national authorities, whether individuals or public agencies, should not be recognized if the basis for the decision is anything but the child's best interests. Actions based on national interests, rather than those of each individual child, such as automatic repatriation for all children, should thus not be honored by the country in which the child is situated. Not only should such decisions be viewed as "contrary to public policy," but they can also be regarded as outside the category of measures "for the protection of the child's person" which the Hague Convention seeks to enforce among party States.

So, too, State usurpation of parental powers or guardianship rights without valid justification must be considered illegal in current international law. In the wake of the Hungarian uprising in 1956, for example, Hungary amended its Act concerning Family and Guardianship to State:[172]

> The consent of the guardianship authorities is necessary for every declaration of the parents residing in Hungary, the object of which is the travel abroad of the child, or the determination of the place of residence abroad of the child.

Such legislation is contrary both to considerations of the child's best interests and to guardianship rights of the parents which have been well recognized in national and international law.

No State has the legal right to insist on the return or "repatriation" of any of its citizens, including unaccompanied children, except through criminal extradition treaties.[173] Furthermore, both the Universal Declaration of Human Rights (1948) and the International Covenant on Civil and Political Rights (1966) guarantee "everyone the right to leave any country including his own."[174] For children of insufficient maturity this right is exercised by the parents. When, because of parental absence, the State acts as guardian, it must exercise control of the child's residence under the same best interests principle.

In sum, then, there is no such thing as a valid "national claim" for displaced unaccompanied children in contemporary international law. In certain circumstances, national authorities may act as guardians for their unaccompanied children living abroad. But guardianship by the child's *national authorities* is not the same as guardianship conducted in the *national interest*. The only national or other guardianship entitled to recognition by the country of the child's residence is one that acts in the child's best interests and for no other purpose.

ADOPTION

In legal terms, adoption "creates between adopter and adopted the same bond as that between parent and child;"[175] in other words, "a full reciprocal parent and child relationship."[176] As is stated in the Draft Declaration on Foster Placement and Adoption (1982): "The primary purpose of adoption is to provide a permanent family for a child who cannot be cared for by his/her biological family."[177]

Adoption is governed in the first instance by national law. Not all countries recognize adoption: a United Nations survey in 1973 found that forty of the fifty-one countries responding had provisions for the legal adoption of children.[178] The trend, however, is clearly in favor of legal recognition of adoption. Though Islamic law, for example, does not traditionally provide for adoption,[179] in recent years Bahrain,

Kuwait, and Oman have legalized it, and Iraq has provided for adoption by "affiliation" of abandoned children, orphans, or children under nine whose parents are unknown.[180] In India, although adoption was quite limited under traditional Hindu law, the Hindu Adoptions and Maintenance Act of 1956 expanded permissible adoptees to include girls and orphans.[181] The Draft Convention on the Rights of the Child (1985) supports this trend, stating: "The States Parties to the present Convention shall undertake measures, where appropriate, to facilitate the process of adoption of the child."[182] Of course, even among countries where adoption exists, frequency of use varies widely.[183]

Conditions and Effects of Adoption

National adoption legislation determines the conditions under which adoptions may take place and the effects of the adoption. There is substantial variation among national laws about requirements concerning the age of adopters, their marital status, whether the adopters may already have children, the age of the person to be adopted (the adoptee), the required age difference between adopter and adoptee, and the nature of the necessary consent to adoption.[184]

There is also a variety of ways in which the effects of adoption are structured. This can be seen in the distinctions between what are often called simple and full (or complete) adoptions. In general, full adoption terminates all ties with the natural family and the child takes the adopter's name. With simple adoption, though, the natural family may retain certain rights of contact with the child and rights to inherit and receive support from the child.[185] Some countries permit only simple or only full adoption; others permit either form, at the option of the parties.[186] In both simple and full adoption, however, the adoptive parents usually have both parental authority over the child and the primary duty of support.

In some countries, adoption is revocable under certain circumstances, although usually only with judicial permission. Simple adoption in France, for example, may be revoked for "serious reasons" on application of close bood relatives or the child himself if he is over sixteen years old.[187] The European Convention on Adoption of Children (1967) provides that adoption may be revoked before the adopted person comes of age only by decision of a judicial or administrative authority on serious grounds, and only if revocation on that ground is permitted by law.[188] Revocation of adoption usually revives the legal guardianship between the child and his natural family.[189]

Almost all countries permitting adoption, and international instruments on the subject, require that there be a finding that a proposed adoption is in the child's best interests and that there be formal judicial or administrative approval of the adoption.[190] There is an increasing tendency to see adoption as a means for protecting children's welfare rather than for perpetuation of the adopters' family line, and therefore, to relax the restrictions on who may adopt whom. What has been said about legal developments in Europe probably now applies worldwide: modification of adoption law since 1945 "is directed on the one hand towards the alleviation of the formal conditions required for adoptions, but on the other hand, it is directed towards a more stringent control by the Courts or other adoption institutions in order to give effect to the principle that adoption should take place in the interest of the child."[191]

Legal Aspects of Intercountry Adoption

Because unaccompanied children in emergencies have often been adopted or moved outside of their countries of origin, their adoption has frequently involved a foreign element—so-called intercountry adoption. As a legal matter, such adoptions involve questions of jurisdiction to grant the adoption, the choice of law governing the conditions and effects of adoption described above, and recognition of the adoption in other countries.[192] Many countries have legal rules governing these questions, but the multiplicity of these rules produces a certain complexity and confusion. It must be recognized, however, that these problems exist with intercountry adoptions in general and are not unique to those concerning children separated from their families by war, natural disaster, or refugee situations. Nor do the legal complications prevent thousands of intercountry adoptions from taking place every year.

In recent years, several attempts have been made to bring some international consistency to the legal treatment of intercountry adoptions. The European Convention on the Adoption of Children, concluded in 1967, contains a set of substantive provisions to which Party States agree to conform their own law.[193] Under this Convention, uniformity of intercountry adoption is achieved through national adherence to common principles. Although, somewhat surprisingly, there is no provision obliging the States Parties to mutually recognize each other's adoptions made according to these principles, the application of normal comity or recognition rules would presumably produce that result. Thirteen European countries have so far acceded to this convention.

The Hague Convention on Adoptions of 1965, in contrast, concerns itself with jurisdiction, choice of law, and mutual recognition of adoptions.[194] Parties to the Hague Convention agree not to grant an adoption "unless it will be in the interest of the child" and has been preceded by "a thorough inquiry relating to the adopter or adopters, the child and his family."[195] They also bind themselves to particular rules about jurisdiction and choice of law in granting an adoption. Adoptions governed by the Convention and effected under its jurisdictional rules "shall be recognized without further formality in all Contracting States."[196] Only three countries, however, have ratified the Hague Convention on Adoption.[197]

The Declaration on Foster Placement and Adoption (1982), existing only in draft form, advances a number of principles regarding intercountry adoption. These include the use of authorized and competent adoption agencies, a prohibition on proxy adoptions, and insistence that consents to the adoption be legally valid in both countries, and that the child be able to immigrate to the adopters' country and obtain their nationality.[198] The Draft Convention on the Rights of the Child (1985) contains similar directives for intercountry adoption on the use of competent agencies and assurance of the legal validity of the adoption in the countries involved.[199]

Consent to Adoption and Adoption Without Parental Consent

All countries which authorize adoption require the consent of the child's parent(s) or guardian, if fit, and if the child is old enough (most commonly fourteen or above), the consent of the child himself. As discussed in greater detail above, such consent must ordinarily be knowing, intelligent, and voluntary.[200] Consent to termination of parental rights and adoption is usually given by a written instrument, executed and witnessed according to the relevant national law.[201] Because of the seriousness of such a step, there have been international recommendations "that a parent, regardless of social and legal status, should have the opportunity for full consideration of what is involved, including legal and psychological consequences, before a decision is made that adoption is the best plan for the child."[202]

For children made unaccompanied by war, natural disaster, and population movements, however, formal parental consent to adoption may be impossible to obtain. The parents may be dead, absent, or impossible to locate, incapable of being contacted if alive, incompetent, or unable for political or other reasons to freely give their consent. In some rare instances, the very identity of the parents may be unknown. In others, the parents may unreasonably withhold their consent to adoption although they are unable or unwilling to care for the child. These problems, by and large, are not peculiar to children separated from their parents in emergencies, but the incidence of children with parents who cannot be located or freely contacted is undoubtedly higher among this group than with other prospective adoptees.

How does the law attempt to resolve the issue of parental consent to adoption in these circumstances? There are two general approaches: (a) declaration of the parents' death or absence, and (b) dispensing with the usual requirement of parental consent to adoption.

Declaration of the Parents' Death or Absence

Obviously, if the child's parents are known to be dead, adoption can proceed without their consent. There are legal procedures in most countries to declare dead or absent persons who have disappeared but cannot be proven dead—a substitution, in effect, for the usual death certificate. These procedures are designed mainly to resolve questions of inheritance, insurance, and dissolution of marriage, but may affect the status of children as well. Common law countries generally have a presumption of death after seven years' unexplained absence.[203] Civil law countries have special proceedings for declaration of absence or death, most often after periods of three, five, or ten years.[204] Some of these declarations are provisional, and in at least one country, the Federal Republic of Germany, a person mistakenly declared dead may later regain his parental authority.[205]

Both civil and common law systems permit a finding a death or absence after shorter time periods if the missing person was exposed to an enhanced danger, including military combat.[206] After World War II, several countries permitted declarations of death or absence for those participating in war activities after six months or a year.[207] In Poland, a special provision of one year applied to those who, during war activities, "were deprived of their freedom by the authorities of a foreign country and were confined to a place where their life was exposed to particular danger" (*i.e.,* concentration camps).[208] In addition, persons forcibly deported from Poland could be declared dead after three years from the end of the calendar year in which they were last known to be alive, but not less than two years from the end of the calendar year in which the war activities ended.[209] This last proviso was meant to prevent the declaration of death of those who were "missing" or not heard from solely because of difficulties in travel or communication.

On the international level, the carnage of World War II produced the United

Nations Convention on the Declaration of Death of Missing Persons of 1950.[210] This Convention provided for the "declarations of death of persons whose last residence was in Europe, Asia, or Africa who have disappeared in the years 1939–45, under circumstances affording reasonable ground to infer that they have died in consequence of events of war or of racial, religious, political or national persecution."[211] The Convention contained uniform procedures for the declaration of death, recognition of foreign declarations, and the establishment within the framework of the United Nations of an International Bureau for Declarations of Death, including a central registry. The Convention required that a period of at least five years had to have elapsed "since the last known date on which the missing person was probably alive" before a declaration of death could issue.[212] Significantly, the definition of those eligible to bring such proceedings included "persons whose personal status may be affected by the survival or death of the missing person" and "by persons desirous of adopting the minor children of the missing person"[213] This Convention did not supplant existing national procedures for the declaration of death; rather, it supplemented them.[214] It did not gain wide acceptance, however. Only seven countries ever acceded to it, and after two extensions, the Convention itself died a natural death in 1972.[215]

In general, declaration of the parents' death is not a legal mechanism that is particularly well adapted to the problem of consent to adoption of unaccompanied children in emergencies. Many of the parents of these children are not dead, even if they cannot be located. Some may be known to be alive but cannot be communicated with; others may not be free to express their wishes or to resume their parental role. And for those children whose parents are, in fact, likely to be deceased, the statutory period of many years' absence may be too long a period to delay permanent placement. For these reasons, in part, countries with legal adoption will in certain circumstances permit adoption without the consent of living or missing parents.

Dispensing with Parental Consent to Adoption

National laws allow the parents' consent to the adoption to be dispensed with on a variety of grounds. These grounds are phrased differently in the various national laws, but can be described generally as abandonment, persistent neglect, persistent ill-treatment, parents' whereabouts unknown, and parents incapable of giving consent. The first three grounds mentioned may form the basis of a separate proceeding, preliminary to adoption, to terminate parental rights. Parents whose parental rights have already been terminated generally have no say in the subsequent adoption of their child.[216]

Abandonment—desertion of the child with no intention of reclaiming him—is discussed above.[217] Short-term dereliction of parental duties can result in temporary State removal of the child from parental control;[218] long-term neglect or ill-treatment can be grounds for terminating parental rights permanently. In the Federal Republic of Germany, for example, parental rights may be lost in cases of "continuing gross neglect of parental duty" or of "malicious ill-treatment."[219] England permits parental consent to be dispensed with if the parent "has persistently failed without reasonable cause to discharge the parental duties in relation to the child" or has "persistently ill-treated the child."[220]

More important, perhaps, for unaccompanied children in emergencies are provisions dispensing with parental consent if the parents cannot be located. These are

quite common. In Spain, a judge may authorize adoption if parents are unable to appear;[221] in England if the "parent cannot be found;"[222] in Israel if "there is no reasonable opportunity to identify or locate the parent or to determine his or her view."[223] Other countries permit adoption without parental consent if the parents are "unknown" or the child is a foundling.[224] By the same token, many countries will proceed with adoption if the parent is "incapable"of consent.[225] In England, at least, it is also possible to dispense with parental consent that has been "unreasonably withheld."[226]

The working—or perhaps stretching—of these doctrines with respect to a refugee child can be seen in the English case *Re R.*[227] R, twenty years old at the time of the proposed adoption, had illegally left a "totalitarian regime" where his parents still resided. The parents "were not sympathetically regarded by the authorities there," but were at liberty when R left. British citizens wished to adopt R, to which he agreed. The chancellor held that since there was no practicable means of communicating with the parents, they "could not be found," and furthermore, that they were "not capable" of giving their consent where "the probability is that R's parents would not be permitted by the authorities in their country freely to give their consent to the proposed adoption."[228] Parental consent to the adoption, and even notice of its pendency, were therefore deemed unnecessary.

Adoption without parental consent is not explicitly regulated on the international level, though the fact that some children cannot be raised by their natural parents is well recognized.[229] Rather, the circumstances in which the requirement of parental consent may be dispensed with are left to national law. Even the European Convention on the Adoption of Children (1967), which insists on several specific principles for national adoption law, simply states that "the competent authority shall not dispense with [parental consent] save on exceptional grounds determined by law."[230]

17

Unaccompanied Children in the Law of Armed Conflict, Natural Disasters, and Population Movements

There is one international principle that applies to all emergency situations: "The child shall in all circumstances be among the first to receive protection and relief." This principle is now contained in the Declaration of the Rights of the Child,[1] and in the law of at least one State.[2] It was first expressed in the Declaration of Geneva of 1924, also known as the Declaration of the Rights of the Child of 1924, which was adopted by the Assembly of the League of Nations.[3] That Declaration stated: "The child must be the first to receive relief in times of distress."[4] The reformulation contained in the Declaration of the Rights of the Child in 1959 puts the child "among the first" to receive both protection and relief "in all circumstances." In view of the directive in the Declaration of the Rights of the Child that "[s]ociety and the public authorities shall have the duty to extend particular care to children without a family,"[5] unaccompanied children may be considered the "first among the first" to receive protection and relief in all emergencies.

Aside from this one principle, the law of "emergencies" must be treated in its various components. There is no international law of natural disasters which bears directly on the situation of unaccompanied children,[6] but there are important conventions concerning war and refugees, and national law and practice concerning the immigration and resettlement of unaccompanied children in emergencies. This chapter will therefore describe the salient doctrines regarding child/family separation and unaccompanied children in the law of (1) armed conflict, (2) immigration, and (3) refugee status.

ARMED CONFLICT

The major sources of the law of armed conflict as it affects unaccompanied children at present are the Geneva Conventions of August 12, 1949, particularly the Fourth Convention Relative to the Protection of Civilian Persons in Time of War and the

Additional Protocols I and II.[7] These instruments provide special protection for civilians, for all children under fifteen years of age, and for children under fifteen who are orphaned or separated from their families. In addition, there are provisions promoting family integrity and family reunion. As of May 13, 1985, 161 States were parties to the Geneva Conventions, making them the most widely accepted international conventions. Fifty-one States were parties to Protocol I and forty-four States were parties to Protocol II on that date.

All children would ordinarily receive the benefits accruing to civilians under the Conventions and Protocols,[8] so long as they were not acting as combatants or members of the armed forces, militias, or other volunteer corps or organized resistance movements as defined in the Convention.[9] The general protections for civilians will not be discussed here, except as they bear on the questions of family separation and reunion. In addition, since the Convention and Protocol I mainly concern international armed conflict ("armed conflict between two or more High Contracting Parties") the discussion here will refer to international conflicts unless otherwise noted.

The provisions of the Convention and Protocols concerning issues relevant to the question of child/family separation in time of war can be grouped into five categories: (a) protection for all children; (b) promotion of family integrity; (c) permission for evacuation, separation, or removal of children; (d) promotion of family reunion; and (e) special protection for unaccompanied children.

Protection for All Children

1. *Newborn babies:* Newborns are treated as wounded or sick under Protocol I and thus "shall receive, to the fullest extent possible and with the least possible delay, the medical care and attention required by their condition," and shall be "respected and protected," in addition to all other protections benefitting children under fifteen.[10]

2. *Special protection for children:* Parties to Protocols I and II agree to provide children with the care and aid they require, to make them "the object of special respect," and to protect them against indecent assault, whether or not they take part in hostilities.[11]

3. *Relief supplies:* All Parties shall, subject to certain exceptions, allow for the free passage of "essential foodstuffs, clothing, and tonics intended for children under fifteen, expectant mothers, and maternity cases."[12] These persons shall receive priority in the distribution of relief consignments.[13]

4. *Military recruitment and imprisonment:* Under Protocol I parties shall refrain from recruiting children under fifteen into their armed forces, and shall endeavor to recruit "oldest first" from those fifteen to eighteen.[14] Protocol II prohibits recruitment of children under fifteen in internal conflicts. These provisions, of course, operate to prevent parent/child separations by reducing those caused by military service. Also, those children under fifteen who are arrested, detained, or imprisoned for reasons related to the armed conflict must be confined separately from adults unless they are held with their parents.[15]

5. *Death penalty:* Children who were under eighteen at the time of a capital offense may not be sentenced to death or executed.[16] Under Protocols I and II pregnant women and mothers of dependent children are also exempt from execution.[17]

6. *Alien children in the territory of a party to a conflict:* "Shall benefit from any preferential treatment to the same extent as the nationals of the State concerned."[18]

7. *Children in occupied territories:* The occupying power must "facilitate the proper working of all institutions devoted to the care and education of children."[19]

8. *Interned children:* A Detaining Power is obligated to support the dependents of internees if the dependents cannot support themselves.[20] Interned children under fifteen years of age shall be given additional food in proportion to their physiological needs.[21] The education of interned children and young people must be ensured.[22]

Promotion of Family Integrity

1. *General:* The "family rights," "religious convictions and practices," and "manners and customs" of civilians must be protected at all times by all Parties.[23]

2. *No change in personal status of children:* Occupying Powers may not change the personal status of children (i.e., family or guardianship status), nor enlist them in formations or organizations subordinate to it.[24]

3. *Limitations on deportations, transfer, and evacuations:* Individual or mass forcible transfers and deportations of all civilians from occupied territory to any other country are prohibited by the Fourth Convention.[25] Evacuation of particular areas may take place only if the security of the population or imperative military reasons so demand.[26] Such restrictions contribute to family integrity by reducing the number of family separations that transfers, deportations, and evacuations inevitably produce. The Convention on the Prevention and Punishment of the Crime of Genocide defines as genocide the forcible transfer of children of one group to another group at any time, when done with intent to destroy, in whole or part, a national, ethnical, racial, or religious group as such.[27]

4. *Parents and children are not to be separated* during forcible transfers or evacuations by Occupying Powers,[28] internment,[29] or arrest and detention.[30]

Measures Permitting Evacuation, Separation, or Removal of Children

1. *By their own State:* (a) *during internal conflicts:* "measures shall be taken, if necessary and whenever possible with the consent of their parents or persons who by law or custom are primarily responsible for their care, to remove children temporarily from the area in which hostilities are taking place to a safer area within the country and ensure that they are accompanied by persons responsible for their safety and well-being";[31] (b) *during international conflicts:* All civilians and noncombatants may be sheltered in *neutralized zones;*[32] children under fifteen (along with the wounded, sick, aged, expectant mothers, and mothers of children under seven) may be placed in *hospital and safety zones.*[33] Both neutralized zones and hospital and safety zones cannot be the object of attack.[34] In addition, children under fifteen and the others listed above may be removed from besieged and encircled areas under agreements of the Parties.[35] There are also special provisions concerning the evacuation of orphans and other children separated from their families, discussed below.

2. *By a foreign State:* The circumstances in which a foreign State may evacuate the children of another nationality to a foreign country are much more limited. Article 78(1) of Protocol I addresses this question:

> No party to the conflict shall arrange for the evacuation of children, other than its own nationals, to a foreign country except for a temporary evacuation where compelling reasons of health or medical treatment of the children, or, except in occupied territory, their safety so requires.

Thus, under this article a State may ordinarily evacuate foreign children to another country (including its own) only for compelling reasons of health or medical treatment or safety and then only with the permission of the parents, guardians, or other persons responsible for the children in law or custom. Furthermore, such evacuation can take place only if "the medical care required for their cure or convalescence cannot be given on the spot or in their country."[36] Even then, the evacuation must be temporary, lasting no longer than the reasons of health or medical treatment require.

Occupying Powers may not arrange the evacuation of children at all for claimed reasons of safety. As was stated during the deliberations on Article 78:[37]

> [T]he limitations to evacuate for compelling reasons of health or medical treatment where the evacuation is to be from occupied territory reflects a deep-seated concern . . . that the dangers of Occupying Powers abusing their discretion are greater than the dangers of prohibiting evacuation for reasons of safety.

The Article leaves open the possibility of legal evacuation of children for purposes of safety by their own State, of course. But given the possibility of puppet governments, even the requirement of the consent of the Party State of which the child was a national was considered inadequate.[38]

As a result, Article 78 demands the written consent of the parents or legal guardians to any evacuations for reasons of health, medical treatment, or safety. If the parents or guardians cannot be found, the written consent of persons "who by law or custom are primarily responsible for the care of the children" will meet the written consent requirement. The term "primarily responsible" was explained during the deliberations of the Diplomatic Conference by the Nigerian representative, at whose instance the word "primarily" was included:[39]

> [T]he provision [was] intended to cover situations provided for in customary law. As a number of persons might be responsible for children to varying degrees, it was essential to indicate that person who was considered to have primary responsibility for the child. That could only be done by retaining the word 'primarily.'

By including "primarily" before "responsible," the delegates thus clearly intended to preclude the consent of those with temporary responsibility for the child, such as teachers, from meeting the Article's demand for written permission for evacuation.[40] *A fortiori* this would apply to child welfare or other government authorities as well, unless they had previously lawfully assumed legal guardianship. In sum, then, there can be no evacuation at all under Article 78, no matter how compelling the reasons, without the prior written consent of the parent, guardian, or other person most responsible for the child.

For legal evacuations of children Article 78 also specifies requirements for protection ("all Parties to the conflict shall take all feasible precautions to avoid endangering the evacuation"), and education ("provided . . . with the greatest possible continuity"), including religious and moral upbringing ("as his parents desire").[41] Further, "with a view to facilitating the return to families of children evacuated," Article 78 directs that the sending (and, if necessary, the receiving) Party send to the Central Tracing Agency of the ICRC a card for each evacuated child with a photograph and, whenever possible and whenever it involves "no risk of harm to the child," the following information:[42]

a. surname(s) of the child;
b. the child's first name(s);
c. the child's sex;
d. the place and date of birth (or, if that date is not known, the approximate age);
e. the father's full name;
f. the mother's full name and her maiden name;
g. the child's next of kin;
h. the child's nationality;
i. the child's native language, and any other languages he speaks;
j. the address of the child's family;
k. any identification number for the child;
l. the child's state of health;
m. the child's blood group;
n. any distinguishing features;
o. the date on which and the place where the child was found;
p. the date on which and the place from which the child left the country;
q. the child's religion, if any;
r. the child's present address in the receiving country;
s. should the child die before his return, the date, place, and circumstances of death and place of interment.

It is important to note that this Article and its associated precautions apply only to evacuations by a State of children who are not its own nationals.[43] The protections in the Conventions and Protocols do not by their terms apply to evacuations by a State of its own children during an international armed conflict, but there is nothing to prevent a State from following these salutary measures of its own accord.

Measures Promoting Family Reunion

Provisions promoting family reunion in the Convention and Protocols include those (1) encouraging identification of children, (2) facilitating communication with relatives, (3) providing for release of civilian internees, and (4) specifically requiring the parties to facilitate family reunion.

1. *Identification of children:* The Fourth Convention directs the Parties to "endeavor to arrange for all children under twelve to be identified by the wearing of identity discs, or by some other means."[44] Occupying Powers are under a special duty to identify children and register their parentage; as mentioned above, they are prohibited from changing the children's personal status.[45]

2. *Communication with relatives:* The Fourth Convention mandates that all persons in the territory of a Party to a conflict, including internees, be able to communicate with family members.[46] Parties to the conflict are obligated to facilitate inquiries by members of families dispersed because of the war.[47] The Convention also directs the Parties to set up national information bureaus concerning civilians and to forward information about them to a Central Information Agency for transmission to relatives.[48]

3. *Release of internees:* The Parties must, at the close of hostilities or occupation, endeavor to ensure the return of all internees to their last place of residence or to facilitate their repatriation.[49] In addition, during the course of hostilities, the Parties shall endeavor to release from internment children, pregnant women, and mothers with infants and young children.[50]

4. *Family reunion:* Article 74 of Protocol I states:

> The High Contracting Parties and the Parties to the conflict shall facilitate in every possible way the reunion of families dispersed as a result of armed conflicts and shall encourage in particular the work of the humanitarian organizations engaged in this task. . . .

There is a similar provision in Protocol II concerning internal armed conflicts.[51]

Special Protection for Unaccompanied Children

Article 24 of the Fourth Convention sets forth the basic obligations of all Parties with respect to unaccompanied children:

> The Parties to the Conflict shall take the necessary measure to ensure that children under fifteen, who are orphaned or are separated from their families as a result of the war, are not left to their own resources, and that their maintenance, the exercise of their religion and their education are facilitated in all circumstances. Their education shall, as far as possible, be entrusted to persons of a similar cultural tradition.

This Article thus directs that unaccompanied children be given assistance and protection and that their care include continuity of education, religion, and culture. In addition, the Convention also requires Occupying Powers to "make arrangements for the maintenance and education, if possible, by persons of their own nationality, language, and religion, of children who are orphaned or separated from their parents as a result of the war and who cannot be adequately cared for by a near relative or friend."[52]

Provisions addressing the issue of whether or not to remove unaccompanied children from their own milieu to a neutral country for the duration of the conflict have undergone an evolution since 1949. Article 24 of the Fourth Geneva Convention of that year, written in the wake of World War II, actively encouraged such evacuation, stating:[53]

> The Parties to the conflict shall facilitate the reception of such children in a neutral country for the duration of the conflict with the consent of the Protecting Power, if any, and under due safeguards for the observance of the principles stated in the first paragraph.

International thinking changed in the following two decades, however, and by 1973, the International Committee of the Red Cross, after consulting national viewpoints and international experts, proposed the following draft provision on evacuation of all children:[54]

> If their conditions necessitate their evacuation for reasons of health, in particular to obtain medical treatment or to hasten convalescence, children may be transferred to a foreign country. Where they have not been separated by circumstances from their parents or guardians, the latters' consent must be obtained. In the case of evacuation to a foreign country, the operation shall be supervised or directed by the Protecting Power, in agreement with the Parties to the conflict concerned.

(The draft article also contained protection for continuity of education, language, and culture, as well as documentation.) The commentary and deliberation on this proposal leave no doubt that it was intended to supercede the encouragement of evacuation of unaccompanied children contained in Article 24 of the Fourth 1949 Convention. The official commentary accompanying the draft Article stated:[55]

> It was found necessary to make Article 24 of the Fourth Convention more precise, by introducing a restriction in its second paragraph [concerning evacuation]; it appears desirable to prevent children from being removed from their environment abusively and unnecessarily.

Similarly, introducing this draft Article before Committee III of the Diplomatic Conference considering the proposals in 1977, the ICRC representative stated that the article[56]

> was designed to cover as fully as possible the question of the evacuation of children to a foreign country during armed conflict. The guiding principle was that evacuation must remain the exception. . . . As far as possible children should not be removed unnecessarily from their natural environments, since even though it might be beneficial medically, it often has undesirable psychological effects. . . . *Paragraph 1 [of the draft Article] thus restricted the scope of the second paragraph of Article 24 of the fourth Geneva Convention* (Emphasis added.)

Although draft Article 69 was revised in some significant respects before emerging as the present Article 78 of Protocol I, discussed at length above,[57] its history amply demonstrates that it was intended to modify and restrict the provisions on evacuation of unaccompanied children contained in Article 24 of the Fourth Convention. For Parties to both the Fourth Convention and Protocol I, the binding principles on evacuation of foreign unaccompanied children are thus contained in Protocol I, rather than the Fourth Convention. Those States which are not parties to Protocol I for reasons totally unrelated to its provisions about children should nevertheless accept Article 78 of that Protocol as the most current international principle on the evacuation of unaccompanied children to other countries in international armed conflicts.

IMMIGRATION LAW AND PRACTICE

In the past forty years, many unaccompanied children have been displaced or moved outside of their countries of origin. The admission and/or retention of aliens, including unaccompanied children, is largely governed by the domestic law and practice of each sovereign State. There are some international instruments bearing on these questions—most notably the 1951 Convention and 1967 Protocol Relating to the Status of Refugees—which bind States Party to certain principles and practices with respect to aliens. These will be discussed in the following section. With the exception of these obligations and others imposed by international law not relevant here, immigration law and practice are issues of national law and ultimately of national policy. Immigration law and practice concerning unaccompanied children will thus reflect national priorities about such issues as immigration control, refugee relief, child welfare, and intercountry adoption.[58] Furthermore, this law and practice may change as thinking about these issues and the balance among them changes.

Categories of Unaccompanied Children in Immigration Law

In general, the immigration status of foreign unaccompanied children in the territory of another ("second") country, or resettled in a "third" country, can be classified into several major categories. The category into which any individual child falls will depend on both the relevant immigration law and the particular circumstances of the child. Both of these can change over time, of course, so a child may move from one category to another.

1. *Refugee.* In legal terms, a refugee is most often defined as a person who, owing to a well-founded fear of being persecuted for reasons of race, religion, nationality, membership of a particular social group or political opinion, is unable or unwilling to return to his country of nationality or former habitual residence. Refugee status for unaccompanied minors is discussed in detail below.

2. *Asylee.* Certain aliens who are not recognized as refugees under the 1951 Convention may nevertheless be granted asylum: the right to stay in the territory of the second or third country with protection against *refoulement* (return). Persons who receive this treatment are often called *de facto* refugees. "Such modalities of protection may, for instance, apply to aliens whose applications for refugee status have been rejected for what governments consider lack of well-founded fear of persecution, but where a certain risk to life and freedom in case of return cannot be excluded."[59] They may also apply to persons who have asked for refugee status, but whom the authorities, for diplomatic or other reasons, have not wanted to recognize as refugees. Other types of asylees include those given temporary residence while claims for refugee status or other forms of asylum are being decided. Unaccompanied children can benefit from any of these types of asylum.

3. *Unaccompanied children settled or resettled as part of a group of refugees or displaced persons.* Unaccompanied children have been resettled in certain countries as part of larger populations which receive group or "quota" refugee status. This is the method by which Indochinese unaccompanied children were resettled in the Federal Republic of Germany, Netherlands, Sweden, and the United States in the late

1970s and early 1980s.[60] The United States' Refugee Act of 1980, for example, provides for the admission of unaccompanied children among the annual admissions of "refugees of special humanitarian concern to the United States."[61] A May 1983 Presidential Directive specifically mandated the inclusion of unaccompanied children among those for whom group refugee status under this Act should be considered.[62]

4. *Special admissions for unaccompanied children.* Since World War II, several countries have set up special admissions categories for unaccompanied children, variously defined, outside of normal immigrant or refugee quotas. These admissions categories have usually been designated for children from a particular emergency and for a limited period of time. The United States' Displaced Persons Act of 1948, for example, permitted the admission on a nonquota basis of 3,000 orphans under the age of sixteen displaced by World War II.[63] For Indochinese refugees, special admissions for unaccompanied children were created by Belguim (200) and France (600) in the late 1970s and early 1980s.

5. *Unaccompanied children admitted for purposes of adoption or as already adopted.* Unaccompanied children from emergencies have also been admitted outside of normal immigration quotas for purposes of adoption or as already adopted children. This practice has been used most frequently by the United States, which enacted four temporary laws permitting the entry of alien orphans for adoption in 1953, 1957, and 1960.[64] In 1961 the United States Immigration and Nationality Act was permanently amended to admit as a nonquota "immediate relative" of citizens:

> a child, under the age of fourteen ... who is an orphan because of the death or disappearance of, abandonment or desertion by, or separation or loss from, both parents, or for whom the sole or surviving parent is incapable of providing the proper care and has in writing irrevocably released the child for emigration for adoption.

To qualify, these children must be adopted abroad by a U.S. citizen, or the prospective parent must have complied with the preadoption requirements of the state of the child's proposed residence.[65] Such "orphans" may not enter "unless a valid homestudy has been favorably recommended by an agency of the State of the child's proposed residence."[66] These unaccompanied children benefit from the unlimited "immediate relative" admission in U.S. law only if they are adopted or about to be adopted by U.S. citizens. Similar exceptions to immigration restrictions for adopted children exist in other national laws.

6. *Unaccompanied children admitted for family reunion.* Children whose parents are already living in a second or third country are often eligible for admission to those countries as "immediate relatives," "dependents," or for purposes of family reunion if their parents are citizens or lawful residents of the receiving country.[67] The admission of the *parents* of unaccompanied children who have been admitted first is more problematic, and is discussed below.

7. *Other lawful residence.* Unaccompanied children are often eligible for other lawful forms of residence in a second or third country, including student visas or immigrant status. In the Federal Republic of Germany, for example, children under the age of sixteen do not need a residence permit.[68] The range of legal grounds for presence in a country is limited only by the terms of that country's immigration law and practice.

8. *Unaccompanied children as illegal aliens.* Unaccompanied children on foreign territory who do not come within one of the legal grounds for presence are illegal aliens, like similarly situated adults. From an immigration law point of view, they are subject to the general measures applied to all illegal aliens. As has been discussed above, there is an obligation in most national and international law to provide temporary care and protection to such unacccompanied children. By extension, it can also be argued that national authorities should not act under immigration law in such a way as to put these children in danger of harm to their person or health. Within these constraints, however, unaccompanied children who are found to be illegal aliens are subject to the normal range of legal measures, including deportation or other forms of exclusion.

Questions of Settlement and Resettlement Policy in Immigration Law and Practice

The main issues concerning unaccompanied children are (1) when should settlement and resettlement opportunities be created for unaccompanied children, and (2) whether to admit the parents, siblings, or other family members of those earlier admitted as unaccompanied children.

With respect to the first issue—when to create settlement or resettlement opportunities for unaccompanied children—there are questions concerning the effects of such settlement on both the receiving country and the children and their families. The potential receiving country must consider the question of unaccompanied children within the context of its policies on immigration, refugee relief, child welfare, and intercountry adoption. It must recognize, in addition, that existence of resettlement opportunities for unaccompanied children in developed countries—where most resettlement has recently taken place—has a profound effect on second- and third-world populations. The chance of legal, expense-paid immigration to developed countries can cause children in less-developed areas to leave their families, or be sent by them, in quest of better education, employment, and material standards of living. In short, resettlement from less-developed to more-developed countries can itself cause child-family separation. This is particularly true when opportunities for later legal migration are limited. Resettlement as an unaccompanied child may truly be a once-in-a-lifetime opportunity. On the other hand, resettlement may be the only practicable means of providing foster family care for unaccompanied children. These factors must be carefully included in any consideration of the creation of resettlement places for unaccompanied children in emergencies.

The issue of later reunion with the parents raises the same concerns. All countries recently involved in resettlement of unaccompanied children have permitted the child to rejoin the parents in the *country of origin* if both desire. Automatic reunion with the parents or other relatives in the country of resettlement, however, may encourage what is sometimes called the "anchor syndrome": the sending of a child abroad to be settled or resettled as unaccompanied and the later use of the child as an "anchor" for immigration by other family members. This ploy may be especially attractive if the family as a whole would not receive as favorable an initial chance for resettlement as would the child on his own. Such a possibility obviously encourages families to send children off by themselves. On the other hand, denial of the possibility of later reunion in the resettlement country can eventually force the child to choose between his new country of residence and his natural family.

This problem has been especially vexing with respect to unaccompanied children from Southeast Asia resettled in the past ten years. Many of these children, especially those from Vietnam, have parents or other close relatives remaining in the country of origin.[69] Most resettlement countries have given immigration preference to close family members of resettled Southeast Asian children, but often only within existing quota limits. The tension between general restrictions on immigration and the goal of family reunion is well illustrated by recent experience in the United Kingdom:[70]

> The United Kingdom Home Office family reunion criteria have changed three times since 1975. Initially it was agreed that existing criteria should be relaxed specifically for the Vietnamese, and that *any* refugee having *any* family link in the United Kingdom should be allowed to enter the country. This was subsequently restricted to include parents, grandparents, and unmarried siblings of any age, in addition to spouses and minor children. In May 1981, the criteria returned to those which had existed prior to 1975 and refugee family reunions were restricted to admitting only the spouse and minor children of those already in the United Kingdom. Other relatives are considered on an individual basis. Thus present criteria do not enable family reunion to occur in the case of unaccompanied refugee minors, though it is known that the Home Office granted visas to 188 close relatives of refugee children already here. Such cases are currently being decided on an individual and discretionary basis. The concerned agencies are taking initiatives to try to change this approach, so that unaccompanied refugee minors can be reunited with their families if at all possible, as a matter of course.

The Netherlands, too, faced with a potentially large influx of relatives of unaccompanied Vietnamese children resettled there, has decided that not all family members will be granted admission. For Cambodian children resettled since 1979, however, the United Nations High Commissioner for Refugees has insisted on, and received, assurances from receiving governments "that they will promptly accept the child's remaining family members once they have been located and have expressed the wish to join the child."[71]

REFUGEE STATUS

This section addresses the "refugee" status of displaced unaccompanied children. Refugee status under any of the major conventions concerning refugees entails many legal rights and benefits, the most important of which is protection against expulsion or return to the territory where the conditions of persecution or danger exist. In addition, a refugee has the right to recognition of his personal status, and significant economic and social rights. Though the term "refugee" is often applied broadly to all people in flight or otherwise displaced, its legal meaning is narrower and more specific.

The preeminent international instruments governing refugee status are the 1951

United Nations Convention and 1967 Protocol Relating to the Status of Refugees.[72] These instruments define a refugee as "any person" who[73]

> owing to well-founded fear of being persecuted for reasons of race, religion, nationality, membership of a particular social group or political opinion, is outside the country of his nationality and is unable or, owing to such fear, is unwilling to avail himself of the protection of that country, or who, not having a nationality and being outside the country of his former habitual residence as a result of such events, is unable, or owing to such fear, is unwilling to return to it.

This definition is virtually identical with that in the Statute of the United Nations High Commissioner for Refugees. The 1969 Organization of African Unity Convention Governing the Specific Aspects of Refugee Problems in Africa (OAU Refugee Convention) adds to the United Nations Refugee Convention definition as follows:[74]

> The term "refugee" shall also apply to every person who, owing to external aggression, occupation, foreign domination or events seriously disturbing public order in either part or whole of his country of origin or nationality, is compelled to leave his place of habitual residence in order to seek refuge in another place outside his country of origin or nationality.

States which are party to these instruments are free, of course, to apply an even broader definition of "refugee" in their own legislation.

Accompanied Children

Most countries adhere to the rule that if the head of a family meets the criteria of the [refugee] definition, his dependents are normally granted refugee status according to the principle of family unity.[75] Natural children are thus assimilated to the status of the head of their family, even when they were temporarily separated during flight.[76] Since adopted children stand in the same legal relation to parents as natural children, the principle of family unity should apply to them equally. As for foster children, Grahl-Madsen suggests that the "principal of family unity" might be applied—by analogy—to foster children who are living together with the refugee."[77] There is, of course, a possible objection that children can be placed with refugee families solely for the purpose of acquiring the attendant legal benefits.[78] This danger can be minimized, however, by insisting that the guardianship or foster arrangement be certified or attested to by a competent authority,[79] and by refusing "dependent status" to children whose circumstances clearly evidence a bogus arrangement.

Unaccompanied Children

After World War II, the International Refugee Organization (IRO) constitution included within its definition of refugees "unaccompanied minors who are war orphans or whose parents have disappeared, and who are outside their country of origin."[80] This made unaccompanied minors under sixteen years of age eligible for the full range of IRO care and assistance, protection and resettlement. The IRO definitions scheme, however, differed greatly from the present United Nations Convention and Protocol by defining refugees and displaced persons largely by past experience rather than present fear of persecution.

"There is no special provision in the 1951 Convention [or the 1967 Protocol] regarding the refugee status of persons under age."[81] This is also the case with the Organization of African Unity Refugee Convention. Unaccompanied children, then, like all other people, must have a well-founded fear of persecution on one of the enumerated grounds, or, in the African case, meet the other objective criteria in order to obtain refugee status. By the nature of their situation, unaccompanied minors are unlikely to be assimilated to the status of a family head like most other children. At least initially, they have no responsible adult to represent them or act on their behalf in the process for acquiring refugee status, and this can be a problem for some children. In practice, however, national authorities have often treated such children as refugees whenever adults in similar circumstances would have been so treated. Experience since 1951 has also shown that it is often less important for an unaccompanied child to obtain formal refugee status than it would be for a comparably situated adult because of the preferential treatment often accorded these children in national immigration law and practice.

The UNHCR Handbook, written for the guidance of Governments, makes several recommendations for handling the determination of refugee status of unaccompanied minors. Claiming that "[t]he question of whether an unaccompanied minor may qualify for refugee status must be determined in the first instance according to the degree of mental development and maturity," UNHCR urges the enrollment of "the services of experts conversant with child mentality."[82] It also recommends the appointment, "if appropriate," of a guardian "whose task it would be to promote a decision that will be in the minor's best interest.

The Handbook goes on to say: "If there is reason to believe that the parents wish their child to be outside the country of origin on grounds of well-founded fear of persecution, the child himself may be presumed to have such fear." Otherwise, "[m]inors under sixteen years may normally be assumed not to be sufficiently mature" to have such a fear, but this must be assessed "in light of his personal, family and cultural background." Stress should be placed on the child's objective situation, including that of his family in his country of origin. The Handbook concludes: "If the will of the parents cannot be ascertained or if such will is in doubt or in conflict with the will of the child, then the examiner, in cooperation with the experts assisting him, will have to come to a decision as to the well-foundedness of the minor's fear on the basis of all the known circumstances, which may call for a liberal application of the benefit of the doubt."[83]

Determination of refugee status for unaccompanied children should focus on the objective situation of the child. As Grahl-Madsen has pointed out, persecution on grounds of race, ethnic group, religion, nationality, and membership in a particular social group is essentially beyond the control of the individual. No individual has a choice of race, and few children have any meaningful say in their nationality, religion, or social group membership. If the persecution is severe enough to give adult members of the relevant group a well-founded fear of persecution, then the guardian or others responsible for the child should have the duty to assert a claim to refugee status on behalf of that child whether or not the child has a "fear of persecution" or even an awareness of his membership in the group. The classic, if extreme, example here would be an infant or young child found outside a country of which he was a national in which all people of his race were being exterminated. The inquiry in cases of this kind would concern the objective circumstances in the child's country of origin, and not the child's state of mind. This would also be true under the OAU refugee definition in determining whether the child was compelled to leave due to external aggression, occupation, foreign domination, or events seriously disturbing public order.

Can measures aimed specifically at children be considered persecution for reasons of membership in a particular social group? As Goodwin-Gill has noted, "[j]urisprudence on the interpretation of 'social group' is sparce."[84] Yet on the face of the language there seems to be no reason not to regard children as a social group. Children, of course, share common minority status in the eyes of the law and are treated as a distinct social group in a wide range of social and governmental matters. Measures such as mass abduction or removal of children from their homes constitute clear persecution of children as such. Given the fact that children are deserving of special protection under general international law, as discussed above, when acts which constitute persecution are specifically directed against children, the measures should be construed to come within the terms of the Refugee Convention and Protocol's protection against persecution on the basis of social group membership.

Persecution for the expression of political opinion, or a politically based act or refusal to act, unlike persecution on the basis of group membership, is premised on individual behavior. Minors are certainly capable of such behavior as history shows. In most, though not all instances, any minor who is capable of expressing political opinion or acting on political motives is also capable of entertaining a reasonable fear of persecution for his actions. Thus, an unaccompanied minor invoking fear of persecution on these grounds should initially be presumed to be sufficiently mature to have such a fear; the only issue, as for all refugees, would be whether the fear is well-founded.

As a matter of experience, many older unaccompanied minors claim "political persecution" after fleeing to evade conscription, desertion from the military, or unauthorized departure and absence from their home country. There would seem to be no reason not to apply the same substantive standard to minors as is used for adults in those situations, provided of course, that the minors have adequate representation in presenting their claims. It is generally agreed that evasion of the draft, desertion, or departure from a country with prohibitions on unauthorized departure or absence are not by themselves sufficient bases for refugee status.[85] Some authorities have required the individual to demonstrate that the nature and severity of the punishment he is likely to receive for the offense will be harsher because of his race, religion, nationality, social group, or political opinion.[86] Other authorities have found a political (or religious, etc.) motivation for the offense to be a sufficient basis for refugee status.[87] Either way, there needs to be a nexus between the offense and the reasons enumerated in the Convention as a basis of persecution for those seeking refugee status after evading conscription, desertion, or leaving a country with prohibitions on departure.

A word needs to be said, too, about what might be called "derivative persecution"; persecution of the child because of the political activities of his parents or relatives.[88] Here the relevant facts concern not only the child's fear and his own situation, but also that of the relevant family members and the parents' fears for the child's well-being. The child himself may not be old enough to grasp the situation. The probable reason for the parent's decision to send the child outside the country of origin is a factor that must be taken into consideration.[89]

Even where a minor does not qualify for refugee status, there may be compelling reasons of a humanitarian nature not to return him to his country of origin. For example, deserters may be subject to extreme sanctions such as capital punishment. Many countries grant *de facto* refugee status or some form of temporary leave to remain to persons displaced outside of their country of origin by war or other serious disturbances of public order.

The Need for a Guardian for Unaccompanied Children
Potentially Eligible for Refugee Status

Even a brief examination of the substantive issues of refugee status for unaccompanied children points to their need for an adult representative in the process of such determination. Procedures for deciding whether a particular person comes within the definition of refugee are not specified in the Convention or Protocol, and differ greatly among the various States. The question of refugee status, however, is an individual one and requires some kind of factual inquiry. UNHCR has recommended some basic requirements for the procedure, which include that "the applicant should be given the necessary facilities, including the services of a competent interpreter, for submitting his case to the authorities concerned."[90] However, "[t]he relevant facts of the individual case will have to be furnished in the first place by the applicant himself."[91] This can be difficult for any refugee without assistance. "An applicant for refugee status is normally in a particularly vulnerable situation; he finds himself in an alien environment and may experience serious difficulties, technical and psychological, in submitting his case to the authorities of a foreign country, often in language not his own."[92]

These difficulties, of course, are compounded for children on their own, who will usually have less education, experience, and familiarity with legal procedures than older refugee applicants. Their situation presents a particular example of the general principle of the need for adult assistance for children separated from their families. There can be little dispute, then, about the need for adult assistance for unaccompanied children whose status as refugees is in question. This need can be filled by a general guardian[93] or by a guardian *ad litem*—one who is appointed only for the particular eligibility proceeding. In view of the especially vulnerable status of these children and the frequent technicality of the refugee determination process it is hard to conceive how unaccompanied children can have their status under refugee law determined with even minimal fairness without such a guardian.[94]

Automatic appointment of a guardian for unaccompanied minors potentially eligible for refugee status finds strong support in the recommendations of the Final Act of the Conference which adopted the 1951 Refugee Convention. The signatories urged Governments " to take the necessary measures for the protection of the refugee's family, especially with a view to . . . the protection of refugees who are minors, in particular unaccompanied children and girls with special reference to guardianship and adoption."[95] Such appointment is also supported by Article II *bis* of the Draft Convention on the Rights of the Child (1985):

The States Parties to the present Convention shall take appropriate measures to ensure that a child who is seeking refugee status or who is considered a refugee in accordance with applicable international or domestic law and procedures shall, whether unaccompanied or accompanied by his parents, legal guardians or close relatives, *receive appropriate protection and humanitarian assistance in the enjoyment of applicable rights set forth in this Convention and other international human rights or humanitarian instruments to which the said States are Parties.* In view of the important functions performed in refugee protection and assistance matters by the United Nations and other competent intergovernmental and nongovernmental organizations, cooperation in any efforts by these organizations to protect and assist such a child and to trace the parents or other close relatives of an unaccompanied refugee child in order to obtain information necessary for reunification with his family. In cases where no parents, legal guardians or close relatives can be found, the child shall be accorded the same protection as any other child permanently or temporarily deprived of his family environment for any reason, as set forth in the present Convention. (Emphasis added.)

In addition, it must be recognized that a child's ability to assemble proof and express his fears will usually be less than an adult's, which, in the words of UNHCR, "may call for a liberal application of the benefit of the doubt."[96]

18

Jurisdiction and Choice of Law
for Unaccompanied Children

Jurisdiction is the legal power to act—in this case, to act on and for unaccompanied children. The question of jurisdiction concerns which State or other entity has the legal right to take actions for any particular unaccompanied child. The entity with jurisdiction over the child has the legal right to treat the child under its general laws of the family, child protection, emergencies, refugees, and aliens.[1]

In virtually all cases the State in which the child is located has jurisdiction over him. As will be explained in greater detail later, however, a State asserting jurisdiction over a child who is not a national of that State may choose to apply the law of the State of the child's nationality. This issue of what law to use is called here "choice of law." Where a State's choice-of-law rules direct it to apply the child's national law, then the law applicable to the child will involve both the law of the State acting on him and that of his national State.

Because of the differences in choice of law depending upon whether a child (1) is found unaccompanied in his national State and those in which he is (2) found in a foreign State, or (3) settled in a foreign State, the discussion here will distinguish between the three situations. At the risk of a certain amount of oversimplification, however, it may be said in general that the State in which an accompanied child is located has primary jurisdiction over him and that the law which applies to unaccompanied children is most often the law of that State. This is probably all the general reader needs to know about jurisdiction and choice of law. The three subsections set forth the legal bases of these general conclusions for readers who want a more technical analysis.

"IN-COUNTRY" SITUATION: CHILDREN FOUND UNACCOMPANIED IN THEIR NATIVE COUNTRY

In cases where unaccompanied children are found in the country of which they are nationals, the question of jurisdiction is simple: the legal institutions of that State will have jurisdiction over those children. A State is sovereign in its own territory, and, clearly, the care and protection of children without parents will fall within its general domestic jurisdiction. As recently as December 1983, the United Nations General

Assembly noted "the sovereign right of Governments to define their national and international policies as regards the protection and welfare of children, including placement and adoption."[2] Every State, then, has the legal right to establish and implement its own family and child welfare legislation. Furthermore, without the State's permission, outside intervenors cannot enter its territory on behalf of the children there.[3]

In several instances in the past, governments have been supplanted during or immediately after the emergency which caused the separation of parents and children and the new authorities have then asserted jurisdiction over those children. After the defeat of the Third Reich in 1945, for example, the Allied High Commissions in Germany and Austria took charge of the unaccompanied children within those territories. The federal government of Nigeria reasserted its authority over children from the eastern states after the collapse of the secessionist government of Biafra in 1970.

There have also been occasions when national authority has been weakened to the point of breakdown, and children separated from their families by the crisis have been left in a legal void, as well as a social and material one. The very emergency which caused their separation may have also incapacitated the institutions, customary or formal, that would have otherwise assumed responsibility for their care. Such was the case in Korea during the 1950–53 War. The collapse of the relevant national legal structure—and the creation of a legal void—is always a danger in the kinds of emergencies which produce unaccompanied children. That a State as a practical matter is unable to exercise its legal authority, however, in no way derogates from the basic principle that it has jurisdiction over its own unaccompanied children within its boundaries.

"CROSS-BORDER" SITUATIONS: CHILDREN FOUND UNACCOMPANIED IN A FOREIGN COUNTRY

In a "cross-border" situation, children from one country are found unaccompanied in the territory of another, "second," country. In recent years, with mass movements of millions of people across national frontiers,[4] unaccompanied children in "cross-border" situations have become a common phenomenon.

The starting point on the question of jurisdiction over these children is the "universal maxim of jurisprudence" that a sovereign "has exclusive jurisdiction over everybody and everything within [its] territory and over every transaction that is there effected."[5] The authorities of the "second country" clearly have power to take any necessary measures for the care and protection of unaccompanied alien children. Under this territorial principle, the foreign children could ordinarily be included within the ambit of the second country's family and child protective laws and agencies.

In the *Boll* case in 1958, the International Court of Justice upheld a nation's right to apply its own child protective legislation to foreign children even where the nation at issue (Sweden) had agreed in a prior convention to apply the child's national law.[6] Marie Elizabeth Boll was a child of Dutch nationality born in and residing in Sweden. After the death of her mother, the appropriate Dutch court appointed a guardian for her. The Netherlands and Sweden were both party to the 1902 Convention Governing the Guardianship of Infants,[7] the first attempt at an international agreement on the choice of law to be applied on the question of the guardianship of foreign infants. That Convention states, in Article 1, "The guardianship of an infant is governed by his national law." Nevertheless, under the Swedish Law of 1924 on Protective Upbringing, the Swedish courts gave custody of the child to the local Swedish Child Welfare Board, which in turn gave the custody of the child to her maternal grandfather. The Dutch guardian—her father and his designee—were denied custody by the Swedish courts. The judgment of the International Court of Justice found that the Swedish Law was "designed" to protect society against dangers resulting from improper upbringing, inadequate hygiene, or moral corruption of young people and that "to achieve the aim of the social guarantee which it is the purpose of the Swedish Law on the protection of children and young persons to provide, it is necessary that it should apply to all young people living in Sweden."[8]

The words of the Court in that case are especially applicable to children in a "cross-border" situation:[9]

> To arrive at a solution which would put an obstacle in the way of the application of the Swedish Law on the protection of children and young persons to a foreign infant living in Sweden would be to misconceive the social purpose of that law, a purpose of which the importance was felt in many countries, particularly after the signature of the 1902 Convention. The social problem of delinquent or even of merely misdirected young people, and of children whose health, mental state or moral development is threatened; in short, of those ill-adapted to social life, has often arisen; laws such as the Swedish Law now in question were enacted in several countries to meet the problem. The Court could not readily subscribe to any construction which would make the 1902 Convention an obstacle on this point to social progress.

The territorial principle serves the purpose of unequivocally specifying who is immediately responsible for foreign children and what law of protection should be applied. It thus avoids confusion and uncertainty over who is responsible for immediate care and protection. The necessity of territorial jurisdiction in cases of urgency is recognized in the 1961 Hague Convention on the Protection of Infants.[10]

> In all cases of urgency, the authorities of any contracting State in whose territory the infant or his property is, may take any necessary measures of protection.

Clearly, children found unaccompanied in a "second country," often without even the most minimal material support, fall into this category of urgency. The Convention on the Protection of Infants thus empowers, and as was discussed above, national and international children's legislation directs, the authorities of any "second" country to assume full immediate responsibility for children in their territory. This assertion of jurisdiction should continue, at the least, until the situation of the child is no longer "urgent."

SETTLEMENT: CHILDREN SETTLED IN A "SECOND COUNTRY" OR RESETTLED IN A "THIRD COUNTRY"

After unaccompanied children are identified, and, as we have seen above, been taken within the jurisdiction of the child welfare system of the second country, there are three possible options for their future: (1) repatriation (return to their State of nationality or previous habitual residence), (2) local settlement (in the "second country"), or (3) resettlement (settlement in a "third country"). Repatriation, of course, means that the child will return to the State of his nationality or previous habitual residence. Once he is there, the national courts and agencies will treat him under their own laws, as with the "in-country" situation described previously.

"Settlement" here means any form of care and protection that extends beyond a few days' duration; in other words, all situations in which a child is not promptly repatriated. In "settlement" situations, the courts and agencies of the State of settlement would ordinarily have *jurisdiction* over the child, but there may be a question as to *choice of law:* whether to apply the law of the child's nationality or the law of the State of settlement. This question only arises in those civil law countries which have traditionally applied the child's "national law" to issues of personal status such as age of majority, guardianship, and parental authority.[11] This civil law approach finds expression, in modified form, in the Hague Convention on the Protection of Infants. In countries with common law legal systems inspired by that of England, the territorial principle is usually followed on all questions of the status of children within the territory; common law countries would thus generally follow their own domestic law with settled or resettled children.[12]

Even in civil law countries, however, for children who are refugees under the 1951 Convention and 1967 Protocol Relating to the Status of Refugees, or who are treated as such by the countries in which they are settled, the choice of law issue is decided in favor of the domestic law of the country of settlement by Article 12(1) of the Refugee Convention:[13] "The personal status of a refugee shall be governed by the law of the country of his domicile, or, if he has no domicile, by the law of the country of his residence." Rights previously acquired by a refugee and dependent on personal status, however, shall be respected by the State where he is if they would have been recognized by the law of that State had he not become a refugee.[14] The law applied to children treated as refugees will thus ordinarily be the law of the State in which they are settled or resettled.

Therefore, it is only for children who are not treated as refugees under the 1951 Convention *and* who are settled or resettled in countries that usually look to the child's national law in matters concerning personal status that there is any question about what law to apply to a child settled in a second or third country.

The international convention with the greatest bearing on the solution to this problem is the Hague Convention on the Protection of Infants.[15] This Convention gives the authorities of the State of the infants "habitual residence" primary jurisdiction to take measures directed to the protection of his person or property.[16] The authorities of that State "shall take the measures provided by their domestic law, and that law shall govern the initiation, modification, and termination of the measures, as well as their effect on relations between the infant and persons or institutions responsible for his care and in respect of third persons.[17] The State of habitual residence, however, must recognize a "relationship subjecting the infant to authority which arises directly from the domestic law of the State of the infant's nationality.[18] The exclusive jurisdiction granted to the State of the infant's habitual residence by the Convention continues unless and until the authorities of the State of the infant's nationality take measures for his protection according to their own law.[19] Once the

State of the infant's nationality takes such protective measures, they shall be recognized in all contracting States.[20]

However, "the authorities of the State of the infant's habitual residence may take measures of protection insofar as the infant is threatened by *serious danger to his person or property*" regardless of any measures of protection or relationships subjecting the infant to authority arising under the infant's national laws.[21] Moreover, in cases where the infant is not in serious danger and the measures of the authorities of the infant's national State would otherwise take precedence, the State in which enforcement is sought can refuse enforcement "if such application is manifestly contrary to public policy" (ordre public).[22]

There can be little dispute that a child "settled"—that is, identified and cared for for any substantial period of time—in a second or third country is "habitually resident" there. Cheshire and North have stated that "[n]o more than a present intention to reside should be necessary for habitual residence."[23] Blom has noted that "[t]he connecting factor is his habitual residence at the present time, not the place where he has been habitually resident in the past."[24] As Blom goes on to say, "habitual residence ought to be defined along the lines of a man's principal home in fact or the one country where he can be said to have settled headquarters at this moment."[25] Wahler has stated: "It is generally understood by the courts as meaning: that State where the child has his actual and substantial center of living at the time of the decision."[26] Under any of these definitions, a child who is still in a second country or has been moved to a third country after the period of urgency has passed, should be found to have his habitual residence in that country of settlement or resettlement.

There are practical reasons, as well, to consider the settlement country to be the "habitual residence" of unaccompanied minors. From the moment of his arrival, the minor is faced with questions of his status under the country's immigration, asylum, and refugee laws; often in addition to very real problems of food, shelter, medical care, and protection from exploitation. The immediate and longer-term aid required in meeting these needs can usually come only from the local authorities. In almost all cases, the authorities of his natural State will not be able to observe or assess the child's situation. Nor will they ordinarily have the means to provide assistance; in the case of refugees or displaced persons they may be totally unwilling to do so. Not to consider the settlement country as the minor's "habitual residence" would often be to deny him aid from the only sources actually capable of providing it.

Thus, under the Convention on the Protection of Infants the law applied to children settled in second and third countries should ordinarily be that of the country of settlement, unless and until the authorities of the national State take measures for their protection. In all cases of children who are settled in common law countries or who are refugees (or are treated as such), the children's law of the settlement country would apply as a matter of course. For those civil law countries applying neither the Hague Convention on the Protection of Infants nor the 1951 Refugee Convention, considerations of child welfare should, in any event, prompt a substantially similar choice of law.[27]

19

The Role of International
and Voluntary Organizations

As we have seen, sovereign States have primary jurisdiction over unaccompanied children in their territory. In the kinds of emergencies in which children are often found unaccompanied, however, States may experience practical difficulties in filling this role. The structure of the State may be weakened or destroyed by the emergency, creating a legal or administrative void. In other instances, there may be so many other demands on national resources and so few means to meet those demands that the authorities find it impossible to identify and care for children left unaccompanied by the emergency. In addition, the necessary experience or expertise among local personnel may be lacking.

International or voluntary organizations can help meet all of these needs, and many States have used such outside assistance in the years since World War I.[1] From a legal viewpoint, this means the local sovereign may delegate some or all of its authority over unaccompanied children to the outside organizations. This practice, of course, is part of a more general pattern of international assistance with social, economic, and development matters which has become common in the last few decades.

The international organizations most active with unaccompanied children in the past four decades have been the United Nations Relief and Rehabilitation Agency (UNRRA) and the International Refugee Organization (IRO) in the years immediately after World War II, and the United Nations High Commissioner for Refugees in the years since. Voluntary agencies, of course, have worked with unaccompanied children in a variety of contexts, sometimes under the coordination of international organizations and sometimes independently. Among the many voluntary organizations active in this field have been the various Save the Children organizations and International Social Service. The following analysis will focus on the international organizations, but the legal questions are basically the same for voluntary organizations as well.

In examining the legal relationships between sovereigns and international and voluntary organizations on the one hand, and these organizations and the unaccompanied children on the other, several common themes can be discerned. The first is one of *authorization:* the scope of the organization's own mandate. In what contexts is the given organization authorized under its own terms of operation (mandate) to intervene to assist unaccompanied children? To what extent is this mandate accepted by sovereign States? The second theme is *delegation:* to what degree has jurisdiction over the children been delegated to the organization by the relevant sovereign? The third theme concerns the *policies and procedures* of the organization in caring for and placing the children. To what degree do the organizations fill the legal void with a body of their own law? An examination of these themes follows.

AUTHORIZATION

An international or voluntary organization must be authorized by the terms of its own charter to act in an emergency on behalf of unaccompanied children if it is to act legally. Its charter or other mandate may impose limits on the kinds of emergencies the organization may involve itself with or the categories of children it may assist. Furthermore, the primary mission of the organization, as spelled out in its basic documents, may also influence the policies and procedures it follows in caring for and placing the children. The terms of the organization's mandate thus bear on the questions of whether the organization may intervene at all, when and where it may intervene, and how it will act if it does intervene. An examination of the authorization of several international organizations with respect to unaccompanied children in emergencies follows.

International Committee of the Red Cross (ICRC)

The International Committee of the Red Cross (ICRC) has as its general mandate "to endeavour to ensure at all times that the military and civilian victims of [war, civil war or internal strife] and of their direct results receive protection and assistance, and to serve in humanitarian matters, as an intermediary between the parties."[2] The Statutes of the International Red Cross define the International Committee of the Red Cross as follows:[3]

> As a neutral institution whose humanitarian work is carried out particularly in time of war, civil war or internal strife, it endeavours at all times to ensure the protection of and assistance to military and civilian victims of such conflicts and other direct results. . . . It takes any humanitarian initiative which comes within its role as a specifically neutral and independent institution and intermediary and considers any question requiring examination by such an institution.

The Fourth Geneva Convention concerning the protection of civilians in time of war recognizes "the humanitarian activities which the International Committee of the Red Cross or any other impartial humanitarian organization" may undertake,[4] and obliges occupying and detaining powers to allow national Red Cross societies and other relief societies to pursue humanitarian activities.[5] The Fourth Geneva Convention also authorizes ICRC to offer its services as an impartial humanitarian body to the parties to armed conflict not of an international character.[6] In the First Protocol the parties bind themselves to "facilitate in every possible way the reunion of families dispersed as a result of armed conflicts"[7] and to "encourage in particular the work of the humanitarian organizations engaged in this task."[8] The Fourth Geneva Convention also provides for a Central Information Agency for civilians,[9] a service ICRC and national Red Cross organizations have traditionally provided through tracing agencies.

In general then, ICRC has a broad mandate in the protection of civilians, internees, and prisoners of war. Its sphere of action, however, is limited both under its statutes and the Geneva Conventions to international armed conflict, civil war, or internal strife. Although ICRC has performed a range of protection functions, traced for family members, and facilitated family reunion, it has not taken legal responsibility for unaccompanied children.

The United Nations High Commissioner for Refugees (UNHCR)

The basic authority of UNHCR is stated in Paragraph 1(1) of its Statute:[10]

> The United Nations High Commissioner for Refugees, acting under the authority of the General Assembly, shall assume the function of providing international protection to refugees who fall within the scope of the present Statute and of seeking permanent solutions for the problem of refugees by assisting Governments and, subject to approval of the Governments concerned, private organizations to facilitate the voluntary repatriation of such refugees, or their assimilation within new national communities.

Since its establishment in 1951, UNHCR has been requested by the General Assembly and ECOSOC to assist particular groups of refugees and displaced persons and persons displaced outside their countries of origin as a result of "man-made" disasters, bringing these persons within the terms of the statutory mandate.[11]

For these groups, UNHCR performs protection functions, seeks durable solutions to their plight, and often provides material assistance.[12] The legal bases for its actions are found in its Statute, particularly Paragraph 8, which empowers the High Commissioner to promote the execution of any measures calculated to improve the situation of refugees and to reduce the number requiring protection; to assist governmental and private efforts to promote voluntary repatriation or assimilation within new national communities; and to promote the admission of refugees to the territories of States.[13]

Material assistance and legal protection for unaccompanied refugees and displaced children clearly falls within the general mandate and the "humanitarian and social character" of the High Commissioner's work.[14] Protection of unaccompanied children and reunification of refugee families have been part of UNHCR's efforts since the subjects were mentioned in the Final Act of the United Nations Conference on the Status of Refugees and Stateless Persons, adopted in July 1951.[15] Most recently, the General Assembly explicitly approved the High Commissioner's assistance of unaccompanied children in Resolution 35/187 of December 15, 1980. In that resolution, after "stressing that the problem of refugee children is especially anguishing and considering the disturbing situation of children who have not yet been settled, many of whom have lost all the members of their immediate families," the General Assembly expressed its gratitude to the United Nations High Commissioner for Refugees

> for the action which he has already taken to assist refugee and displaced children, and requested him to intensify his efforts in that respect, endeavoring to ensure as

far as possible that the cultural and family identity of the minors settled is preserved.

For refugee children and children in refugee-like situations, then, UNHCR clearly has competence to take necessary measures of care and protection. It has not, however, taken on the formal legal guardianship of unaccompanied children whom it has assisted in past emergencies.[16]

United Nations Children's Fund (UNICEF)

UNICEF was created in 1946 as the International Children's Emergency Fund to provide "supplies, material, services, and technical assistance" "for the benefit of children and adolescents of countries which were victims of aggression" and "for child health purposes generally."[17] In its first few years, UNICEF devoted most of its resources to meeting the emergency needs for food, drugs, and clothing of children in Europe.[18] But when the Fund's existence was extended in 1950, the General Assembly expanded UNICEF's mandate:[19]

> Recognizing the necessity for continued action to relieve the sufferings of children, particularly in underdeveloped countries and countries that have been subjected to the devastation of war and to other calamities. . . .
>
> Decides . . . that the Board, in accordance with such principles as may be laid down by the Economic and Social Council . . . shall . . . allocate the resources of the Fund for the purpose of meeting, through the provision of supplies, training and advice, emergency and long-range needs of children and their continuing needs particularly in underdeveloped countries, with a view to strengthening, whenever this may be appropriate, the permanent child health and child welfare programmes of the countries receiving assistance. . . .

This direction was reaffirmed and made permanent in resolution 802 (VIII) in 1953.

UNICEF has focused its efforts since then on longer-range programs benefiting children within the context of national development programs. In emergencies in which unaccompanied children have been present, it has tended to concentrate on material assistance, leaving care and placement responsibilities to other organizations. Nevertheless, its legal bases make clear that UNICEF remains authorized to assist unaccompanied and other children in emergencies arising out of wars and other calamities.

DELEGATION

International or voluntary organizations only have such authority over unaccompanied children as is delegated by the sovereign authority in the territory in which they operate. The extent of their role will therefore depend on the terms of their agreements with the local sovereign,[20] or the terms of international agreements to which the State may be a party. It is not uncommon for the sovereign to grant some responsibility for unaccompanied children to an international or voluntary organization while withholding other important prerogatives.

The very presence of an international or voluntary organization at the scene of an emergency depends on consent of the local sovereign.[21] Ordinarily, an international organization will thus have no involvement with unaccompanied children unless invited by the local authorities. UNICEF, for example, can provide assistance in emergencies only with consent of the government concerned.[22]

The positions of UNHCR and ICRC are somewhat different. Article 35 of the 1951 Refugee Convention and Article II of its 1967 Protocol obligate party States:

> to co-operate with the Office of the United Nations High Commissioner for Refugees, or any other agency of the United Nations which may succeed it, in the exercise of its functions, and shall in particular facilitate its duty of supervising the application of the provisions of this Convention.

By this agreement States are bound, at the least, to cooperate with UNHCR in its functions of international protection and of supervision of the well-being of persons in reception and other refugee centers.

As mentioned above, countries adhering to the Fourth Geneva Convention and its First Protocol agree in advance of any particular conflict to permit certain ICRC activities. Article 81 of the First Protocol, for example, states:[23]

> The Parties to the conflict shall grant to the International Committee of the Red Cross all facilities within their power so as to enable it to carry out the humanitarian functions assigned to it by the Conventions in order to ensure protection and assistance to the victims of conflicts.

Furthermore, in the same Article, the Parties agree "to grant to their respective Red Cross . . . organizations the facilities necessary for carrying out their humanitarian activities" and to "facilitate in every possible way the assistance with Red Cross organizations and the League of Red Cross Societies extend to the victims of conflicts" in accordance with the Geneva Conventions and Protocols.[24] Thus, by adhering to the Geneva Conventions and Protocols, States agree in advance to permit the presence of ICRC, local Red Cross Societies, and, in some contexts, other humanitarian organizations.[25]

Because the jurisdiction of international and voluntary organizations depends on the terms of agreements with the sovereign authority, the nature of their involvement with unaccompanied children can vary widely from one emergency to another. The agreement between UNHCR and the United States relating to UNHCR assistance with Cuban refugees in 1980, for example, merely requested UNHCR to make a representative available "to assist" the United States' authorities with the reunification of family members.[26] In Korea during the war, at the other extreme, voluntary agencies constituted themselves as the *de facto* guardians of the children. In Thailand in 1980–82, UNHCR was granted broad authority to designate unaccompanied children for resettlement, but its decisions were subject to approval by the Thai government.[27]

Some of the complexities of the division of jurisdiction between a sovereign and international organizations assisting it with unaccompanied children can be seen in the relationship between the Allied powers and the international relief organizations in Europe in the years after World War II. UNRRA was initially designated to take charge of unaccompanied children, but from the outset its responsibility was limited to the terms of its agreements with the Allied forces.[28] It was made clear that "primary authority and responsibility for displaced children found in enemy territory is vested in the military or other control authorities pending repatriation."[29] The guardianship of children of United Nations' (that is, Allied countries) nationalities was the responsibility of the respective governments, while authority over stateless children rested with the military.[30] Therefore, UNRRA could not and did not assume legal guardianship.[31]

Similarly, when the Preparatory Commission for the Refugee Organization (PCIRO) succeeded UNRRA, its authority over the unaccompanied children in its care was circumscribed by its agreements with the various Allied Commissions. In the American zone in Austria, for example, PCIRO performed its functions of reuniting families and providing care, maintenance, repatriation, and resettlement of unaccompanied children "subject to the authority of the U.S. High Commissioner."[32] The U.S. High Commissioner remained "the sole authority to decide on matters involving the extradition or removal of refugees from the U.S. zone,"[33] and all refugees had such legal status and were subject to the jurisdiction of such courts as the High Commission prescribed.[34]

In its agreement with the British Element of the Allied High Commission for Austria, ACA(BE), PCIRO recognized that the High Commissioner "has sovereign authority in the Zone and full and complete responsibility in respect of law, order, and all aspects of security," and, agreed that ACA(BE) was ultimately responsible for the care and maintenance of refugees, though PCIRO would administer the camps and repatriation and resettlement programs.[35] While the agreement gave PCIRO responsibility for "plans for the priority assistance of unaccompanied orphan children and for the reuniting of separated families," it also specified the principles that had to underlie those plans:[36]

> ... since plans will be formulated solely in the humanitarian interest of the persons concerned and without regard to considerations of a political nature. The general principle to be applied in the case of unaccompanied orphan children is that children under the age at which they can be reasonably expected to make their own decisions will be repatriated when their nationality is known and when there is assurance that they will be properly cared for upon their arrival in the country to which they are being repatriated.

In Germany the Allied Commissions specified that the competent authority established by the Occupying Powers would review and decide on the children's final disposition on the basis of IRO recommendations.

From even this brief summary it can be seen that the possible relationships between international and voluntary organizations and local sovereign authorities are virtually limitless. The main factors determining the extent and nature of an organization's jurisdiction are the specific agreement between it and the relevant sovereign and the terms of any international agreements to which the State may be bound.

POLICIES AND PROCEDURES OF THE INTERNATIONAL ORGANIZATION

An international or voluntary organization which has been delegated any substantial control over decisions for unaccompanied children must establish policies and principles to guide its actions. These amount, in effect, to a body of family and child welfare law, the kind of procedures and rules which are normally propagated by the legislature, courts, and the administrative organs of States. There may come into being legal systems of the international or voluntary organizations which operate geographically within, but often in addition to or in place of, the legal systems of the States. To observe this, of course, is neither to praise nor condemn it. It is merely to note that the *de facto* legal and administrative structures of the international organizations must be recognized as one part of the legal framework which has come to influence the treatment of unaccompanied children.

A brief look at recent policies of the ICRC and UNHCR pertaining to unaccompanied children illustrates these points. As described above, the Fourth Geneva Convention and its Protocols provide a number of protections for civilians and children in general, and unaccompanied children in particular. In its work to implement these documents, the ICRC has attempted to embody their principles in certain substantive rules for the treatment of unaccompanied minors. Its Rules to be Observed Regarding the Problem of Unaccompanied Children,[37] used with Cambodian children in 1980, for example, set forth several substantive policies, including:

1. the children's welfare overrides all other considerations;
2. the unity of the family is a fundamental objective;
3. children who have been separated from their families should be kept in their own cultural environment;
4. as long as there is any chance that inquiries might lead to reuniting of the members of a family, no change in the situation of unaccompanied children which might prevent such an event should be contemplated.

In addition, ICRC has been compelled to develop a body of rules under its family reunification mandate to address such issues as who may reunite with unaccompanied children and the nature of the proof necessary to establish the family relationship. For example, its insistence that reunion of Cambodian unaccompanied minors along the Thia-Cambodian border in 1980–81 take place only with parents and not with other close family members had the effect in some cases of delaying beneficial reunions and forcing the parents to undertake a long trek by foot from distant villages, in contravention, some contended, of established Cambodian cultural practices of care for children by members of the extended family. Such rules, and the others mentioned above are, of course, the functional equivalent of substantive legal doctrines.

Similarly, when faced with substantial numbers of displaced unaccompanied children in Southeast Asia beginning in 1979, UNHCR was forced to create a childcare and placement system which became, in effect, a children's welfare and legal system. Acting through various voluntary agencies, UNHCR established children's centers, group homes, and foster placement, as well as a tracing program. It approved children for resettlement but initially prohibited their adoption, and insisted that receiving countries give the children the option of reuniting with close family members if they were subsequently located.[38]

A review committee was established to pass on long-term placement decisions. The committee was composed of specialists in resettlement and law from the UNHCR staff, representatives from ICRC and UNICEF, and an observer from the Thai Red Cross.[39] The committee considered the child's family and personal history, the tracing of his family members, medical background, assessments by child-care workers and a statement of the minor's expectations for the future. There was not, however, an advocate to present the child's point of view. The review committee attempted to meet the individual needs of the child within the available options. These options changed over time for reasons related to larger policy concerns about resettlement on the part of the Thai government, UNHCR, and the resettlement countries. The review committee acted, in effect, as a court deciding on the placement of the children who came before it. In addition, UNHCR was compelled to replicate other legal forms, such as agreements for and certification of foster placements.

Policies for unaccompanied children were further developed by UNHCR in the Conclusions of Family Reunification of its Executive Committee in its 1981 meeting. The Executive Committee concluded:[40]

> Every effort should be made to trace the parents or other close relatives of unaccompanied minors before their resettlement. Efforts to clarify their family situation with sufficient certainty should also be continued after resettlement. Such efforts are of particular importance before an adoption—involving a severance of links with the natural family—is decided upon.

The same Conclusions urged support for the reunification of separated refugee families with the least possible delay, including the application of liberal entrance and exit policies and the relaxation of requirements for documentation of family relationships.[41]

Introducing these recommendations before the Subcommittee of the Whole on International Protection, the Director of International Protection gave a somewhat more comprehensive statement of UNHCR's placement policies for unaccompanied minors in countries of first asylum:[42]

> The point of departure should always be to seek to ensure that their best interests were fully protected. This called in the first place for tracing action to ascertain the whereabouts of parents and close family members with whom reunification might be appropriate. If such tracing efforts prove unsuccessful, the solution of resettlement could usefully be envisaged. Tracing should moreover be continued after resettlement has taken place. Only when it became reasonably certain that the possibility of finding parents or other close relatives did not exist could adoption be considered.

UNHCR policies for unaccompanied children in Southeast Asia were elaborated further in its "Guidelines for the Promotion of Durable Solutions for Unaccompanied Minor Children in Southeast Asia" in 1982.[43] These guidelines are meant to assist in discharging the child welfare responsibilities that have devolved on UNHCR; in other words, to establish the substantive rules to apply in effecting UNHCR's authority. The guidelines permit unaccompanied children fifteen years of age or older to choose their own "durable solution" (*i.e.,* placement). Children between ten and fifteen who are of "mature judgment" may also determine their own future. For all others, UNHCR undertakes "to determine which durable solution is likely to be in the minor's best interest" with a preference for return of an "immature" child to his parents where the parents were not in agreement with his departure. A number of factors to be considered in determining the child's best interests are mentioned, including the situation of both the parents and the child *vis-à-vis* the authorities of the country of origin. The guidelines also recommend consideration of the establishment of a review committee "to assist in determining the age and maturity of unaccompanied minors and to advise on the most appropriate durable solution."

Without evaluating the content of either the ICRC or UNHCR policies, it can be seen in even a cursory review that they constitute a broad body of substantive rules for decision on the issue of care and placement of the unaccompanied children falling within the agencies' jurisdiction. UNHCR includes, as well, a decision-making body—the review committee—and other procedural devices. Understandably, in both cases, the organization's policies reflect its primary mission: for ICRC, the protection of vulnerable civilians in armed conflict under the Geneva Conventions; for UNHCR, the promotion of durable solutions for refugees and displaced persons. The acceptance of jurisdiction over unaccompanied children in emergencies, then, has led these international organizations to create a *de facto* legal system through which to discharge the resulting responsibilities for care and placement. Voluntary agencies, too, have been compelled to establish policies and procedures for unaccompanied children with whom they work.

In sum, then, the role of international and voluntary organizations depends primarily on the terms of their mandate and on the extent of responsibility delegated to them by the States in which they operate. Any significant degree of involvement will necessitate the creation of policies and procedures to guide their action. International and voluntary organizations must thus be included among the institutional actors involved with unaccompanied children, and their policies and procedures considered as part of the legal framework for their care.

Abbreviations

ACHR: American Convention on Human Rights (1969), OAS Official Records, OEA/Sec. K/ XVI/1.1.

ADRDM: American Declaration of the Rights and Duties of Man (1984).

DCRC: Draft Convention on the Rights of the Child, 1984 draft, E/CN. 4/1984/71, February 23, 1984.

DDFPA: "Draft Declaration on Foster Placement and Adoption"; full title: Draft Declaration on Social and Legal Principles Relating to the Protection and Welfare of Children, with Special Reference to Foster Placement and Adoption Nationally and Internationally, A/ Res/36/167, Feb. 3, 1982.

Draft Declaration on Foster Placement and Adoption: see DDFPA above.

DRC: Declaration of the Rights of the Child, G.A. Res 14/1386, Nov. 20, 1959.

ECHR: European Convention for the Protection of Human Rights and Fundamental Freedoms, Council of Europe Treaty Series, N. 5, November 4, 1950.

ECOSOC: Economic and Social Council of the United Nations.

Fourth Geneva Convention: Geneva Convention Relative to the Protection of Civilian Persons in Time of War of August 12, 1949.

Hague Convention on Adoption: Convention on Jurisdiction, Applicable Law and Recognition of Decrees Relating to Adoptions, Hague Conference on Private International Law (1965).

Hague Convention on Child Abduction: Convention on the Civil Aspects of International Child Abduction, Hague Conference on Private International Law, October 25, 1980.

Hague Convention on Guardianship: Convention Governing the Guardianship of Infants (Convention pour Régler la Tutelle des Mineurs), Hague Conference on Private International Law (1902).

Hague Convention on the Protection of Infants: Convention Concerning the Powers of Authorities and the Law Applicable in Respect of the Protection of Infants, Hague Conference on Private International Law, October 5, 1961.

ICCPR: International Convenant on Civil and Political Rights, G.A. Res. 2200A (XXI), December 16, 1966.

ICRC: International Committee of the Red Cross.

IECL: International Encyclopedia of Comparative Law.

ICESCR: International Covenant on Economic, Social, and Cultural Rights, G.A. Res. 2200A (XXI), December 16, 1966.

ICJ: International Court of Justice.

IRO: International Refugee Organization.

OAU Refugee Convention: Organization of African Unity Convention Governing the Specific Aspects of Refugee Problems in Africa, U.N.T.S. No. 14691, September 10, 1969.

PCIRO: Preparatory Commission for the International Refugee Organization.

1951 Refugee Convention: Convention Relating to the Status of Refugees, 189 U.N.T.S. 137, July 28, 1951.

1967 Refugee Protocol: Protocol Relating to the Status of Refugees, 606 U.N.T.S. 267, October 4, 1967.

UDHR: Universal Declaration of Human Rights, G.A. Res. 217A (III), December 10, 1948.

UNHCR: United Nations High Commissioner for Refugees.

UNICEF: United Nations Children's Fund.

UNITAR Report: Law and the Status of the Child (A.M. Pappas ed.) (United Nations Institute for Training and Research: 1983) also published in 13 Col. Human Rights L. Rev. No. 1 and 2 (1981–82).

UNRRA: United Nations Relief and Rehabilitation Administration.

UNTS: United Nations Treaty Series.

Sources of International Law

This chapter draws on the following sources: international conventions, protocols, and declarations (and their draft versions); the law and practice of nation States and international organizations; and scholarly writings.[44] A brief definition for the layman of each of these sources, and discussion of their legal effect, follows:

1. *Convention.* A convention is a multilateral treaty, an agreement of a contractual character among States, creating legal rights and obligations among the parties with respect to a particular subject.[45] Probably the best-known existing conventions are the four Geneva Conventions of August 12, 1949 for the protection of victims of war.

The provisions of a convention are paramount over everything else regarding the rights and duties of the States party to it, as among themselves.[46] They may or may not, however, create legal rights for individuals. Conventions are binding only on those States which consent to be bound, usually by ratification or accession. "A treaty does not create either obligations or rights for a third State without its consent."[47] The Geneva Convention Relative to the Protection of Civilian Persons in Time of War (the Fourth Geneva Convention), for example, does not protect nationals of a State not bound by it.[48] It may, however, apply vis-à-vis a State which "accepts and applies the provisions thereof," even without formal ratification."[49] Some conventions and some countries require the passage of national legislation before a convention has any legal effect.

Though a convention is an agreement among States, its aim may be the adoption by the various Parties of definite rules for the treatment of individuals, creating individual rights and obligations enforceable by national courts.[50] Thus, not only does a convention create obligations between the States party to it, but it may also establish legal rights for individuals in relation to all the contracting States, including their own.

Under the doctrine of *pacta sunt servanda,* "every treaty in force is binding upon the parties to it and must be performed in good faith."[51] If one of the parties nevertheless materially breaches a multilateral convention, the other parties can suspend or terminate the operation of the convention in their relations with that State, or suspend or terminate the convention as a whole.[52] Some conventions provide for the arbitration of differences among States Parties or the referral of disputes to a judicial body such as the International Court of Justice.[53] The Fourth Geneva Convention obligates its parties to search for and prosecute persons who have committed "grave breaches of the convention," but this kind of sanction is exceptional in conventional law.[54]

In general, however, the possibilities of meaningful enforcement of conventions are often few. Among party States, a breach by one can often lead only to termination by others, which hardly furthers the objectives of the convention. Furthermore, individuals aggrieved by violation of the terms of a convention may often have no realistic legal remedy. Even where an

278

international tribunal has found a State in violation of a convention or complaint of another State, there is usually no international enforcement mechanism to compel compliance. Nevertheless, conventions will, at minimum, contribute to the development of recognized international standards.

2. *Protocol.* A protocol is a treaty that amends or supplements another treaty (for example, the Protocols additional to the Geneva Conventions). A protocol has the same legal effect as a convention among party States.

3. *Declaration.* A declaration is a statement of principles endorsed and recommended by several States acting together. Among countries in the Western Hemisphere, declarations and resolutions of certain inter-American conferences are regarded as creating legally binding obligations without further action.[55] With respect to declarations and resolutions of the General Assembly of the United Nations, there is a range of legal opinion about the legal effect. Brownlie describes their legal effect as follows:[56]

> In general these resolutions are not binding on member states, but, when they are concerned with general norms of international law, then acceptance by a majority vote constitutes *evidence* of the opinions of governments in the widest forum for the expression of such opinions. Even when they are framed as general principles, resolutions of this kind provide a basis for the progressive development of the law and the speedy consolidation of customary rules. . . . Resolutions on new legal problems provide a means of corralling and defining the quickly growing practice of states, whilst remaining horatory in form.

4. *Draft conventions, protocols, and declarations.* Drafts of the above instruments obviously have no legal effect at all. The drafting of these documents, however, brings together experts and representatives from many States, and the process is often long and detailed. As a result, the product (a draft convention, protocol, or declaration) is often excellent evidence of the consensus of international thinking on the particular issue, involving as it does persons with many national viewpoints rather than just one.[57]

5. *Law and practice of States.* The law and practice of sovereign States includes their statutes or codes, regulations, judicial decisions, and administrative practice, as well as customary law of those societies operating without written codes. National law and practice is the law that usually applies in all contexts, including questions concerning unaccompanied children in emergencies. The examination of national law and practice is also necessary to learn the extent of agreement on legal doctrines in formulating an international legal framework on any issue. In turn, international norms may aim to influence national law and practice. National practice can, in some instances, become so convergent and widely accepted among countries that the practice becomes part of customary international law, which may be binding on other States.[58]

6. *Law and practice of international organizations.* The law (i.e., rules and regulations) and practice of international organizations, particularly organs of international institutions, may, in areas within their competence, have the effect of international law.[59] In other instances, they may contribute to the evolution of customary rules, or simply give evidence of widely accepted international principles.

7. *Scholarly writings.* This term is used here to refer to "the teachings of the most highly qualified publicists of the various nations" to which the International Court of Justice is directed to refer.[60] Scholarly writings are of importance in the field of international law simply as evidence of what the law and practice are, since, in contrast to national law, less of the law is codified. It is often left to scholars to discern trends, particularly commonalities, or recommend new approaches.

It should be clear from this brief discussion of sources that only conventions unequivocally establish binding legal obligations among the parties, and even with these instruments, enforcement possibilities may be quite limited. For the other sources noted above, such issues as the weight to assign a given source or combination of sources, when a practice has crystallized into customary international law, or a declaration is to be considered to be widely enough accepted in practice to represent a binding international obligation, are matters of great dispute among scholars, both as questions of general theory and in particular instances.[61]

IV

RECOMMENDATIONS FOR THE PROTECTION, CARE, AND PLACEMENT OF UNACCOMPANIED CHILDREN IN EMERGENCIES

The three preceding Parts have looked at unaccompanied children from historical, psychological, and legal perspectives. This Part sets forth recommendations for future action taken on behalf of these children. Admittedly, no set of general recommendations can anticipate all aspects of what would or would not be best for children and adolescents displaced from their families during whatever natural disasters, wars, and mass population movements may come. Indeed, what the fate of any one child will or should be in any given emergency will always remain a question of probability. Yet, as we have seen, unaccompanied children have existed in past emergencies and they are likely to exist in future ones. This fact alone necessitates that national and international agencies be well prepared to offer swift assistance to these especially vulnerable human beings.

In this regard, all three authors share the belief that this assistance be informed by what is known about children's developmental needs in general, unaccompanied children's needs in particular, and the dynamics of relief intervention in emergencies. Although the care and placement of unaccompanied children has been subject to debate over the years, today law and psychology have converged in recognition that the "best interests of the child" is the paramount consideration. This consensus is contained in the child welfare legislation of the majority of States and in all recent international legislation concerning children. Given the widespread global acceptance of this principle, and its accord with international human rights law and with what is known about children and their development, the best interests standard should guide all actions undertaken on behalf of unaccompanied children in emergencies. Thus, the recommendations presented in the following sections are organized around the basic premise that it should be the child's individual well-being—not a government's, an agency's, or a private citizen's—that determines the protection, care, and placement of children unaccompanied during emergencies.

THE BEST INTERESTS OF THE CHILD

The best interests standard has three main implications for States, agencies, and individuals who act on behalf of unaccompanied children in emergencies. These intervenors must: (1) protect and assist the child at all times, (2) put the child's welfare ahead of all other considerations and, (3) meet the child's developmental needs.

Protection and Assistance

Unfortunately, the protection of unaccompanied children in emergencies has too often been ignored or given a position of secondary importance. As a result, children have been abducted from their families, unlawfully recruited into armed forces, used as slave laborers, raped, sold into prostitution, abandoned, and illegally moved or adopted. Similar failures have been true with respect to assistance. In a number of emergencies, unaccompanied children have been left without food, medical care, shelter, or emotional and social support. In these and other instances, relevant national and international law has been ignored and violated by those who have acted or should have acted upon the children.

Most States have child welfare and protective laws which by their terms apply to unaccompanied children within their territory, as well as jurisdiction and choice of law rules for foreign children. These laws will often provide the necessary protection to children made unaccompanied by emergencies. Important protections for unaccompanied children are also found in a number of international conventions, especially in the Fourth Geneva Convention and the Convention Relating to the Status of Refugees and their respective protocols. States party to the Geneva Conventions agree to provide special protection for children in general, and unaccompanied children in particular, in armed conflicts. The Refugee Convention includes important protections for people, not excluding children, who fear persecution on the grounds of race, religion, nationality, social group, or political opinion. The rights of unaccompanied children under these and other conventions must be respected by all Party States which deal with these children.

Unaccompanied children, like all human beings, have civil and human rights under international and national law. These rights should be respected by those who intervene to assist them and by all other States, organizations, and individuals. One of the most important functions of those who assist unaccompanied children is to see that their legal rights are ensured.

In addition to relevant national law, there are a number of international instruments prescribing human, civil, social, and cultural rights which bear on the treatment of unaccompanied children in emergencies. Most notable of these are the Universal Declaration of Human Rights, the Declaration of the Rights of the Child, and the International Covenants on Economic, Social, and Cultural Rights and Civil and Political Rights, as well as various similar regional instruments. A Convention on the Rights of the Child and a Declaration of Social and Legal Principles Relative to the Promotion and Welfare of Children, with Special Reference to Placement and Adoption Nationally and Internationally, are both still in draft form. These instruments are discussed in Part III. Suffice to say here that international human and civil rights law has an important contribution to make to the protection, care, and placement of unaccompanied children and should be the basis of any intervention on their behalf by States or other organizations.

In sum, unaccompanied children, like all children, are in a privileged position vis-à-vis adult society: they must be among the first to receive aid in any emergency. When parents or other family members are unable to fill this role, other adults are legally and morally obligated to do so. At base, this means that States, agencies, and individuals must protect unaccompanied children against all forms of cruelty and exploitation and provide them the best material, educational, and social support possible under the circumstances.

The Child's Welfare

The best interests standard also directs the appropriate authorities to put what is best for the child above all other considerations. This begins with respecting and promoting the child's legal and human rights, and involves an assessment of each child's circumstances, as well as an individualized placement decision. Determining what is truly best for the child may also require his or her participation. Participation in decisions meant to be in their best interests, in fact, is a right to which children are entitled simply by virtue of their humanity.

The best interests standard dictates that the unaccompanied child must be treated as an individual and that his or her particular circumstances and needs are the most important factors to be considered. Solely national, political, religious, or agency bases for treating unaccompanied children are inconsistent with a determination of what is best calculated to "enable the [child] to develop physically, mentally, morally, spiritually, and socially in a healthy and normal manner," in the words of the Declaration of the Rights of the Child. Individual considerations and individualized placement decisions are therefore required; what is truly best for a child cannot be determined by a general formula. What is best for one child will not necessarily be best for another. The touchstone is what is best for the individual child in his or her particular circumstance.

Finally, choices of placement, custody, and, in the case of children outside their country of origin, residency (local settlement, repatriation, or resettlement), are important issues for unaccompanied children. By definition, these children do not have a parent or guardian to make decisions for them. Though it will be recommended that whenever possible unaccompanied children receive a guardian or representative to advocate on their behalf, even this procedure does not eliminate the question of the child's role in decisions about placement and residency. For all unaccompanied children these issues involve serious decisions about their future, and whether or not newly appointed guardians, representatives, and other community members or authorities also participate, the child's involvement in the decision remains an important question.

Cultures and nations vary considerably in the degree of importance they give a child's wishes in decisions about him or her. National laws and societal customs usually concern issues of placement and custody which for children who cross national boundaries may not be applicable to issues such as repatriation, settlement, and resettlement. International legislation bears on the child's role in decisions in two ways:

through the general provision of participation and through special mention of children's rights to participate in decisions about them. In general, many international instruments provide that all human beings are born free and equal in dignity and rights, have the right to recognition before the law, are entitled to a fair hearing by an independent and impartial tribunal in the determination of their rights and responsibilities, and have the right to freedom of movement. In addition, the Draft Convention on the Rights of the Child requires that the views of the child be heard in all matters, and his or her wishes be "given weight in accordance with his age and maturity." Similarly, the Hague Convention on Civil Aspects of International Child Abduction permits courts to follow the wishes of the child who "has attained an age and degree of maturity at which it is appropriate to take account of its views."

The extent to which unaccompanied children should participate in decisions that directly affect them will thus depend upon the child's developmental maturity and the degree to which the child's culture considers such participation appropriate. Of particular importance, too, is the actual decision being contemplated: interim placement, permanent custody or place of residence. Specific guidelines for the unaccompanied child's participation in these decisions will therefore be made in the appropriate Interim Placement, Long-Term Placement, and Unaccompanied Children Outside Their Country of Origin sections. Suffice to say here that participation in important decisions about their future is a right to which unaccompanied children are entitled, and which should be exercised to the maximum extent allowed by their maturity, culture, and circumstance.

Meet the Child's Developmental Needs

While the implementation of the best interests standard begins with physical protection, material assistance, and respect for the child's individuality, it also requires satisfaction of the unaccompanied child's developmental needs. Since parental separation or loss has already placed unaccompanied children at risk, this aspect of the best interests standard often demands actions that will prevent further developmental harm. In general, this is accomplished by protecting whatever psychological, social, and cultural ties the unaccompanied child may still possess, thereby minimizing further stresses, traumas, and deprivations. It also involves providing unaccompanied children sustained and age-appropriate adult care.

First, unaccompanied children need the continuation of existing relationships with people to whom they are psychologically attached. This guideline rests on the understanding that an unaccompanied child is helped to cope with the emotional distress of parental separation or loss when there are no further disruptions in surviving relationships with brothers, sisters, extended family members, other familiar adults, and peers within their community. Separation from other people to whom the child is attached exacerbates the initial trauma of parental separation or loss, and consequently, is likely to result in more serious developmental harm. Therefore, from the onset of intervention through all subsequent placement decisions, responsible authorities and relief agencies should recognize, support, and strengthen unaccompanied children's existing relationships with people to whom they are attached.

Secondly, along with preserving existing attachments to important people, the continuation of ties to the child's own community and cultural group is a central aspect of protecting and promoting unaccompanied children's best interests. In the short run, the familiarity of known social, linguistic, cultural, and religious practices can provide the firm identifications, stable values, and enduring goals necessary in helping to offset the inherent turmoil that results from family separation or loss. In the long run, this sameness is important for the continued development of personal identity. Identity formation is a dynamic process involving the merging of both psychological and social components. During each stage of development, the child's evolving sense of self gains strength through the constant awareness that his or her own way of mastering experiences is in accordance with the basic demands and values of a given society. Disconnecting unaccompanied children from their community often exacerbates the inherent vulnerability caused by family separation or loss, and can lead in later life to identity confusion: a sense of not knowing who one is or with whom one belongs.

Unaccompanied children's need for this type of continuity is maximized when they are cared for in their own community, which should be the first priority. There are emergencies, however, in which unaccompanied children are displaced from and cannot return to their native communities and homelands. In such cases, placements with members of their own cultural group, while usually more difficult to arrange, are also desirable. Without parents of their own, unaccompanied children are clearly dependent upon other people of similar backgrounds to provide them with this needed continuity.

Most importantly, unaccompanied children need to be provided sustained and age-appropriate adult care. This requirement rests on the knowledge that every child's immediate and future developmental well-being is dependent upon establishing and maintaining close attachments with adults. Indeed, it is the disruption or loss of these essential attachments that place unaccompanied children at increased risk in the first place. Yet, equally important if not more so, in determining the developmental fate of each unaccompanied child, will be the kind and quality of adult care that follows the original separation or loss. When intervention efforts provide an unaccompanied child sustained and age-appropriate adult care, the likelihood of more serious developmental harm is reduced. When however, adult care is absent or given in a generalized or largely impersonal manner, the probability of more serious harm increases dramatically. The speedy provision of sustained and age-appropriate adult care is thus a minimum requirement for preventing further developmental harm, and should not be sacrificed for long-term objectives, even if one of these objectives is eventual reunion with natural parents.

RECOMMENDATIONS AND CONSIDERATIONS

The best interests of the child principle offers a unifying standard by which actions undertaken in the protection, care, and placement of unaccompanied children may be measured. In many instances, however, this principle does not indicate in precise terms what actions should be taken and by whom. Without more detailed directives the determination of what is in an unaccompanied child's best interest at any given time in an emergency may become subject to ideological, political, and other kinds of arguments or debates. In essence, there is the danger that decisions about the "best interests" of these children might be made in name only.

In an effort to eliminate this potential confusion the following sections offer recommendations and more detailed considerations which take the "best interests" principle one step further by translating its basic premises into specific actions, policies, and programs for unaccompanied children in emergencies. These suggestions represent both a distillation and an integration of information and analysis offered in the previous historical, psychological, and legal sections. They are arranged into six categories—preparedness, prevention, assistance, interim placement, long-term placement, and unaccompanied children outside their country of origin—which address the major issues intervenors (governmental and community authorities as well as international and voluntary agencies) will confront while offering protection and assistance to unaccompanied children during and after an emergency.

Since what is initially done or not done in the beginning phase of an emergency is often linked to later care and placement measures, and consequently, to the children's immediate and future well-being, these recommendations are designed as a whole and one section cannot be fully understood in isolation from other sections. Readers are therefore encouraged to consider these guidelines consecutively, beginning with the preparedness recommendations and continuing on through the ones in the subsequent sections.

One final explanatory note is in order. The greater portion of the information, observations, and research on unaccompanied children presented in this book is based on work undertaken by individuals from Western nations. Indeed, much of the knowledge of children's development as a whole, and many of the principal tenets of international law can be said to encompass a western view of the individual. The possibility of Western bias is heightened further by the fact that this book is written by three Westerners employing a Western language and Western terminology.

Yet, we believe this background does not negate the fact that all children unaccompanied during past emergencies—children from different nations and different cultures—have encountered many of the same problems and have shared many of the same basic needs. Protection from exploitation, the necessity for material assistance and human attachments, and support from their own community are examples of what may be rightfully conceived of as being basic needs of unaccompanied children all over the world. It thus seems not only possible but also timely to establish a set of protection, care, and placement recommendations which can be generally applied to unaccompanied children in all emergencies. Nonetheless, an awareness of the customs, beliefs, and child-care practices of any affected community is critical to informed intervention efforts. Thus, specific references to important cultural considerations are made in a number of relevant recommendations and guidelines which follow.

An earlier draft of these recommendations was discussed by representatives of nongovernmental and international organizations at a June 1985 seminar on the care and protection of unaccompanied children in emergencies in Halvorsbøle, Norway, sponsored by Redd Barna. The recommendations which follow benefited from the comments of the seminar participants and others. As of September 1986 these recommendations had been adopted as the basis for emergency action by UNICEF in its resource handbook for field staff, *Assisting in Emergencies,* and by the United Nations High Commissioner for Refugees in the *UNHCR Handbook for Social Services.*

AN UNACCOMPANIED CHILD

It is suggested that an unaccompanied child be defined as a person under the age of majority not accompanied by a parent, guardian, or other adult who by law or custom is responsible for him or her. The need for a somewhat flexible age component in this definition stems from the fact that there is no one internationally recognized age of majority. While all definitions include children up to the age of fifteen, many regard children eighteen years of age and older as minors. Consequently, whether or not an individual is under the age of majority will depend on national law, including the laws of receiving nations for children outside their country of nationality.

The second component of this definition—unaccompaniedness—focuses on the absence of an adult with firm legal or customary responsibilities for the child. Again, there is not one universally accepted definition of the "unaccompanied" aspect of children in need of special protection. However, the formulation offered here is consistent with the thrust of most international instruments.

For further discussion of the definition, or absence of definition, of the term "unaccompanied minor" in various legal systems, see the discussion in Part III. The recommendations in this Part, however, do not depend on the definition proposed here; rather, they can be applied to unaccompanied children however defined by the relevant law or custom. In the absence of a different existing definition though, it is proposed that the recommendations here be followed for all persons under the age of majority not accompanied by a parent, guardian, or other adult who by law or custom is responsible for them.

20

Preparedness

An emergency is not the appropriate time to experiment and test methods of furnishing assistance to children. Preparedness is required to ensure that child welfare services are available to meet the needs of unaccompanied children. Preparedness means being in readiness; it consists of actions taken prior to the onset of an emergency to organize and facilitate services which will attempt to prevent the separation of children from families and ensure that unaccompanied children are cared for and protected. Preparedness should be attempted by all concerned parties, including those with mandated responsibilities in times of emergency to care for and protect unaccompanied children, and by all others likely to provide services to children, such as local and national services, and international agencies and organizations.

PREPAREDNESS: RECOMMENDATION 1

**All concerned agencies should be prepared to address the problems
of family-child separation in emergencies.**

Preparation should be based on informed decision making, the preparation of personnel for the responsibilities they assume, establishment of policies and procedures to guide future action, interagency coordination, and research and evaluation.

Informed Decision Making

A substantial body of knowledge exists about the developmental needs of children, the consequences of separation and trauma on their health and well-being, the phenomena of family separation, the circumstances of unaccompanied children in emergencies, effective provision for relief services, and the legal framework within which services are provided. Knowledge about such issues is essential for planning and carrying out effective programs in future emergencies. The effectiveness of preparedness actions is dependent in large part upon the accuracy of the planning assumptions—based on these bodies of knowledge and about human needs and behavior in emergencies—those of parents, children, persons within the community, and outside intervenors. Inaccurate assumptions lead to inappropriate programs, a fact substantiated in many past programs for unaccompanied children.

Assistance to unaccompanied children in emergencies requires specialized programs, for example, to identify unaccompanied children, to offer legal and physical protection, to document their situation and assess their needs, to care for them in ways that meet their developmental needs, and to trace their families. The services typically provided to adults and families in emergencies, such as those ensuring the availability of food, clothing, shelter, medical care, and protection, do not usually address the special needs or circumstances of unaccompanied children.

Therefore, as the beginning point in any preparedness effort or emergency program, it is crucial that planners and program staff fully understand the dynamics of family separation in emergencies, the common characteristics and circumstances of unaccompanied children, the developmental needs of children and how those needs are best met, the lessons learned from past experience about the care and placement of unaccompanied children, and the legal framework within which children are to be cared for. A mastery of these considerations is necessary as a precursor to the establishment of sound policies, the training of staff, and the formulation of effective emergency programs. The first three parts of this book provide an overview of these issues.

Staff Recruitment and Training

Helping unaccompanied children in emergencies requires special skills, including those of working with children in providing the specialized services required in emergencies. Identification and documentation of unaccompanied children, evaluation of their developmental and special needs, provision of guardianship, placement of children, and assessment of family situations are examples of routinely required tasks related to the care of unaccompanied children for which experience and expertise are helpful.

To assure that persons with appropriate knowledge and skills are available when needed, agencies serving unaccompanied children must establish special recruitment practices and train persons who will assume responsibilities for the care of unaccompanied children. All persons who may plan policies or carry out programs for unaccompanied children should be trained to understand the needs of children, the most effective ways to help, the legal implications, and lessons learned from the past. For standing emergency organizations and other agencies that routinely assist unaccompanied children in emergencies, there is no adequate substitute for expertise on staff. Other agencies should adopt recruitment practices which assure that trained and competent people are chosen to implement programs. In many past emergencies this was not done, and many problems arose because people without relevant child welfare experience or expertise were made responsible for the care of unaccompanied children.

Policies and Procedures

All intervenors who are likely to encounter unaccompanied children or provide emergency services to families should adopt policies and procedures to guide their actions within the operational mandates of each agency. Policies and procedures should ensure that services are provided from the onset of an emergency concerning initial, interim, and long-term care and placement. All emergency plans and procedures should include provisions for unaccompanied children.

There are many layers of intervention: national governments, direct service organizations, and international organizations. In a world composed of sovereign States, national governmental authorities constitute the first line of responsibility towards unaccompanied children in emergencies. It is therefore essential that national and local policies and procedures exist to protect family unity and to ensure that unaccompanied children are cared for. No amount of exhortation or assistance from concerned observers can substitute for resolute and informed action by the governmental authorities responsible.

Direct service organizations, as the second line of responsibility, such as hospitals and clinics, standing emergency organizations, social welfare services, religious institutions, orphanages and other special facilities, and the police and the military are likely to encounter unaccompanied children. It is essential that these agencies have appropriate policies and procedures to deal with children.

Because of their unique roles, international organizations occupy a third line of responsibility towards unaccompanied children in emergencies. They often supplement and complement government actions, especially in emergencies, when there may be other demands on public services. International organizations should adopt policies and procedures to provide effective assistance to vulnerable families and unaccompanied children or support for programs that do.

The policies and procedures adopted by each intervenor must correspond with the role and responsibilities assumed by that emergency. For example, national governmental authorities may allocate local responsibility for programs and set national policy. Police and soldiers may refer to certain agencies unaccompanied children whom they locate. Social welfare services might assist vulnerable families and establish placement programs for unaccompanied children. Health care services may ensure that children are not separated from their families by rescue procedures or faulty record-keeping practices and are properly referred. Therefore, although based on the same set of child welfare principles, policy and procedures will differ among agencies. The children's interests are best served when such policies are complementary and are based on the principle of the best interest of the child. The recommendations and guidelines proposed in this book should assist intervenors in establishing policies and procedures for their own agencies.

Interagency Coordination

Many of the responsibilities involved in the care and protection of unaccompanied children are likely to be shared. For example, one agency may assume responsibility for the care and placement of unaccompanied children, while another may assist in tracing the families. A coordinated response system can increase the effectiveness of all services and avoid the confusion caused by conflicting programs. Experience has shown that collaboration before the onset of the emergency improves cooperation during emergencies. Therefore, interagency coordination on all matters related to unaccompanied children is integral to preparedness. There are two essential aspects of coordination—definition of roles and responsibilities and the sharing of information. Preparedness should guarantee the free and continual flow of information about unaccompanied children among agencies and should ensure that needs are met through the roles and responsibilities assumed by intervenors.

Finally, among international voluntary agencies it would be useful if one were to make unaccompanied children a permanent concern and become actively committed to their plight. This agency would document information, conduct training, and serve as a resource and advocate for unaccompanied children, giving special attention to those in emergencies. This one agency's coordination would improve the participation of all others and ensure that the care of unaccompanied children was not inadvertently overlooked. Assumption of this institutional responsibility would, we believe, contribute to the development of comprehensive and sensitive treatment of children separated from their families. At the time of this writing, various international and voluntary agencies are implementing programs for unaccompanied children around the world, but the coordinating agency, as proposed herein, does not yet exist.

Evaluation and Research

Developing organizational competence and staff expertise requires ongoing assessments, study, and research to determine how services can be provided more effectively. While there is convergence of thought about the needs of unaccompanied children and ways to meet those needs, much is still not known. The situation of children in emergencies is constantly changing; thus continual monitoring and evaluation are necessary.

21

Prevention

In most cases children are best off in their own families, for the family is the fundamental unit most capable of providing for the numerous physical and psychological needs of children during infancy, childhood, and adolescence. Most children are fortunate enough to be a member of a family which provides them with life's physical necessities and offers them love and acceptance. In such a family (whether it is nuclear or extended, or consists of biological or adoptive members), the child is afforded an unmatched opportunity to first establish and then maintain close psychological attachments. These attachments—products of the mutual exchange of ongoing affection, social involvement, and personal commitment between the child, the parents, and in some cases, other close family adults—are as essential for healthy growth and development as the provision of physical protection and material assistance.

During emergencies, which place unusual demand on their physical and emotional resources, children have an even greater need for family attachments. Family adults, most often parents, remain the primary source of security and protection for children, and anything that signals the threat of separation is likely to intensify anxiety reactions in children of all ages. Observations indicate that of all the potential determinants of children's responses to emergencies, those in the category of the parent-child relationship are most significant. Children are usually able to endure the stress and disruption of most emergencies if they remain with their families and if their parents are able to continue offering them adequate care, comfort, and protection. Conversely, events that break or prevent the formation of secure family attachments endanger the child's developmental well-being and place him at increased psychological risk. Thus, family attachments, which are essential for children's growth and development during normal circumstances, often take on increased importance in emergencies and must be preserved.

The law recognizes the importance of the family and the parent-child relationship. International law unequivocally describes the family as the basic group unit of society. The law confirms the right of parents and children to remain together and upholds the initial and presumptive right of parents to raise, educate, and care for their own children. These rights yield only when the parents are unable or unwilling to fulfill their responsibilities, and it thus becomes necessary to remove the child from parental control. Concomitantly, parents have the duty to provide food, shelter, care, education, discipline, and representation of the child's interests. International and national law are virtually unanimous on this approach to parent-child relations.

Since they have the right to control the education, care, and upbringing of their child, parents may entrust their child to other adults for any or all of these functions. They may also, if they so desire, surrender all of their parental rights to the child, although in most cultures it is usually subject to judicial or other governmental approval. They may, in addition, grant the child his independence, leaving him free to choose his own residence, or in fact send the child away as an emigrant.

PREVENTION: RECOMMENDATION 1

In emergencies, efforts should be made to prevent the separation of children from their families.

In virtually all wars, mass population movements, famines, and natural disasters, some children become separated from the adults responsible for their well-being. Some separations are intentional on the part of parents, such as abandonments and voluntary surrenders of children; other separations are unintentional, such as abduction, accidental separation, or death. Both types of separation can be prevented to some degree.

Protect the Integrity of the Family

Protection of family integrity implies firstly, noninterference with satisfactory parent-child relations. In emergencies, children have often been involuntarily separated against the wishes of the parents, through kidnapping, illegal recruitments, or other forms of abduction. The forcible separation of children from their families is legally prohibited at all times, including during wars, natural disasters, and mass population movements. Parents may voluntarily place their children with others or surrender all parental rights, but legally they may not be coerced or deceived into doing so. Nor, under legal norms, can parents and children be separated for political, racial, religious, or other ideological reasons. Indeed, the forcible transfer of children of one group to another group constitutes the crime of genocide when done with intent to destroy a national, ethnic, racial, or religious group. War and refugee situations, particularly, should be monitored for forced separation of children from their parents. Diplomatic and public advocacy, even on an international scale, may be required to stop such separations, especially when carried out by political bodies and military groups.

Administrative practices, such as organized population movements, should be designed and implemented to minimize family separation and maximize the integrity of parent-child relationships. In the field of immigration and asylum law, family unity can be protected by recognizing foster and adoptive relations under general family and dependency status rules, and by permitting immigration or asylum by sibling and family groups. Law and practice, in other words, should in all respects prevent the separation of children from parents and other close family members.

Provide Assistance to Families Vulnerable to Separation

Problems within the families themselves are a major cause of parent-child separations. In some cases, these problems result from adversities stemming from the emergency itself, such as forced migration, the death of a parent, the absence of the father because of military conscription. In other cases, there are chronic problems that predate the emergency, as for example, when families are struggling with divorce, alcoholism, unwanted pregnancies, or poverty. Most often existing problems are exacerbated by emergencies. The poor become poorer. Poverty, especially, appears to be the largest cause of the parent-child separation.

Preventive assistance thus requires that a social welfare framework be established through which individual, social, and economic problems can be more effectively dealt with. Material assistance, cash-generating schemes, skills training, medical care, family-planning services, counseling, and conflict resolution systems can all be essential in protecting and promoting family unity.

The first task in offering these services is to identify vulnerable families. This vulnerability will not always be evident: often parents in such families are either unaware of the available services or reluctant to ask for them. Preventive social services must therefore be organized on an "outreach" basis, paying close attention to single-parent families. Often, women and children constitute the majority of displaced persons in war and refugee situations, and whether widowed, abandoned, divorced, or unmarried, these women must shoulder the multiple burdens of coping with a different and often precarious environment while supporting their children emotionally and economically.

Special services may also be required for families who have children with exceptional physical or psychological needs, and who, in some cultures, are more likely to be abandoned than other children. Again it is always best when these required services are brought to the families in the affected area rather than sending the children to a far-away community to receive them. While in extreme situations, the movement of children for specialized services may indeed be required, the principle of family integrity still applies: children should be accompanied by at least one parent.

Avoid Relief Policies and Practices Which May Inadvertently Cause Parent-Child Separations

In past emergencies, relief policies and practices have sometimes inadvertently contributed to or caused parent-child separations. This has occurred, for example, when relief workers, assuming children to be unaccompanied, have removed them from an endangered area without ascertaining the whereabouts of the parents or informing other members of the affected community about their plans for the children. In various emergencies, there have been reports of young children whose names and origins have been lost due to poor documentation when they were, for example, taken for emergency medical treatment or moved from one location to another for other reasons. Other separations have resulted from poor record-keeping and poorly organized transfer of records to and from hospitals, clinics, or other emergency facilities.

Family separations have also resulted from relief assistance focused only on children, thereby ignoring the needs of families and negating their traditional responsibilities in the care of their children. When emergency feeding has been limited to children, for example, infants have been abandoned at the feeding centers.

In many past emergencies the way in which assistance has been provided to accompanied children has contributed to additional family separation, particularly when separation was the only means by which parents could avail their children of the perceived benefits. For example, orphanages, children's centers, and other institutions for children alone, set up for emergency purposes, have led to significant increases in separations, especially when unaccompanied children received more food and material assistance, better shelter, protection, and educational opportunities than children with their families. In the same way, in cross-border situations, preferential resettlement status has caused intentional family-child separations. The "magnet" effect of these practices has been apparent in many emergencies, including the Korean war, the drought in Ethiopia, the so-called Vietnamese "babylift," and the continuing flow of unaccompanied children from Kampuchea into Thailand and from Vietnam to many resettlement countries.

Separations caused by relief policy and practice can to a large extent be avoided. Accidental separations caused by relief workers are preventable through adequate preparation of relief personnel and through the establishment of special provisions for the care and protection of unaccompanied children in all emergency systems. The adoption of a family-oriented assistance approach that supports and enhances the capabilities of family members to meet the needs of their children can eliminate those separations caused by relief measures that tend to encourage family separations between parent and child as a precondition to receipt of service; aid is best provided to children through their families. Services to unaccompanied children should be provided in ways that minimize the "magnet" effect which cause additional separations of parents and children, a point discussed in greater detail later in these Recommendations.

PREVENTION: RECOMMENDATION 2

At least one parent should accompany any child removed from an emergency by organized evacuation.

The evacuation of children from disaster-affected areas to locations where they are safer and may have greater access to goods and services is an option to be considered in every emergency, and in past emergencies various evacuation schemes for the removal of children have been conceived and implemented. However, experience confirms that the separation of children from their families for evacuation may actually be more harmful than helpful to the children. Firstly, evacuation programs are typically plagued by unanticipated complications and consequences. For example, parent-child separations from past evacuation schemes have been much longer in duration than anticipated by parents, children, and organizers. A significant percentage of evacuated children were never returned to their natural parents. In past evacuations, difficulties consistently arose in providing adequate care for the children after they were separated from their families, in spite of the good intentions of evacuation organizers. Evacuations have also led to changes in the ethnic identity, religion, and language of the children, which complicated or prevented later reunion with their families. In the final analysis, the occurrence of family reunions has been determined more by the outcomes of war and resulting political circumstances than by the intentions or supposedly well-formulated plans of any evacuation scheme.

Secondly, comparative studies of evacuated and nonevacuated children clearly suggest that separation from families during most emergencies was more traumatic than exposure to war, bombardments, and reductions in food rations. Distress reactions among evacuated children were usually more intense and long-lasting than among nonevacuated children. These emotional disturbances and age-related behavioral problems are most prevalent during what Anna Freud and Dorothy Burlingham have described as "the no man's land of affection": the period of time after old attachments have been broken and before new ones have been formed. In this regard, children evacuated with a parent or cared for in new settings by familiar adults have fared substantially better than children sent alone or in sibling groups to strange caretakers and communities.

Thirdly, some evacuated children have not returned to their families because family reunions after lengthy separations have not necessarily been in the children's best interests. Children have a limited ability to sustain psychological attachments to absent parents depending on age and corresponding developmental capacities. The younger the child, the more quickly separation is experienced as permanent loss, accompanied by profound feelings of deprivation and despair. After lengthy separations, evacuated children have often formed new and more important attachments with substitute caretakers.

Furthermore, for many the emotional scars that resulted from the initial separation from natural parents were further aggravated by the breaking of these second psychological attachments. Evacuated children have often been returned to biological parents whom they have become emotionally estranged from, or in the case of infants and young children, had completely forgotten. Family reunions have been particularly stressful and problematic for children who were cared for in different sociocultural settings, especially when new languages had replaced old ones.

For all of the above reasons, children should not be intentionally separated from their families through evacuation schemes. There are, however, situations in which compelling reasons do exist for the movement of children away from areas of danger, particularly when their safety and physical health is in imminent danger, and when adequate protection and assistance is neither present nor forthcoming. In such situations, if the children are to be moved, parents or other adult family members should be evacuated with the children.

Parents should not be encouraged or enticed to separate from their children by sending them in evacuation schemes. However, parents do have the right to place their children with others or send them to other locations. This parental right must be respected, for parents are likely to be in the best position to know what actions are required to care for and protect their children, and, in any event, have the right to choose their child's place of residence unless their choice is clearly inappropriate.

In rare situations children may need to be moved for compelling reasons, where the parents may not be able to accompany the child. In such conditions, parental consent should be secured and the evacuation should be temporary in nature. These children should be provided care and protection as unaccompanied children in accordance with the recommendations which follow.

22

Assistance

Children, by virtue of their developmental immaturity, are more vulnerable than adults to the misfortunes and deprivations that arise during natural disasters, wars, and mass population movements. Physically, children are especially susceptible to infections, illnesses, and diseases that can result from changes in home environment, diet and water supply, and from temporary placements in overcrowded, unsanitary living quarters. Even more serious are the dangers imposed upon the child's physical, cognitive, and personality development by malnutrition or undernutrition.

Mentally and emotionally, children also differ from adults in ways that place them at increased risk during emergencies. Normally, adults are able to perceive and respond to changing circumstances with reason and intellect. Children's perceptions and reactions to these very same circumstances, however, are likely to be clouded by irrational throughts and based on impulsive emotions. Young children, for example, often respond to events in an "egocentric" or self-centered manner. Thus, the need to flee one's home may be experienced as the result of some past misdeed or angry thought, or the accidental separation from a loved one as an intentional abandonment. In situations which demand increased coping abilities, these reactions which distort or deny reality and reverse or displace feelings, may place children even more at the mercy of events.

In emergencies as in peaceful times, children's physical and psychological well-being is dependent upon the constant care and protection of adults. Normally, parents assume responsibility for their children's welfare by tending to and nurturing their bodily needs, alerting them to potential risks in the environment, helping them understand and organize their thoughts and perceptions, serving as essential emotional attachments, and acting as principal guardians of their immediate and future well-being. Most organized relief efforts, in fact, reflect this expectation of parental guardianship by directing food, shelter, medical, and other needed assistance to children via the family unit as a whole. The assumption is, of course, that parents will utilize these resources to the advantage of their children.

By definition, however, unaccompanied children are without parents or legal guardians. Thus, in addition to the external emergency, these children must also contend with the internal emergency resulting from family separation or loss. Their double vulnerability of being children and children without parents or guardians necessitates swift and informed intervention on their behalf. The prognosis for unaccompanied children is vastly improved when other adults assume responsibility for their care and protection. When, however, adequate protection is absent and developmental needs go unmet, the immediate and future well-being of unaccompanied children is greatly jeopardized. Therefore, intervention is necessary to fill to the maximum extent possible the parent's role vis-à-vis the child.

Under international law, children must be among the first to receive aid in any emergency, and unaccompanied children, by virtue of their separation from their families, must be the first among the first to receive aid. When parents or other family members are unable to fill this role other adults are legally obligated to do so. At base this means that other adults must provide food, clothing, shelter, medical assistance, and direct human care to enable the child "to develop physically, mentally, morally, spiritually, and socially in a healthy and normal manner," in the words of the Declaration of the Rights of the Child. The authorities must, therefore, provide the best material, educational, and psychological support possible under the circumstances. They must also protect unaccompanied children against all forms of cruelty and exploitation.

In past emergencies, unaccompanied children have often gone without basic goods and services. It is incumbent on intervenors, therefore, to ensure that unaccompanied children are provided the assistance normally given by the absent family members. This means that the situation must be assessed and unaccompanied children identified, protected, and documented. Second, aid must include physical protection, food, shelter, clothing, medical care, education, and recreation. In addition, due to the greater possibility of psychological and emotional trauma among these children, unaccompanied children in emergencies should receive special evaluation, and where necessary, treatment services. Because they are not in the care of a person legally responsible for them, a guardian or representative should be designated. In short, assistance to unaccompanied children entails protection, the provision of such special services

ASSISTANCE: RECOMMENDATION 1

The appropriate parties should ensure that unaccompanied children are receiving care and protection.

The Appropriate Governmental Authorities Should Intervene

The obligations for the care and protection of unaccompanied children fall in the first instance to the authorities of the State where the children are located. In national and international law, the country in whose territory unaccompanied children are found is responsible for emergency measures of care and protection. Preexisting child welfare agencies should be used wherever possible; new local systems should be created if needed. National agencies are thus the first resort for care and protection of unaccompanied children in emergencies. In some countries the appropriate authorities will be determined more through custom than through law and will exist on the local rather than national level.

Other Concerned Agencies Should Offer Their Assistance

National authorities may fulfill their duty by inviting an international or voluntary organization to assume full or partial responsibility for the care, protection, and placement of the children. When State legal and administrative structures break down, international organizations should offer their services to fill the void which unaccompanied children would otherwise face. Whatever legal or administrative form is used, however, the appropriate authorities clearly have the duty to see that unaccompanied children are provided adequate care and protection.

The Community's Own Efforts to Care for Its Unaccompanied Children Should Be Supported

Studies and direct observation clearly indicate that community members within an emergency situation do, on their own initiative, provide assistance to unaccompanied children. In all communities, laws and customs exist for the care of children not with parents or guardians. In every emergency reviewed, people opened their homes, established programs, and provided goods and services to needy children, even in the face of extreme hardships and, sometimes, at great danger to themselves. These findings have two important implications for outside intervenors.

First, the "spontaneous" care typically provided unaccompanied children in emergencies is evidence of the potential resources for taking care of these children within the affected community and should be supported. Specifically, caring relationships between unaccompanied children and substitute families should be respected and not interfered with except to support the relationship or to offer assistance in locating separate family members. Secondly, unaccompanied children can be helped through a wide spectrum of preexisting local and national services, such as community hospitals, schools, churches, and social service centers which can often adapt themselves to help meet the needs of these children.

However, the effectiveness of community actions must be judged by whether or not the unaccompanied children's needs are being met. In many past situations, the needs of unaccompanied children have not been fully satisfied by local efforts. Examples abound of unaccompanied children living unattended on the streets, in deplorable institutions, with little or no legal protection, or without access to organized tracing programs or other essential services. Nor can traditions or local child care practices be supported when they violate the basic rights or physical or mental wellbeing of the child. The lack of effective, child-oriented, community services should serve as a stimulus for strengthening and enhancing local efforts rather than as a justification for independent actions which will further weaken community response. Thus "support the community's own efforts" requires that outside intervenors be cognizant of what is already being done locally for unaccompanied children and, when needed, devise supplementary assistance programs that will enhace, rather than undermine or replace, the interest and willingness of community members and organizations to care for these vulnerable children.

The care and protection of unaccompanied children has often involved various diverse parties including governmental and nongovernmental agencies, long-established child welfare organizations, and recently created ad hoc groups, and even persons acting individually in their own capacity. They have come from the local community, from within the region or nation, and from outside the country. Not uncommonly, each party has adopted its own program philosophy and method, resulting in some cases in disagreement as to the most appropriate intervention approach and role of each intervenor, particularly those from outside the victim community. By reason of their autonomy from local structures and their access to significantly greater financial resources than the local community, outside intervenors many times have tended to establish independent programs, even bringing in outside personnel, without adequate consideration of local values and traditions in the care of unaccompanied children or the efforts and capabilities of the local community. These problems can be avoided by ensuring that from the onset intervention assists the affected community to care for and protect its own children, because children are best provided for within the local community in accordance with traditional approaches of child care. Emergency assistance should benefit and enhance, rather than replace or diminish, a community's ability to care for its own members, including its unaccompanied children.

ASSISTANCE: RECOMMENDATION 2

The first steps of assistance should include a situational assessment, and the identification, documentation, and immediate care and protection of unaccompanied children.

Assistance to individual children is usually a three-phase effort—immediate assistance, interim placement, then long-term placement. Interim and long-term placement are discussed in the following two chapters. Here we are dealing with the initial actions that must be taken in providing assistance to unaccompanied children to ensure that they receive adequate care and protection. First, in order to determine what assistance is required a situational assessment is necessary. Second, the children must be identified, for only by locating them can their individual circumstances be determined or assistance given. Third, any children found in need of immediate care and protection should be provided such without delay. Fourth, as a precondition to subsequent assistance, the circumstances of individual children should be assessed through documentation.

Situational Assessment

The first intervention step in any emergency is to determine to what extent unaccompanied children do exist and why. A determination of the scope and cause of the problem is necessary to plan a program response that will meet the needs of the children. Intervenors in the past have often acted in haste, solely on assumptions which have later been proven to be incorrect. Time and time again it was wrongly assumed that unaccompanied children did not exist or that special intervention was not required.

Thus basic services for unaccompanied children have often been unnecessarily late or have not been provided at all. The lack of an analysis of the problem has also resulted in inappropriate programs being instituted. Orphanages have been established or children adopted abroad, for example, only to learn too late that their families existed nearby. Along with helping to determine an appropriate course of action for currently unaccompanied children, an assessment of how and why these children have been displaced from their families can provide the information necessary to help prevent further separations. Initial intervention has long-term consequences for both the unaccompanied children and their separated family members. It is essential that policy decisions and program actions be planned and implemented on the basis of careful analysis rather than on conjecture.

The following categories of information about unaccompanied children are required in a situational assessment:

1. The number of unaccompanied children;
2. Information about the children and absent family members;
3. Current physical, emotional, and social needs of the children;
4. Causes of the child-family separations;
5. Local child welfare practices, laws, and customs which may bear on the children's situation;
6. Current actions being taken on behalf of these children;
7. Available local resources for taking care of unaccompanied children;

Identification of Unaccompanied Children

Following a situational assessment indicating that unaccompanied children may exist, an "active" search should be carried out to locate such children. It cannot be assumed, as has happened in the past, that the absence of highly visible unaccompanied children or lack of a direct request for assistance to them means that they are not present. They are likely to be overlooked unless an active search is undertaken, as a matter of course, to locate them. The search is to locate them; whether they are getting care and protection comes next.

Locating unaccompanied children during and after an emergency requires an active search of the many places in which they are likely to exist. A majority of unaccompanied children will probably be found in homes: with neighbors, family friends, or with other adults to whom they had no previous connection. Unaccompanied children also turn up at feeding stations, social services offices, medical facilities, or other emergency centers. Moreover, unaccompanied children will often be in orphanages, jails, and prisons, or living on the streets without any support other than that afforded by peers. Unaccompanied children have also been found in factories or other businesses where they were being exploited as cheap labor; or with armies as "mascots."

Since child-family separations have been known to occur months, if not years, after the initial crisis, the search for unaccompanied children should be conducted on an ongoing basis rather than as a single survey. During World War II, for example, newly unaccompanied children continued to be found for more than six years following the end of armed conflict. In fact, in most emergencies, family separations continued to occur for as long as the affected community or nation was plagued with social upheaval and poverty. Therefore, for an ongoing process of separation, it is usually necessary to establish an intercommunity referral and consultation system to ensure that unaccompanied children who are identified are referred to appropriate authorities. As a practical measure it is usually necessary to designate a person or service to which the children will be referred.

It is also important to note that the identification of unaccompanied children is not itself grounds for removing them from their present living arrangements. What should happen to any unaccompanied children after identification is discussed in subsequent sections.

Immediate Care and Protection

Once unaccompanied children are identified, the first responsibility of intervenors is to ensure that every child is protected from harm and exploitation and is receiving at least minimum life-sustaining amenities—food, shelter, clothing, and medical care. A search for unaccompanied children is likely to result in the identification of some children who are in need of such basic amenities and some who are not. Among those who may not require immediate emergency assistance are those who, although separated from parents or guardians, are being cared for by other adults, such as neighbors, family friends, or others. For those children who are without any adult care and in need of immediate assistance, intervenors must as a first responsibility ensure that aid to them is immediately provided. In the most extreme situations, the absence of immediate assistance can result in death or severe deprivation of children, which has repeatedly occurred in past emergencies.

The provision of immediately required life-sustaining food, clothing, shelter, and medical care, is best considered as a short-term emergency measure only. At the moment unaccompanied children are identified it is usually impossible to immediately arrange satisfactory interim or long-term care through which immediate emergency needs can be met. The arrangement of such care, discussed in the next chapters, requires the careful, individualized consideration of the needs and circumstances of each child and arrangement of whichever placement would most appropriately meet individual needs. Immediate care should not be provided by methods that lock the child into an arrangement which would preclude a considered interim or long-term placement of the child.

The method of providing immediate emergency care to children will be determined in large part by the circumstances and options that exist within the emergency situation. The basic aim is to put unaccompanied children in the care of an adult. Common options include families, existing community services, and child welfare services newly established specifically for this purpose. It is often possible to find families willing to care for unaccompanied children. If families cannot be found, unaccompanied children may be put in existing community services such as hospitals or temporary shelters. Again, it must be emphasized that these are temporary measures which should be limited to a matter of days. Subsequent care of unaccompanied children is to be guided by the recommendations of interim and long-term placement which follow in later sections.

The way in which immediate goods and services are provided is of critical importance because it often influences subsequent programs for these children. Indeed, a major problem arising in past emergencies has been that the indiscriminate gathering together of children in moments of necessity foreclosed subsequent placements based on individual evaluations of each child's needs. Orphanages, for example, are a particularly poor form of short-term care, and once established, often perpetuate themselves.

Documentation

Once an unaccompanied child is located and his immediate needs taken care of, he should be documented. Documentation is a compilation of all information relevant to both the child's immediate care and placement and to a search for missing family members. An assessment of the children's current situation and the original causes of child-family separations is necessary in order for planning a future course of action. Separated children should be documented as soon as possible. Experience suggests that there are certain basic categories of information needed for this documentation process:*

1. Personal data on the child;
2. Assessment of the child's current care and placement;
3. Medical and health status;
4. Psychosocial assessment;
5. Family history/tree and other information relevant to the determination of the child's legal status and for tracing purposes;
6. Children's own intentions and future plans.

As a practical matter, it is seldom possible to assemble all information about an unaccompanied child when he is first identified. Thus a "registration"—a partial documentation—of the child may be necessary to ensure that certain facts are known. At the very minimum, the child's name (including nicknames and alias) and current whereabouts should be recorded in the written language of both the child and the interviewer when there is a difference. If the child has no known name, or if others have renamed him, this should be clearly noted in his record. Otherwise, later tracing efforts may be seriously hampered. Photographs are especially important for family tracing, and consequently, a snapshot of each child should be taken during this initial interview and attached to that child's file. Any other pertinent information that a child may be able to offer verbally or possesses in the form of family letters, photographs, etc., which would be helpful in locating separated family members should be recorded, preferably without taking them away from the child. Finally, children, as well as their current caretakers, should be instructed to report any changes in status or residence to a designated person.

Specially designed forms and a coordinated information system are likely to be required. Much of the same information is required for legal protection, tracing, and placement. The information system should be integrated to minimize duplication and to avoid having to interview children repeatedly to obtain the same information,

*See Appendix A for more detailed questions within each of these basic categories.

which, of course, can be harmful to a child as well as a waste of time. An up-to-date registry of unaccompanied children is also essential to guarantee that the whereabouts of each child is known and that he is receiving proper attention.

Children themselves are a primary source of information for documentation. This has implications for both the interviewing process as well as the skills required of the interviewer. First, while nearly all children will be able to provide clues as to their identities and needs, the approach required to obtain this information needs to vary in accordance with each child's age, emotional state, and circumstance. Children below the age of five or six, for example, will often be able to offer descriptive accounts of their families that may prove useful. However, their grasp of factual data about their parents, brothers, sisters, extended family members, and previous place of residency is likely to be limited. Older children, on the other hand, are normally able to recall family details more precisely. Yet, here too, children's recollection of these details may be vague or unreliable, depending upon the length of separation. Many unaccompanied children, for instance, have reported that their parents are dead based on their emotional experience of separation rather than upon actual fact. Conversely, children and adolescents may intentionally give false or misleading information, especially if they have come from countries with repressive or authoritarian governments, suffered persecution in the past, or possess future aspirations such as resettlement to another nation that may be hampered by family tracing.

It is therefore essential that the important task of documenting unaccompanied children be entrusted to individuals who are experienced in interviewing children and knowledgeable about their developmental capacities, limitations, and needs. Furthermore, these individuals should, whenever possible, speak the same language and be of the same cultural background as the children.

Since the information provided by an unaccompanied child is likely to be incomplete, or at times, inaccurate, it is important to speak to other individuals as well. This necessitates interviewing current caretakers, adults from the child's community, religious leaders, teachers, accompanying siblings, and anyone else who may be able to shed light on the child's current situation or the location of separated family members.

Because documenting unaccompanied children always runs the risk of being traumatic for the particular child being questioned, as well as disruptive to the current placement, it must be carried out with a great deal of sensitivity. This process must not be permitted to degenerate into an insensitive bureaucratic counting exercise for its own end or to jeopardize the child's current relationship to a care-giver or a family. A survey, for example, that implicitly questions the right and desirability of children remaining with foster families may cause the family or caretaker to conceal the child's true identity. It can also, on the other hand, act as an incentive for abandonments and separations. Therefore, while the search for and documentation of unaccompanied children is imperative, it must be undertaken by individuals who are sensitive to children and sensitive to the various issues involved.

ASSISTANCE: RECOMMENDATION 3

Unaccompanied children should be provided basic goods and services equal to, and part of, those afforded other children within the affected community.

Equality of Basic Goods and Services

Because unaccompanied children are deprived of normal familial aid in securing basic goods and services during or after an emergency, without outside assistance they are likely to have less food, clothing, shelter, medical care, education, and other goods and services than other children. Intervenors should ensure that they are not so deprived. Several international instruments direct assistance for children without families, and notions of fairness and need dictate that unaccompanied children should not receive a lesser amount of material aid than other children in emergencies.

Just as unaccompanied children should not receive less, however, they should not be given more goods and services than other children. In some past emergencies, unaccompanied children have been singled out for exceptional gifts, a higher level of care or education, or different food and clothing. While such actions may be well-intended, they have had two serious consequences. First, these disparities have often isolated unaccompanied children from their community's living patterns, with negative consequences for their psychological and social well-being. Secondly, greater aid for unaccompanied children has encouraged additional child-family separations as parents and other care-givers have sent their children "unaccompanied" to orphanages, institutions, group programs, or other facilities in an effort to help their children obtain food, shelter, or opportunities they themselves could not provide. While these intentional separations were often meant to be temporary, many became permanent. Material or other assistance for unaccompanied children which is significantly greater in quantity or quality than that provided other children in the affected community may thus have a "magnet effect," drawing children away from their families. Equality of assistance helps to avoid this phenomenon.

Assistance as Part of That Afforded Other Children

As well as being equal to that afforded other children in the affected community, assistance for unaccompanied children should not be separate from that community. Thus, unaccompanied children should ordinarily attend the same schools, use the same clinics and hospitals, and enjoy the same play and recreational activities as all other children. In this way, alienation from the community and its attendant psychological and social support can be minimized. Furthermore, new and separate institutional arrangements for unaccompanied children are likely to deliver more or less assistance rather than an equal amount, or at least to be perceived as doing so by the community at large. Integration of assistance for unaccompanied children with that given to the community, valuable in itself, thus also serves to ensure equality of quantity and quality in the assistance provided.

No Removal of Unaccompanied Children Except Temporarily for Compelling Reasons of Health or Safety

The question of whether unaccompanied children should be temporarily evacuated from countries at war has generated considerable debate over the last forty years. The 1949 Fourth Geneva Convention recommended that the parties to an armed conflict "shall facilitate reception of [unaccompanied children under fifteen years of age] in a neutral country for the duration of the conflict." The evacuation of unaccompanied children proved to be problematic, however. Removal is often harmful psychologically because it uproots the child from familiar community members, values, and traditions, thus adding additional distress to the inherent trauma of family loss or separation. Experience has shown, too, that many evacuated children have never returned to their countries of origin. Evacuation to other countries has caused problems even for those children who were eventually reunited with their natural families, especially when new languages had replaced old ones. Recognition of these and other problems with evacuation led to a provision in the 1977 First Protocol to the Geneva Conventions restricting evacuations to only those necessary for compelling reasons of health, medical treatment or safety. Under this provision also, the evacuation must be temporary and consented to by the children's guardian or those who are primarily responsible for him.

In accordance with this law and the general recommendation that assistance for unaccompanied children be part of that given other children in the same community, unaccompanied children should not be removed or evacuated except for compelling reasons of health or safety. Thus, removal or evacuation, if proper for all children in the area, should be considered for unaccompanied children as well. If unaccompanied children were exposed to special danger because of the absence of their families—for example, abuse or exploitation of unaccompanied girls or military recruitment of unaccompanied boys—their temporary separation or removal might be warranted if there were no less drastic means to protect them. Where such exceptional circumstances are not present, however, unaccompanied children should remain in the affected community.

ASSISTANCE: RECOMMENDATION 4

Unaccompanied children should be individually evaluated and, when necessary, given culturally appropriate treatment.

From the onset of assistance, culturally appropriate diagnostic and counseling services are essential for unaccompanied children in emergencies. Acute emotional trauma, withdrawn and regressive behaviors, depression, somatic complications, and other age-related distress reactions are common among children who have been separated from or lost parents and other family members during emergencies. While the provision of age-appropriate adult care and stable substitute placements is the fundamental component of assisting these children, other culturally appropriate treatment services may also be required. This is especially true for unaccompanied children whose separation experiences have been compounded by other trauma or deprivation. Exposure to violence, persecution, or the witnessing of the death of a parent, for example, invariably results in psychological disturbances which will evolve into an even more complex syndrome of disorders if left untreated. For these reasons, evaluation of all unaccompanied children is recommended along with treatment for those who require it. Although these services have been sadly lacking in past emergencies they are not without precedent, and in fact have been of crucial importance for the relatively few unaccompanied children for whom they have been available.

Individual Evaluation

Each child's experience of separation and loss and the adversity of the emergency is unique. Family background, cultural experience, temperamental and personality characteristics and, most importantly, age at the time of separation or loss, will determine how a particular child is affected by, understands, and responds to the stress of family separation or loss. Similarly, experiences of adversity and deprivation due to the emergency can also differ enormously. Some children may be victims of violence themselves or have witnessed death (including that of parents or other family members), or have endured hunger, exploitation, or cruelty. Others may have gone through far less traumatic events before reaching safety. Since individual background and history and the way in which that past is experienced will vary from child to child, individual assessments are essential. The tendency to view unaccompanied children en masse and to offer blanket prescriptions for their care and placement must be avoided. Only through careful assessment can the true needs, capacities, and limitations of each child be ascertained and then be carefully matched to placement and treatment.

Evaluation and Treatment within the Child's Cultural Milieu

Evaluation and treatment of unaccompanied children in emergencies is best accomplished within their own cultural setting by people from or thoroughly familiar with that background. Local staff should be used whenever possible, with outside personnel providing advice and training when needed. Unaccompanied children should have access to traditional healing practices, religious exercise, and other expressive activities such as art, theatre, song, and dance within their cultural tradition. This suggestion stems from the importance of culture, community, and religious beliefs to an understanding of the child's experience and to the treatment of resulting trauma or disorders.

ASSISTANCE: RECOMMENDATION 5

Unaccompanied children should be provided physical and legal protection as their individual circumstances require.

From the time they are identified, unaccompanied children must be given physical protection as part of the assistance provided by intervenors. The intervenors' duty to physically protect unaccompanied children is part of their general obligation to care for children without families contained in national and international law. Physical protection means protection from physical harm and disease, forced labor, sexual abuse, military recruitment, and other forms of exploitation, cruelty, and harm. Such security is a precondition for other measures of assistance and care and must continue through all stages of the treatment of every unaccompanied child.

The second aspect of protection is legal protection, particularly establishment of legal responsibility for each unaccompanied child. The issue of legal responsibility for unaccompanied children in emergencies has been a frequent problem throughout the years. Often there has been no agency able and willing to assume legal responsibility, thus leaving the children to the vagaries of those who happen to take custody of them. Unaccompanied children have often been without even an advocate or representative to act to ensure minimal rights and benefits.

For these reasons, legal responsibility for the children should be clarified as soon as possible after their identification, preferably by designation of a guardian or other legal responsible party. Where for legal or practical reasons creation of a formal guardianship is impossible, we recommend the appointment of a representative to advise and advocate for the children as a group or as individuals. The guardian or representative could be an individual or agency. The guardian's overriding mandate should be to act in the best interests of each individual child as outlined elsewhere in these recommendations. Because the legal rules concerning guardianship and representation will often differ depending on the location of the child, the following guidelines address three of the most common such situations.

Unaccompanied Children in Their Countries of Origin

Children separated from their families in emergencies in their own countries of origin would ordinarily fall within that country's child protective law and administration. Virtually all societies provide by formal or customary law for some form of guardianship of children whose parents are dead or unavailable. In practice, responsibility for such children may be spontaneously assumed by other family members; individual adults may be appointed by a court; or a child welfare or protective agency may be similarly designated. The guardian would then substitute for the absent parents in exercising parental rights and duties, including control of the child's residence, education, and discipline, and responsibility for his care, protection, and representation.

The designation of a guardian ensures that a specific individual or agency will have both the legal right and the legal obligation to assist the child. The legal guardian can protect the child from exploitation by other adults, including interference by others who may wish to act on him. Guardianship provides the child with an adult or agency directly responsible for ensuring his clothing, housing, and education, representation, and for generally overseeing his upbringing. The child is thus not left to his own resources or the vicissitudes of purely charitable aid, which entails no legal obligation and may cease at any time.

Unaccompanied Children Outside Their Countries of Origin

For unaccompanied children outside their countries of origin, the country where the child is found has the initial obligation to give emergency care and protection and to clarify legal responsibility for the child. It may, under private international law, defer to guardians appointed in the child's country of origin, but the country where the child is located has the responsibility for interim supervision and for longer-term measures in the absence of recognition of a guardianship designated by the authorities of the country of origin.

Because unaccompanied children outside their countries of origin often face critical questions under refugee, asylum, and immigration law, as well as issues of entitlement to social assistance in general, an adult guardian may be essential for the assertion of the children's legal claims and the resolution of their legal status. For some displaced unaccompanied children, including those for whom the United Nations High Commissioner for Refugees exercises his protection function, there are legal and practical obstacles to creation of a formal guardianship. In these situations it is recommended that a representative be designated to act as an advocate for the children. This individual or agency would consult with and advise the children and, where necessary, assert claims for assistance, refugee, asylum, and immigration status. This arrangement would mitigate any potential conflict of interest which may arise between an organization seeking to aid and provide permanent solutions for a general population on the one hand, and the special needs of unaccompanied children on the other. The representative could also act to ensure compliance with the children's other civil and human rights under applicable national and international law. Separate representation, by providing an independent advocate for the needs of unaccompanied children, would also contribute to prompter resolution of their displaced status.

Unaccompanied Children Settled or Resettled Outside Their Countries of Origin

In recent years it has not been uncommon for unaccompanied children to move through several countries before acquiring a permanent residence. Those who arrange the movement of these children should ensure that a guardianship is maintained at all times, as the Draft Declaration on Placement and Adoption recommends. This can be facilitated by designating a guardian in the country of reception before a child is moved there. It is irresponsible to arrange the resettlement of a child without including guardianship among the preparations for his arrival. By the same token, children should not be resettled in a country of first asylum without the establishment of legal responsibility for their care.

23

Interim Placement

The term "interim placement" is used to refer to the living arrangements for unaccompanied children after the earliest days of an emergency, when they are provided initial care and protection, and before family reunion or other forms of long-term placement take place. Since these living arrangements will determine the kind and quality of care an unaccompanied child receives on a day-to-day basis, they will affect most aspects of that child's developmental well-being. Individual consideration of these placements is therefore essential.

In attempting to meet the best interests of the child, situational factors must also be considered. First, these placements should be culturally appropriate: emergencies are not occasions to introduce into an already vulnerable community placement alternatives which may alter its traditional way of caring for children. Indeed, intrusion into a community's basic way of organizing its social relationships is often contrary to the best interests of unaccompanied children as it can separate them from the essential support of their community. Secondly, while priorities can be established, the selection of interim placements must be tailored to the actual conditions of the particular emergency. What may be ideal in a stable community may not be possible in one disrupted by a war, natural disaster, or mass population movement. Thirdly, interim placements for unaccompanied children must complement and strengthen rather than replace or weaken the affected community's own efforts to care for its unaccompanied children. Assessment of the community's response to unaccompanied children, cooperation between participating relief agencies, and integration of interim placements into the larger relief effort are essential.

INTERIM PLACEMENT: RECOMMENDATION 1

Family care within the child's own community should be the first placement option considered.

Interim family care refers to the placement of unaccompanied children in nuclear or extended families that already exist. The structure and membership of these family placements will depend upon the customs, practices, and societal arrangement of the particular community and culture involved. However, in general terms, the placed

312

child receives care from and participates in the daily life of an intimate family group consisting of adults and sometimes other children, while receiving assistance and supervision from an outside authority or agency. The basic rights and responsibilities of both parties—the receiving family and the unaccompanied child—will be discussed below.

There are three basic reasons why family care is preferable to various forms of group care for most children who are unaccompanied during emergencies. To begin with, until adolescence is fully attained children are primarily dependent upon their families, and when separations do occur they will usually begin reaching out emotionally to those adults who provide for their bodily care as potential attachment figures. Here, the placement itself plays a central role in determining the strength and quality of these attachments. Obviously, no placement alternative is able to guarantee with absolute certainty that secure attachments will evolve between a child and an adult. Yet, it is known that attachments do evolve through the ongoing interaction between a child and an adult, and that they entail the reciprocal exchange of affection, warmth, and commitment. Normally, a family from the child's own community is more capable than a group home or an institution of providing the everyday security, intimacy, and stability necessary to foster this complex psychological and social exchange.

Furthermore, family placements accord more with the goals set forth in the Assistance Chapter than do other alternatives. As already noted, intentional family separations occur frequently when special programs and structures for unaccompanied children are established while the needs of families, especially vulnerable ones, go unmet. In contrast, the rigorous use of interim family care will usually support the affected community's own efforts to care for its unaccompanied children and can actually help prevent subsequent separations from occurring. It must also be noted that nuclear and extended families, not group homes or institutions, are the basic socializing agents in most societies; consequently, interim family placements are less likely to be culturally disruptive for both the unaccompanied child and the affected community.

Finally, family placements are better than group or institutional placements in meeting the long-term developmental needs of most unaccompanied children. This is critical, as in most emergencies it will be beyond anyone's ability to predict how long separations will last and how many unaccompanied children will eventually be reunited with natural parents or other extended family members. And even if family reunions do occur, it is far better that a child has been provided an opportunity to form close attachments with adults that will have to be broken than to subject that child to a lengthy period of emotional deprivation and, in the case of insitutions, social isolation as well.

Moving from Emergency Care to Family Placements

Emergency care arrangements, as described in the Assistance Section, may be necessary when there are large numbers of unaccompanied children and when conditions are too unstable to immediately implement an effective system of interim family placements. Nonetheless, relief agencies must respond to the unaccompanied child's need for age-appropriate adult care. In all cases, relatives and familiar adults already caring for parentless children within the affected community should be given first priority as interim caretakers. Often widows, single women, and married couples from within the affected community can be enlisted to provide one-to-one care for infants and young children without any support either in their own families or in temporary group facilities. Every effort should be taken to keep together siblings, other related children, and peers who have close relationships.

If the temporary grouping together of children is required agencies should still begin planning for interim family placements. Recruitment of qualified child care staff from within and, when these are unavailable, from outside the affected community should begin at once. Screening of children and potential interim family caretakers should be undertaken as soon as possible. As conditions stabilize and family placements are made, agencies can also begin wait-listing other qualified interim caretakers. In this way, needless and often harmful placements of infants and young children with short-term caretakers can be eliminated.

In a similar way, these temporary group settings can shift from use as care facilities to reception centers for older children and adolescents later identified as being "unaccompanied." Once this shift occurs, the children should not remain in these reception centers longer than is required to document them for tracing purposes and to undertake the kind of evaluations necessary in making reasonably informed decisions about interim placements. With respect to these evaluations, it is important to note that the outcome of interim family placements, good or poor, is often influenced by the atmosphere of the family in terms of it being consonant or dissonant with the child's personality, temperament, and cultural upbringing. Careful matching is therefore essential.

Family Care for Infants and Young Children

Children under the age of five are made especially vulnerable by family separation, loss, and desertion. These are the years when selective attachments are first forming and when children are just beginning to be able to maintain relationships during brief separations from primary care-givers. What is at stake developmentally is the infant's and young child's ability to establish meaningful relationships in the future as well as other aspects of emotional and mental growth, all of which are endangered by social isolation. The unaccompanied infant's or young child's capacity to form selective attachments has already been weakened by the original separation or loss and may have been damaged further by other adversities. A grave responsibiltiy thus rests upon those agencies and individuals who intervene on behalf of these youngest unaccompanied children to ensure that each child is placed within a caring family as soon as possible.

In all cultures and in all emergencies, continuity of adult relationships is crucial for infants and young children. So great are their physical needs, and so limited their own ability to meet them, that they will, within a matter of hours or days at most, begin forming psychological ties with those adults who provide for their bodily care. In this sense, infants and toddlers do not understand the difference between what adults term "temporary" and "long-term" caretakers. Rather, each adult who offers even brief care is a potential attachment figure, and every change in caretaker reduces the infant and young child's ability to form these essential attachments. Even if such a child was eventually returned to an absent parent or other members of their extended family, or placed permanently with another adult, his or her capacity for tenderness and empathy with others will have been seriously impaired by the previous lack of sustained adult care and affection. Therefore, the goal is to ensure that the first placement be a viable placement.

Family Care for School-Aged Children

The speed with which school-aged children need to be placed with local families depends upon the length of time that separation from natural parents will last and the child's understanding of his or her situation. For example, the lack of family care may not prove to be overly harmful for school-aged children who know their parents are alive and well and that they will be returned to them shortly. Children of these ages generally possess cognitive skills which allow them to understand that in such cases separation does not mean permanent loss of the relationship; older children are able to anticipate reunion with family while younger children are not. School-aged children are also able to evoke emotionally-charged memories of their parents which provide a measure of psychological continuity when they are absent.

Group or institutional care, however, is at best a temporary solution for children under about fifteen years of age. In most cultures, these children are still primarily dependent upon families for developmental accomplishments, and when they are likely to be separated from their own parents for more than a month or two, or permanently, they should be provided the opportunity to form close attachments with substitute families. In the absence of natural parents, interim family care can provide these children the security and consistency necessary to continue organizing and consolidating various aspects of their mental, emotional, social, and moral development in a coherent and adaptive manner.

Placement of Siblings and Other Related Children

Preservation of sibling and extended family attachments is a central aspect of protecting and promoting the best interests of unaccompanied children. As a general rule, when there are significant age differences among siblings the selection of an interim placement should be oriented towards meeting the developmental needs of the youngest member. It is likely to be far less harmful for an adolescent to be cared for in a family, for example, than for an infant to be cared for in a group home.

However, problems may arise in finding a family capable of providing for large numbers of brothers, sisters, or other children from an extended family. In such cases, the culture or community's own child care practices should guide alternative solutions. Caring for such children in an existing extended family, creating a new "family" by arranging for a married couple or a single woman to care for the children, placing siblings and other related children in neighboring families, or, in some instances, establishing a small community-based group home, are alternatives which may be more or less appropriate, depending upon the cultural norms of the affected community.

Rights and Responsibilities

In many nations interim family placement is referred to as "foster care" in which the basic rights and responsibilities of the receiving family, the child, and the child's absent parents are specified in legal terms. In other nations, societies, and communities, family care for parentless children may be based more on customary practices than on formal legal mandates. While respecting local customs, the basic rights and responsibilities of all parties must be clarified in advance of placements.

At base, the receiving family is obligated to protect the unaccompanied child from all forms of exploitation and cruelty as well as to provide him or her with the best physical, emotional, and social care possible under the circumstances. This means that the unaccompanied child must be treated as a full member of the family—not as a marginal figure or servant. Any practices which endanger the child's physical or psychological well-being should not be tolerated.

The law protects the unaccompanied child's right to maintain his or her identity, including name, nationality, language, religion, and cultural heritage. No changes in the unaccompanied child's status should be made without the child's consent and the approval of the appropriate authority as required. In the same way, absent parents retain certain rights to reclaim the unaccompanied child (*see* chapter 24).

Follow-up and Counseling Services

Once placements have been made, ongoing supervision is required to ensure that adequate care is provided. In addition to medical screenings, evaluations of the child's adjustment within the current placement must be undertaken on a regular basis. These evaluations should be multidimensional: regular assessments of the child's placement, including interviews with the child, his or her caretakers, teachers, and other involved community members, often yield important information about the child's progress.

Counseling services have proven to be especially important for unaccompanied children cared for in interim family placements. Even when the rights and responsibilities have been clarified, the differing emotional expectations of the child and the caretakers can lead to adjustment problems within the family. Counseling services have been identified as an essential support for both the child and other family members during this period of mutual adjustment.

Finally, unaccompanied children whose separation experiences have been compounded by other trauma, deprivation, or abuse, demand special consideration. Even when received with kindness and affection, such children do not always respond positively to their new adult caretakers. Some of these children may have endured so much and been neglected for so long, that they are unable to regard adults as potentially loving individuals. Seriously traumatized, abused or deprived children need to be brought gradually to accept the trust and security of adults through specialized services which may include: (1) family care with ongoing counseling, or (2) specially staffed and structured transitional homes prior to interim family placements.

INTERIM PLACEMENT: RECOMMENDATION 2

Group care for unaccompanied children should be considered if family placements are not feasible, and may be preferable for certain adolescents.

In group care a number of children live together under adult supervision and participate in the daily life of the community by going to school, holding jobs and making friends outside the group home. The quality of care in any given group home will depend, of course, on the extent of individualized adult care provided, continuity of adult caretakers, as well as by the child-adult caretaker ratio. However, in general, group care relies more heavily on peer relationships as the primary means of socialization than do family placements.

Thus, for infants and most children, group care is not as conducive to healthy emotional and social growth or to the development of an adult identity as is family care. For most unaccompanied children it should therefore be considered only in emergencies in which interim family placements are not feasible. However, in some cases, group care has been a viable or better placement option for children who are displaced from their families during their adolescent years. It must be noted that many unaccompanied adolescents have not adjusted to family placements and have often required alternative care arrangements, especially when cared for outside of their native communities. In such cases the potential advantages of family placements have clearly not been realized.

Why have certain adolescents made adequate adjustments to interim family placements while others have not? Past family and cultural experiences, exposure to sustained trauma and deprivation, temperamental and personality characteristics, and the "goodness of fit" between an adolescent and a receiving family have usually been important factors. In cross-cultural placements, differences in language, communication styles, and expectations also seem to affect outcomes. Moreover, developmentally, a number of adolescents have had a difficult time accepting alternative adults as parental figures, particularly when they have remained in contact with natural parents. Thus, it is better the adolescent's needs and desires, as well as the affected community's own traditional means of caring for unaccompanied children, determine the appropriateness of a family or a group placement, rather than the singular approach of a given relief agency.

Requirements for Group Placements

The kind and quality of care unaccompanied children receive within interim placements is often more important than the actual structure of the placement per se. Indeed, in some emergency settings it has been difficult to make clear distinctions between a large extended family caring for unaccompanied children, for instance, and a small group home which is staffed by a married couple and closely integrated into the community. In either case, the essential requirement is that the interim placement be able to respond flexibly to the developmental needs of the particular children involved.

As a general rule, child-staff ratios within each group home should not exceed eight children per one adult giving direct care: the object is to provide each child individualized adult care. Moreover, children in all cultures need sustained adult care if they are to progress emotionally, mentally, and socially in a healthy and adaptive manner. For most unaccompanied children, individualized and continuous adult care will usually be more important than child care experts who devise programs and then leave. Thus, continuity of adult caretakers and individualized care should be minimum requirements for any group care program.

Group programs should also be structured to accord with the unaccompanied children's need for continuity of personal relationships and cultural and community ties. Siblings, other related children, and close friends should be placed in the same home whenever possible. All the homes, in turn, must be carefully integrated into the community; unaccompanied children need to attend school, participate in cultural and religious activities, make friends, and when appropriate, hold jobs within the community itself. It is also beneficial when adult caretakers within group homes are from the children's own community or cultural background.

Group Care for Adolescents

It is further recommended that group homes for unaccompanied adolescents do not attempt to replicate a family setting per se; instead, the overall goal should be to assist adolescents to become increasingly independent and self-sufficient. While personal involvement, affection, and continuity of staff is indeed essential, adult caretakers should exercise their authority in a mentor capacity rather than as a parental figure. Independent behavior and decision-making skills should be stressed and the adolescent helped to anticipate the consequences of his or her own actions. Intensive and culturally appropriate educational and occupational programs must be a core component of group care for unaccompanied adolescents.

Independent Group Arrangements for Older Adolescents and Young Adults

As the term implies, "independent living arrangements" require that the unaccompanied child be quite self-reliant and capable of functioning with a minimum of adult supervision. Like more traditional forms of group care, independent living arrangements rely upon peer relationships as the primary mode of socialization. Unlike group care, however, adult supervision is provided by social workers, mentors, or neighboring families who often do not live full time with adolescents. Therefore, independent living arrangements should be limited to: (1) older adolescents who require only transitional care before assuming work-related responsibilities, and (2) young adults over eighteen years of age who have been included in the unaccompanied children's program because of inaccurate age assessment or because they were deemed vulnerable and in need of extra assistance. Again, it should be the adolescent's or young adult's needs, desires, and cultural practices that determine whether or not independent living is an appropriate placement option.

Even though candidates for independent living are older and more self-reliant, they still need adult supervision on an ongoing basis. Regular contact, material assistance, and the provision of educational and occupational opportunities are essential. Supervisors or mentors of independent living programs should be able to assist these older adolescents and young adults towards an acceptance of adult responsibilities.

Specific social, educational, and occupational goals need to be established at the onset of placement, and as these goals are reached, adolescents and young adults should be emancipated from the independent living program.

Finally, even when candidates for independent living are from the same community, area, or nation, they may differ from one another in terms of ethnic, linguistic, and religious backgrounds. Special attention thus needs to be paid to selecting adolescents and young adults who share common traditions. Siblings, other related individuals, and close friends should also be placed together whenever possible.

The Adolescent's Choice

In some cases, the needs of adolescents will not differ significantly from those of school-aged children, and the fostering of close adult attachments within families will be desirable. This may be especially true of adolescents reared in cultures in which parents continue serving as the principal decision makers throughout these years. In other cases, adolescents may already be well on their way towards establishing a separate identity, with physical, emotional, social, and moral self-reliance. Such an adolescent may not need nor be willing to tolerate another adult as a parental figure. "Parent-like" attachments entail that the child or adolescent become emotionally and socially dependent upon the adult. This dependency, crucial for infants and children, may actually prove to be harmful for certain adolescents as it can pull them backwards at precisely the point in developmental time they need to push forward towards increasingly independent attitudes and behavior. At this time, adolescents may desire close peer relationships and benefit from group placements in which the adults exercise their authority as mentors rather than as substitute parents.

Because of this developmental diversity, it is recommended that adolescents over the age of fourteen be allowed to choose between interim family or group care if both placements are available. Whether or not both placements should be available will depend upon conditions within the emergency situation as well as the affected community's attitude towards these placements. In a number of cultures, group care may not be viewed as an acceptable method of child care, even for older adolescents. Conversely, in some emergencies group care may not only be a practical way of responding to large numbers of unaccompanied adolescents but it can also help promote continuity of existing psychological, social, and cultural attachments. Again, the touchstone of what is appropriate will be the cultural appropriateness of group care for adolescents and the magnitude of the actual emergency.

INTERIM PLACEMENT: RECOMMENDATION 3

Institutional care should only be used when family and group placements are not available.

Institutional care refers to traditional institutions and orphanages. It may also include other arrangements such as children's villages. As with group homes, the internal dynamics of institutions do vary. However, the distinguishing feature of all institutional forms of care is that they are largely self-contained environments in which an attempt is made to meet most of the children's social, educational, and occupational experiences within the setting itself. Children in institutional forms of care remain separated from the surrounding community more than children in group care.

The advantages claimed for institutions are professional care, clear, unequivocal goals, organization and stability; in reality, however, institutions often fail to provide adequate physical care, intellectual stimulation, and emotional and social support. It must be noted that the concept of psychological deprivation was developed from studies of unaccompanied children in institutions during World War II. As a result of these studies, conditions in many institutions have improved. Still, there is no empirical evidence to suggest that institutions are as capable as family or community-based group care of meeting the developmental needs of children, while there is indeed a consistent body of literature which clearly points out that institutional care is harmful for infants and young children. Moreover, in times of crisis, the quality of care in institutions and orphanages is usually very poor. In the past, child-staff ratios have been exceedingly high, physical conditions impoverished, and adequate social care and intellectual stimulation lacking. Such conditions have resulted in high infant mortality and have jeopardized the physical, emotional, social, moral, and intellectual development of older children.

Somewhat better results are reported from children's villages—a second type of institutional care. In children's villages a relatively small number of children, usually not more than ten, are placed in a home with a maternal caretaker or a couple and together constitute a sort of large "family." Many of these "families" comprise the village. Some villages are homogeneous in that all chidlren come from similar cultural backgrounds. Others are heterogeneous with clusters of children from different regions or nations living in the same village. In both cases the children and their caretakers participate in the daily life of the village, which remains distinct from the neighboring community and host society. Thus, unlike age-group institutions and orphanages, children's villages utilize the socializing influence of a primary intimate group which under normal circumstances is the family. Yet, like other forms of institutional care, children's villages are largely self-contained environments within which efforts are made to meet the children's basic needs.

In one sense then, children's villages can provide unaccompanied children protection, age-appropriate adult care and a measure of culture continuity when locations and caretakers are carefully selected. Finding married couples willing to serve for periods of several years has proven difficult, however, and in recent years the trend has been towards increasing use of homes headed by single women. Yet, even more problematic from the perspective of the long-term developmental outcome, is that by remaining in the self-contained environment of the village, children are not able to form realistic relationships with the surrounding community, society, or nation. Unaccompanied children eventually grow up, and the available evidence indicates that they are not adequately prepared to assume personal, social, and work responsibilities once they leave the village.

Assisting Children in Institutions

Even when intervening agencies have not established new structures, children unaccompanied during emergencies, especially in urban areas, have often been found in existing institutions and orphanages. In such cases, responsible authorities and relief agencies must alert themselves to the greater vulnerability of these infants and children. Immediate attention should be given to these children's needs for sustained and age-appropriate adult and adequate environmental experiences. Financial assistance, child care expertise, and most importantly, recruitment of sufficient caretakers from the community should be provided.

History suggests that many unaccompanied children in institutions and orphanages could have been transferred to more viable community placements had relief agencies worked to create such opportunities. Reunion with parents and other extended families is possible only if institutionalized children are documented and included in tracing programs. Material and financial assistance to the receiving families may also be required to facilitate reunions. Community development projects and attaching financial aid to the child rather than to the institution itself is another means of promoting more viable placements for unaccompanied children. In this way, a community can be prepared to receive these children, and the child's financial support continues regardless of where he or she resides. Such actions need to occur even over the possible protests of orphanage personnel who may see the transfer of children as a threat to their livelihood.

24

Long-Term Placement

Children do not start off as "unaccompanied children" and they should not end up "unaccompanied." Permanent adult relationships and societal ties are essential for healthy psychosocial development. Actions taken on behalf of these children should therefore be aimed at ending their "unaccompanied" status by providing them with a permanent place in the world. Permanent adult relationships are secured through family reunion, other stable age-appropriate placements, or, in the case of older adolescents, emancipation as adults. Permanent societal ties are made available by the stable settlement of unaccompanied children in viable communities.

Reunion of the child with his parents is the first priority among long-term placements. Reunion with other family members is the second priority. Family reunion accords with the most fundamental psychological and emotional needs of both children and adults, and is consistent with the law and practice of virtually all societies. These principles require assistance for unaccompanied children in finding and communicating with their families through an organized tracing program. Other forms of aid may also be necessary to enable reunion to take place once contact is established.

Family reunion may not be possible or desirable for certain unaccompanied children, however. It can be contrary to the child's best interests in some cases; for example, where the child's primary attachment is now with another family, or where the child is likely to be abused or neglected after reunion. In other cases, family members may no longer be living or may be impossible to find or communicate with. For these children, other satisfactory long-term placements need to be arranged. The basic aim is to secure for all children a permanent, stable place within family, community, and society.

LONG-TERM PLACEMENT: RECOMMENDATION 1

Intervenors should assist unaccompanied children to find and communicate with their family members through tracing and other services.

Because family reunion is the first priority for long-term placement, it is essential that unaccompanied children be assisted in locating and communicating with their family members. Without such a process, family reunion is much less likely to occur. Furthermore, even if the family members cannot be found or freely communicated with, and even if reunion cannot or should not be effected once contact is made, tracing serves a useful purpose by clarifying the child's situation vis-à-vis his family. Information about the existence, location, and desires of the child's relatives will help other placement and guardianship arrangements to be made and contribute to the child's accurate assimilation of his situation and sense of identity.

"Tracing" is the term generally used for the process of locating and communicating with missing family members. Tracing can occur in two directions: searching for family members on behalf of the child or for the child on behalf of the parents. Tracing on behalf of unaccompanied children is different than tracing on behalf of adults. First, children are less able than adults to locate family members on their own and thus are dependent upon others to assist them. Second, while tracing between adults is primarily to establish contact, there is more at stake for children, since the results of this search often have a direct bearing on their placement. Third, because children cannot sustain emotional ties to absent family members to the same extent as adults, time is of the essence, and tracing efforts for them must proceed rapidly. Finally, the skills required for interviewing children and assessing the potential benefits of family reunion are clearly different than those required to establish contact between separated adults who then are able to make their own decisions about the reunion. A separate program or, at minimum, a child-oriented approach within a more general program, will therefore be required when organizing and implementing a tracing effort for unaccompanied children.

Many family reunions for unaccompanied children in past emergencies would not have occurred without tracing. After World War II some 5,800 of a larger group of 22,800 displaced children were reunited with relatives through an organized search program; after the Nigerian Civil War almost all of the 10,000 evacuated children were returned to their families as a result of organized tracing; in Ethiopia nearly one-half of the approximately 1,500 children unaccompanied during the 1973 drought and ensuing famine were helped to rejoin separated family members; and through tracing, more than 1,900 unaccompanied Cambodian children displaced in Thailand were returned to their families. Unfortunately, however, in a number of major emergencies tracing programs were either never established, were established too late to be of maximum benefit to the children and their families, or inadequate documentation and search techniques severely limited the results. Moreover, in some emergencies, political and military concerns have obstructed tracing efforts and blocked the return of children to their families. In such cases, opportunities to reunite unaccompanied children with separated family members have been lost.

The importance of family reunion and the steps necessary to facilitate it are well-recognized in international law and practice. The Fourth Geneva Convention of 1949 obligates parties to international conflicts "to facilitate in every possible way the reunion of families dispersed as a result of armed conflicts," to encourage the work of humanitarian organizations engaged in this task, and to ensure communication among separated family members. The United Nations High Commissioner for Refugees, in his work with displaced children, has emphasized the need for tracing and the importance of facilitating family reunion. The Draft Convention on the Rights of the Child contains "the right to maintain personal relations and direct contacts with both parents" for children separated from their parents, as well as a right to information about the whereabouts of parents in detention, imprisonment or exile, deported, or dead as a result of state action.

Which Unaccompanied Children to Include in a Tracing Program

In theory, a search for the families of all unaccompanied children may be required as it is only through this procedure that the situation of each child is confirmed. In any given emergency, for example, children found alone may be lost, stranded, orphaned, abandoned, or separated from their families in other ways as noted in the Intervention Section. The circumstance in which the child is found as well as the child's statement of the cause of his unaccompaniedness are not always conclusive of his status. This latter point was well illustrated in Thailand where, of the first 735 parents located through the tracing program for unaccompanied Cambodian children, nearly twenty percent had been reported by the children to be dead. In practice, however, the whereabouts of parents or other relatives or their intent regarding custody of their children may not be in question when, for instance, documented records exist or abandonment can be clearly established. In such cases, then, tracing may not be required. As a general rule, a search for the families of unaccompanied children should be undertaken whenever there is uncertainty about their whereabouts or their intentions regarding reunion.

There is one important exception to the rule that unaccompanied children should be included in an organized tracing program. This involves emergencies in which it has been established that tracing itself—the asking of questions and the circulation of information—would endanger the child or the family, as might occur in some war or refugee situations. In such cases, the potential benefits of tracing must be weighed against the risks that this procedure would impose upon children and their families.

Interview and Documentation

In-depth interviews must be conducted to obtain the basic documentation necessary for tracing. The kinds of information required have already been discussed in these Recommendations and are more fully set forth in Appendix A. This information is likely to come through interviews with at least five sources: persons who first identified the child as "unaccompanied"; the children themselves; current adult caretakers; accompanying siblings and extended family members; current or former neighbors; and teachers, hospital workers, and other community leaders or residents who are in contact with the child or who knew his family.

Search for Family Members

A thorough search for family members should be carried out. The foremost goal of this effort is to locate the natural parents or whoever else might be responsible for that child by law or custom, but this search should not exclude grandparents, aunts, uncles, and any other extended family members. In all cases, the precise boundaries of the search will be determined by how the given society organizes its familial relationships.

In the same way, the methods required for conducting this search will vary with circumstances. When the home location or address is known, personal visits or letters will be the most direct approach. When the address is not known, the community, or at times, a larger region, will have to be canvassed. Photographs have repeatedly been found to be one of the most important tracing tools. One of the simplest methods of documenting the children is to photograph each child individually holding a sign which includes his name (spelled exactly as it is recorded in his file), along with an assigned case number. Ideally, both a polaroid and a negative film should be taken to provide an immediate photo for the child's file and copies for tracing. Often, a "tracing flyer," which includes a recent photograph and pertinent personal and family data on an individual unaccompanied child can be posted in prominent places or circulated within a community or region. With groups of children, individual pictures can be compiled into large composite posters and displayed on bulletin boards. Search can also be effectively carried out through newspapers, radio, and television broadcasts. In most emergencies several or all of these techniques will be required simultaneously. Central registries or information bureaus which compare names of persons searched for with names and addresses of persons compiled in a file can complement, but not substitute for, the more decentralized and action-oriented search methods required for unaccompanied children.

Duration of Tracing

Active tracing for unaccompanied children should continue at least until one of four events occur: the child is reunited with his family; all reasonable efforts to locate family members have failed; the child is adopted; or the child reaches the age of majority. Under the principles proposed here an unaccompanied child must be provided age-appropriate adult care independent of tracing efforts, and neither these efforts nor the prospects of family reunion is a ground for denying that child an age-appropriate placement.

Communication with Located Family Members

When parents or other adult family members are located they should be contacted by person or in writing and told of the child's whereabouts and situation. They should have the opportunity to communicate with the child and to express freely, privately, and without duress, their desires regarding reunion with the child. This stage is critical to decisions about reunion and determinations of the child's long-term placement and legal status.

Verification of Requests for Reunion by Adult Family Members

When adult family members seek reunion with unaccompanied children, the identity of the claimant(s) and their relationship to the children must be authenticated. (Relatives who are minors and who are not living with adult relatives would themselves be unaccompanied children and their "reunion" with the subject child would be determined by suggestions elsewhere in these recommendations concerning siblings and peers.) The verification process is necessary to ensure that a child is not mistakenly given to a person who has no prior connection with him. In a number of emergencies, multiple claims for the same child have been received. This most often is the result of mistake, although in some cases fraudulent claims have been made.

When official documentation is not available, experience suggests four practical ways to verify a claimant's relationship to a child. First and foremost is the independent exchange of photographs between the child and the claimant which permits each to see whether the other party is who he claims to be. Second is whether the description of the child provided by the claimant matches the actual child. Third is knowledge of events or places that only the child and a related adult would know. Fourth, the comparison of information provided by the claimant about the child's family tree with that offered by the child and others can be helpful in determining the legitimacy of the relationship, although cultural differences in how family titles are used can limit the usefulness of this approach. Finally, an assessment must be made about the desirability of the reunion. Although this determination is not usually considered tracing, it is mentioned here because it is the next logical step in the whole process. In other words, documentation and tracing provide the facts necessary for making the decision about reunion. The next recommendation discusses the basis on which this determination should be made.

LONG-TERM PLACEMENT: RECOMMENDATION 2

Unaccompanied children should be reunited with parents and other family members unless reunion is contrary to their best interests.

Reunion with Parents

In national and international law, parents have the presumptive right to raise their own children. This includes the right to reclaim their children from those who hold them illegally and from those who have assumed custody of the children during a parent-child separation. Reunion of unaccompanied children with their parents implements these well-recognized parental rights, as well as the child's complementary right to be raised by his own parents. Depending on local law and custom, identical or similar rights may also extend to other close relatives. National and international law clearly establish reunion with parents as the first priority among permanent placements for unaccompanied children in emergencies.

Reunion with Other Family Members

In the words of the universal Declaration of Human Rights, "the family is the basic group unit of society." In many societies adult relatives other than the parents are intimately involved in child-rearing, and the child will have established close emotional and psychological ties with them, as for example, in extended families. Many, if not most, societies look to the child's relatives to assume responsibility for children whose natural parents are dead or unavailable. Given the widespread recognition in law and practice of the child's connection with other family members, other family members should be the first alternative if the biological parents cannot care for the child. Again, decisions on reunion with family members should be guided by the relevant local law, including customary law, and practice.

Is Reunion with Parents or Other Family Members Contrary to the Child's Best Interests?

The presumption in favor of reunion with first the natural parents, and second other adult relatives, is not absolute, however. When necessary, it must yield to protect the child's welfare. After a period of separation, the child's psychological ties to his parents (or other previous adult care-givers) can lapse and be replaced by new ties to other care-giving adults. The younger the child at the time of separation, and the longer the time with the new care givers, the greater is the likelihood that this will occur. If this happens—and whether and when it has will depend on the individual child's circumstances—it can be harmful to the child's overall developmental well-being to remove him from his new care-givers. At that point, the best interests standard may dictate that the child stay with his present care-givers and not rejoin his parents or other family members. Harsh as this may appear on occasion to the family members, such a possibility is an inevitable and necessary consequence of any approach which makes the child's interests paramount, as most legal systems currently do.

In the absence of such new, stronger attachments, the parents and extended family members' right of reunion should generally be absolute (except, of course when they are likely to abuse or neglect the child, as discussed below). Thus, family reunion can safely take place if the child has not been with new adult care-givers for a substantial period of time, and this period must be linked with the age of the child at time of placement. When new attachments have been established, the absolute right of family reunion must end at some point, however, and family reunion decisions be determined by the child's best interests.

When, then, does the presumption in favor of family reunion end, and the best interests standard begin to determine whether or not family reunion takes place? It is proposed that this should happen only when the child has been with the new individual adult caretakers (in contrast to group or institutional care) for a substantial period of time:

> a. for a child who was placed with his present care-givers at age four or less: one year with those care-givers;
> b. for a child who was placed with his present care-givers after age four: two years with those care-givers.

Under this recommendation, children who have not been with individual adult care-givers for the specified periods of time would automatically reunite with parents or other appropriate family members on the latters' request. If the child is not currently in the care of individual adult caretakers, reunion would ordinarily take place as a matter of course, unless there were other nonfamilial grounds to withhold reunion, such as the child's eligibility for refugee status, or exceptional circumstances such as likely abuse or neglect. On the other hand, if the child has been in individual care for the specified time periods or longer, the presumption in favor of reunion would yield to an evaluation of whether or not the proposed reunion was in the child's best interests.

The duration of these suggested time periods is primarily based on what is known about children and their differing capacities to sustain psychological ties to absent parents, recognizing that numerical time limits can be only an approximate expression of that knowledge. Given the social upheaval during and after many emergencies and the attendant difficulties in communication and travel, these time periods are weighted somewhat in favor of the parents' right to reclaim their children, but not so much as to cause undue harm to the children. Younger children, it has been found, are not usually able to retain earlier attachments after a year with new adult care-givers. In a majority of cases, a two-year period with individual adult care-givers would clearly exceed most school-age children's ability to hold on to psychological ties to their parents. Children who have been with their parents for three or more consecutive years before separation, however, may maintain attachments to parents for a longer time. In any event, the time periods proposed merely attempt to set a general point at which the presumption in favor of family reunion yields to an assessment of whether the proposed reunion meets the child's best interests.

This determination involves an evaluation of the child and his past and current experiences. A number of factors must be considered, including the amount of time he had spent with the claiming relative before their separation, the amount of time they have been separated, the time spent in the care of the substitute family, his adjustment to that family, the substitute family's willingness and ability to provide long-term care, and the child's own wishes—for those over the age of five. Of course, both the natural parents (or other previous care-givers) and the new care-givers must desire to care permanently for the child for such a choice even to arise. If one or the other does not want custody of the child, there is no issue to be decided; the child would normally join (or stay with) whichever adults were willing to raise him.

Determination of the best interests of children under five years old in family reunion contexts involves an evaluation of their overall adjustment. Are they happy, healthy, and emotionally sound in their new setting? Are their needs being met? Does the child appear to be flourishing or floundering? Most important here are the amount of time the child had been with his parents before separation, the amount of time he has been away from them, and the time he has been with the new care-givers. The appropriate authorities are responsible for deciding the reunion question solely on the basis of what is best for the child.

The Desires of Children over Five Years Old

Placements are most likely to be in the child's best interests if they are based on mutuality; that is, if they are desired by both the caretaking adults and the child. Children five years and older who are living with a substitute family which desires to continue caring for them should therefore have the opportunity to express their opinion regarding the proposed reunion; to say whether they want to remain with those caretakers or join their family members. This opinion should be given weight according to the child's age and maturity. For those over the age of fourteen, the child's choice should ordinarily be determinative, since most children of that age are as capable as adults in understanding and assessing choices about their own futures. Respect for the child's wishes, to the maximum extent possible in accordance with his age and maturity, increases the likelihood that his long-term placement will be the product of mutual choice: that the child will live with adults who desire to care for him and with whom he desires to live.

Likelihood of Abuse or Neglect

When there is good reason to believe that the parents or other family members will abuse or neglect the child, reunion can usually be considered as contrary to the child's best interests. Abandonment—desertion of the child with no intention of reunion—is an extreme form of neglect. In addition, if the child has previously been removed from abusive or neglectful parents, or if the child or another informed person has reported previous abuse or neglect, there should first be the kind of social work inquiry about the suitability of returning the child to the familial home that would ordinarily take place in cases of this sort. The child should rejoin his family members only if it is concluded that the previous misbehavior is not likely to recur.

Prior Surrender of Parental Rights

A parental request for reunion with a child can be complicated or barred by the parents' prior surrender of their parental rights. The law permits parents to consent to the termination of their parental rights, usually to permit adoption by others, but normally such a decision is valid only if it has been informed and voluntary. A surrender of parental rights is not valid if done under duress or without an understanding of its consequences. The conditions of confusion, danger, and deprivation common during and after emergencies increase the likelihood that surrenders of parental rights during these periods will be made under duress or without an understanding of the consequences. If parents subsequently challenge their previous surrender of parental rights on these grounds, their claim should be evaluated. Even where the original surrender of the child was not completely knowing or voluntary, however, reunion should not take place if it is clearly contrary to the child's best interests.

Practical Implications

The first implication of this recommendation is the responsibility of intervenors to determine the existence and location of the child's parents and to communicate with them about their desire for reunion. This task involves an investigation of the circumstances of separation, a search for the child's parents, informing them of the child's situation, and discerning their wishes regarding the child. These efforts are likely to produce the following factual situations:

    a. no parent or relative can be located or communicated with;
    b. a parent or relative can be contacted but does not desire reunion or does desire reunion but it is not practical;

    c. a parent or relative desires reunion, but reunion is not appropriate because the
child had been abandoned, surrendered, or is likely to be abused or neglected; or

    d. a parent or relative desires reunion and reunion is not clearly inappropriate.

For children in group (a) and group (c) other forms of permanent placement need to
be sought, while for those in group (a) efforts to locate the parents continue. For those
in group (b), with support services the parents may be in a position to reunite with
the child (see next Recommendation); in other cases they may never desire reunion
or it may never be possible. In the latter event, long-term placement must be arranged.
At the same time, continued contact between parents and child should be facilitated.
In all three cases, efforts to find and reunite the child with other family members may
continue to be necessary.

    For the children in group (d), those whose parents or relative desires reunion and
who have been with their present adult caretakers for less than the prescribed one or
two-year period would automatically rejoin their parents upon request. The wishes of
the new caretakers and of the child himself would ordinarily not be taken into
account; reunion would be a simple return to the previously existing relationship
between parent or relative and the child.

    When the child had been in the individualized care of other adults (as distinct
from group or institutional care) for the prescribed minimum time periods, however,
the question of whether the child should rejoin his parent or relative or stay in his
present placement would be evaluated under the best interests standard as discussed
above. In general, if the present care-givers wished to care permanently for the child,
and a child of sufficient maturity wanted to remain with them, he would do so. If the
caretakers did not want to assume permanent responsibility for the child he would
ordinarily rejoin his parent(s) or relative. For children under five years old, who are
unable to express a meaningful preference, a determination of their best interests
would be made by the responsible authorities.

    Because the approach described here is designed to ensure the child's best inter-
ests above all, the time periods with new care-givers operate independently of tracing
efforts. In other words, if the child has been with the new care-givers for the prescribed
one- or two-year period, the presumption in favor of family reunion would lapse
regardless of whether a reasonable effort had in fact been made to contact the parents.
Of course, timely and thorough tracing for family members is strongly recommended
(see Recommendation 1 above). But in order to protect the child's developmental
well-being, it is important that timing of reunion and other decisions not be inextricably tied to the efficacy of tracing.

LONG-TERM PLACEMENT: RECOMMENDATION 3

Special efforts should be taken to facilitate and support family reunion.

    In addition to the tracing services discussed in a previous recommendation, spe-
cial efforts may be necessary to make family reunion possible and to ensure the well-
being of the child during his reintegration into the family. These measures include the

exchange of information between child and family before reunion, diplomatic and legal efforts to permit reunion involving two or more countries, physical protection during actual transit, and counseling and other preparatory and follow-up social services.

Preparation for Reunion

Children who have been separated from their families during stressful or dangerous situations are likely to have retained or developed fears and concerns about the environment they will be returning to: "Is my family alive and well?" "Is it safe to return to them?" "Do my parents still love and want me?" "Indeed, since they 'lost' me once already, are then even capable of caring for and protecting me?" These and other fears, whether spoken or unspoken, imaginary or real, are likely to exist to varying degrees in children of all ages, causing a great deal of anxiety as the child awaits reunion with the family. Often, this anxiety can be lessened through correspondence with parents or other family members, which should begin as soon as initial claims have been verified. The exchange of letters, photographs, and tape-recorded messages, for example, can provide the child tangible "proof" of his family's existence and give the needed reassurance that he is still loved and wanted. However, when these fears persist and become barriers to reunion efforts, it is usually indicative of more serious problems within the child or between the child and the family members and counseling services are required. In all cases, the responsible authorities must provide unaccompanied children accurate and up-to-date information about their situations and the status of family reunion efforts.

On the other end, family members also need to be prepared for the child's return. Parents or other adult relatives should be provided information about the child's medical health and any emotional or behavioral problems encountered during the period of interim care. When, from the child's perspective of time, separation has been lengthy, family members should be told to expect an often difficult period of mutual adjustment after the reunion takes place. What for the child may begin as emotional distance or demanding and clinging behavior can evolve into more serious and long-lasting disturbances when family members respond to the child's struggles with anger, disappointment, or rejection rather than with understanding, patience, and reassurance. Experience has shown that counseling services both before and after family reunion are often essential. This appears to be especially true when the family's membership has changed through the death of a parent or the presence of a new marriage partner.

Political and Legal Obstacles to Reunion

In addition to personal and interpersonal obstacles, political circumstances can endanger or block family reunions. All too often in the past, military and governmental officials have not cooperated with tracing efforts, or, indeed, have refused to allow unaccompanied children to reunite with their families once they have been found. This has occurred most frequently in cross-border situations in which the host government, the receiving government, or in many cases, both governments, have placed national and political concerns above the rights and well-being of accompanied children and their families by preventing the entry or departure of the child or family members. States have the right to control who enters and resides in their territory, but immigration for purposes of family reunion should receive high priority. As a general matter, countries receiving accompanied children for resettlement should permit close relatives to join them there.

There is strong direction in international law that political and legal barriers be relaxed in order to permit family reunion for unaccompanied children. Protocol I to the Fourth Geneva Convention obligates States Parties to "facilitate in every possible way the reunion of families dispersed as a result of armed conflict." The United Nations High Commissioner for Refugees Executive Committee has called for the application of liberal entrance and exit policies to facilitate family reunion and the relaxation of national requirements for documentation of family relationships. In addition, the Draft Convention on the Rights of the Child would mandate that "applications by a child or his parents to enter or leave a State Party for the purpose of family reunification shall be dealt with by States Parties in a positive, humane, and expeditious manner."

Movement of Unaccompanied Children for Reunion

Reuniting unaccompanied children with their families involves careful planning and coordination between the sending and the receiving parties. The actual means of transportation must be determined in light of what will be the safest and least politically complicated route. If reunion efforts necessitate passage through situations of armed conflict, the children's physical safety must be the paramount concern. Moreover, if there is any doubt that the children will be allowed to remain with their families once reunions occur, or be subjected to persecution for having sought asylum in another country, an agency and personnel qualified to verify and monitor family reunions on an ongoing basis are required.

Postreunion Assistance

Finally, from the perspective of social welfare, simply returning unaccompanied children to their families does not guarantee that their developmental needs will be met. Since poverty and family vulnerability are primary causes of parent-child separations during emergencies, material and cash assistance as well as social services are often required to ensure adequate care for reunited children and to prevent subsequent separations or abandonments. One or more of the following factors indicates potential problems: the returned child is physically or mentally handicapped; the family has been weakened through the death of a parent or through a substantial decline in income or basic household amenities; the community itself is impoverished and undernutrition or malnutrition rates among children are already high. In such cases, follow-up services will be required before as well as after family reunions.

LONG-TERM PLACEMENT: RECOMMENDATION 4

When family reunion is impossible, placements should seek to maximize continuity of relationships and community ties.

Despite the best efforts of intervenors and the children themselves, family reunion will not always be possible. In some instances, there may be no known family members still alive; in others, it may be impossible to locate or communicate with them. Even when living relatives are found and contacted, they or the child may not desire reunion. In addition, for some children, there may be practical or political obstacles to reunion, or reunion may not be appropriate because the child has been abandoned or surrendered, or is likely to be mistreated.

For all of the children in these categories, long-term placements other than family reunion are necessary. Long-term placements are those which will provide continuity and stability consistent with the child's developmental needs until adulthood. They include foster care, adoption, group homes, and independent living, all of which are discussed in the section on Interim Placement. In accordance with the other recommendations on the best interests of the child, these long-term placements should seek to maximize continuity of the child's relationships and community ties, and to provide age-appropriate care.

Continuation of Beneficial Interim Placements

As noted elsewhere in these recommendations, while efforts to accomplish family reunion are in progress, unaccompanied children should be provided age-appropriate interim placements. Where these placements prove beneficial for the child they should be continued in the absence of family reunion. That is, unless there is some good reason to change interim placements, they should be used as permanent placements unless the child's needs dictate otherwise.

Making Existing Relationships Permanent through Adoption

When the likelihood of family reunion is nonexistent or minimal and the unaccompanied child has been in the continuous care of the same adults for a substantial period of time, and a permanent legal relationship with that family would serve his best interests, adoption should be possible. Adoption provides the legal means to give the child a permanent tie to individual adult care-givers. Although national laws differ somewhat in the effects they give adoption, in general, adoption creates the same legal bond between adult(s) and child as that between natural parents and their child—in other words, a full parent and child relationship. In this critical respect adoption differs from foster, group, and institutional care, all of which are legally temporary. Although any of these forms of care can, in fact, continue undisturbed for years, they retain at all times the possibility of termination or modification, giving neither the child nor the adult care-givers a legal expectation of the uninterrupted continuation of the relationship.

Adoption should be possible for children placed below the age of four after they have been with new individual care-givers for a year or more, and for children placed at age four or above after they have been with their new care-givers for two or more years when the need for permanency outweighs the likelihood of family reunion. Adoption should be possible, but need not necessarily take place. The most critical concern is that the adoption be in the child's best interests and the child's own views be taken into account consistent with his age and maturity. The recommendation here is simply that adoption be permissible for children established in new homes for a substantial period of time who would benefit from the additional security that adoption would provide. If necessary, such adoptions can and should dispense with parental consents under applicable national laws which allow adoption to take place when parents cannot be located or are incapable of giving their consent.

Continuity of Community Ties

Where long-term placements cannot provide continuity of relationship between the child and his existing care-givers, they should at least seek to ensure continuity of culture. The stresses of adjustment to a new substitute family or group are compounded when the placement also involves a change in culture. Long-term placements when family reunion is impossible should thus seek to maximize cultural continuity, preferably by using culturally similar caretakers in the child's native locale. Where the child is settled or resettled in a second or third country, members of his ethnic and cultural group are still the first priority as care-givers.

Continued Communication with Family Members

Some unaccompanied children will not be able to rejoin family members whose location is known. Those who have left home with parental consent, particularly, are likely to know the addresses of parents and other family members. In some cases, the parents or child may not desire reunion, while in others, it may not be possible for other reasons. Communication between the child and family members should, nevertheless, be facilitated where the child and his relatives desire to maintain contact.

Intercountry Adoption

Ordinarily, adoption is the legal process by which satisfactory ongoing family foster placements are made permanent. By contrast, intercountry adoption usually involves children and families who have had little or no contact before adoption takes place, therefore producing a much greater risk of serious incompatibility. Furthermore, it also often causes an abrupt change of culture for the child. For these reasons it is generally a very undesirable method of permanent placement for unaccompanied children.

As reviewed in Part II, children who are transculturally adopted face unique psychological and social risks. Initial adjustment problems, such as insomnia; night terrors and nightmares; frequent crying; clinging and fear of separation; and fear and rejection of native food, language, and customs, are all common. Some child care specialists have rightly pointed out that adoptive parents may not be adequately prepared to cope with what is often a long, difficult period of mutual adjustment. However, studies on transcultural adoption emphasize the transience of these problems and the capacity of children to adapt to the linguistic and cultural demands of a new home and community. While older children are likely to have more serious initial adjustment problems, reports indicate equally successful adaptation over time. In the process of identifying themselves as members of a new family and community, many children lose contact with past cultural roots. Struggles between past and current ties and over wanting to be similar and yet looking and feeling different from family and friends can sometimes lead to the emergence of psychological and social confusion in adolescence. How transculturally adopted children come to terms with the issue of identity is not known; there are no follow-up studies of adults who were adopted cross-culturally as children.

Before any intercountry adoption is considered for unaccompanied children, efforts need to be expanded to secure age-appropriate community placements for them. In fact, the necessity of international adoption reflects the failure of both the national and international community to respond to the best interests of unaccompanied children. Nonetheless, when such placements are not available, from the perspective of the individual child, cross-cultural adoption is likely to be less harmful than prolonged placement in a poor-quality institution. On balance, then, the possible benefits of intercountry adoption clearly outweigh the risks only for children whose alternative to intercountry adoption is long-term, low-quality institutional care.

Intercountry adoption for unaccompanied children in emergencies should thus be viewed only as a last resort. Efforts to find satisfactory interim and long-term placements within the affected community should take place before intercountry adoption is even considered. Emergency situations should not be regarded as a source of adoptable children by individuals from other countries. Rather, international efforts should be directed towards ensuring the best interests of children within their own communities. As indicated, only when these efforts have been tried and have failed should intercountry adoption be a possibility for unaccompanied children in low-quality institutional care.

In the rare instances when intercountry adoption is to occur, it would obviously be impossible to require that the children be with prospective adoptive parents any length of time before adoption takes place. Other safeguards of the childrens' interests suggested in recent international recommendations on intercountry adoption should be followed, however. These include the use of authorized and competent adoption agencies, assurance of the legal validity of the adoption in all countries involved, and guarantees that the child be able to immigrate to the adopters' country and obtain their nationality.

25

Unaccompanied Children Outside
Their Countries of Origin

In many recent emergencies, unaccompanied children have been found or displaced outside of their countries of origin (citizenship or habitual residence) to "second" countries. In addition, some children have been resettled from their homelands or second countries to yet other lands ("third" countries), or have been settled permanently in the second country. Unaccompanied children outside their countries of origin thus have composed a substantial subgroup of all unaccompanied children in emergencies.

In addition to the many problems of separation from families during emergency conditions present with all unaccompanied children, for those displaced and settled outside their homelands there are additional and different issues. First, there is the question of their need for special assistance in countries of asylum and resettlement. This includes both basic care and protection and aid in the resolution of their legal status in the second or third country. Second, for the intervenors responsible for these children and for potential receiving countries, there is the question of when settlement or resettlement outside the country of origin should be permitted or encouraged, as well as subsidiary questions of family reunion after such settlement or resettlement. Finally, if settlement or resettlement outside the child's country of origin does take place, there are questions of what methods of reception and placement are most likely to ensure the child's best interests in his new land.

The following recommendations address these issues, which concern the particular problems of unaccompanied children outside their countries of origin. These recommendations are meant to supplement those previously discussed that apply to all unaccompanied children. In other words, unaccompanied children outside their countries of origin should benefit from both the general recommendations and these that specially address their status outside their homeland.

UNACCOMPANIED CHILDREN OUTSIDE
THEIR COUNTRIES OF ORIGIN: RECOMMENDATION 1

Unaccompanied children in countries other than their own are entitled to care, protection, and representation regardless of their legal status.

For unaccompanied children in countries other than their own, the need for basic care and protection—food, shelter, clothing, physical and legal protection, medical care, education—does not change. The legal context in which these needs are to be met is different, however, since unaccompanied children in countries other than their own may face special questions of entitlement to such assistance. In addition, the legal right of some of these alien children to remain in the second country may be in question, necessitating the appointment of a representative or guardian.

Basic Care and Protection of Unaccompanied Children in Countries Other Than Their Own

Depending on their circumstances and the local immigration law, the status of unaccompanied children in countries other than their own may be that of legal temporary or permanent resident, asylum-seeker, or illegal alien. Unaccompanied children are not immune from the consequences of their immigration status simply because they are minors or are not accompanied by a parent or guardian. Regardless of their status under aliens or immigration law, however, unaccompanied children in countries other than their own should be given care and protection on a par with that provided the nationals of the country in which they are then located. As mentioned above, unaccompanied children are especially at risk; their need for care and protection is enormous and their ability to obtain it without adult aid is minimal.

In many countries, the general duty to protect and care for unaccompanied children in national law will extend to alien children as well as the country's own nationals. On the international level, direction for the provision of basic care and protection of alien children arises from the various statements on care and protection of all children without families reviewed in Part III. Furthermore, the Hague Convention on the Protection of Infants, the Fourth Geneva Convention, the Convention and Protocol Relating to the Status of Refugees, and the Draft Convention on the Rights of the Child all contain provisions about protection for children in countries other than their own. The Hague Convention empowers the authorities of the country where the child is found to take any necessary measures of protection in all cases of urgency and when the child is threatened by serious danger of his person or property. The Fourth Geneva Convention directs that alien children in international armed conflicts "shall benefit from any preferential treatment to the same extent as the nationals of the state concerned." For all unaccompanied children seeking refugee status or considered as refugees, the Draft Convention on the Rights of the Child would guarantee the same protection as is provided any other child permanently or temporarily deprived of his family environment. The Refugee Convention and Protocol guarantee equal public assistance to all refugees. Even in those instances in which these instruments do not apply as a legal matter, they provide support for the recommendation that all noncitizen unaccompanied children be given care and protection (both physical and legal) to the same extent as the "host" country's own unaccompanied children.

Representation for Unaccompanied Children Whose Legal Right to Remain Is in Question

For those children whose legal right to remain in the country where they are located is in question, there is a special need for representation in resolving their status under the local aliens, refugee and immigration law. It is well recognized that adults in countries other than their own are often in a particularly vulnerable situation and will experience serious difficulties in submitting their cases to local authorities. The difficulties are obviously even greater for children, who have less education, experience, and familiarity with legal procedures, and are even more likely than adults to be at disadvantage in an alien culture. For these reasons, the Draft Convention on the Rights of the Child includes special insistence that a child who is seeking refugee status or is considered a refugee, whether accompanied or unaccompanied, receives "appropriate protection and humanitarian assistance" in enjoyment of his rights. This point also is made in refugee legislation that is already in effect. If a guardian has been provided to unaccompanied children, as is recommended in the Assistance section, it should be the guardian's duty to advise and advocate for the child regarding his right to remain in the country. In the absence of a guardian willing and able to perform this role, the unaccompanied child should be provided a special representative who will do so.

UNACCOMPANIED CHILDREN OUTSIDE THEIR COUNTRIES OF ORIGIN: RECOMMENDATION 2

Settlement or resettlement of unaccompanied children in emergencies may be appropriate if (1) they are eligible under applicable asylum, refugee, or immigration law, or (2) they are in the care of a family unit which is being settled or resettled, or (3) adequate placements cannot otherwise be arranged.

This recommendation addresses the question of when it is appropriate to settle or resettle an unaccompanied child outside his country of origin, and applies both to those children already displaced and to those for whom removal from the country of origin is being considered. The first condition in which settlement or resettlement may be appropriate recognizes that the law already defines possibilities for settlement or resettlement for which unaccompanied children are eligible. Existing law permits or mandates settlement or resettlement outside the country of origin, for example, in refugee or family reunion cases. The other two conditions in the recommendation address the policy question of when other opportunities for settlement or resettlement should be created and used. Their basic thrust is to limit settlement or resettlement—that is, the raising of the child in a new country—to those cases in which this movement is needed to protect the best interests of the child concerned. Because settlement and resettlement usually involve displacement of the child to a new and different culture, disrupting and complicating the development process, this recommendation treats them as exceptions to the general rule of placing the child within his own community whenever possible.

Eligibility for Settlement or Resettlement under Applicable Asylum, Refugee, and Immigration Law

Unaccompanied children, like adults affected by emergencies, may have legal rights to the opportunity for settlement or resettlement under applicable immigration, refugee, and asylum laws. Perhaps most important of these in emergencies is refugee law, protecting from compelled repatriation those who face persecution, or, in the African definition, those displaced by external aggression, occupation, foreign domination, or events seriously disturbing public order.

In addition, resettlement opportunities for a displaced group as a whole or specially for unaccompanied children within a larger population are also often available through national immigration, asylum or group refugee law or policy. The consequences of resettlement can be profound, and unaccompanied children should not be denied resettlement simply by virtue of their unaccompanied status. Rather, they should be included in the general treatment of the group of displaced persons to which they belong. In other words, where resettlement opportunities already exist under applicable law and practice, unaccompanied children should be free to take advantage of them to the same extent as other members of the affected population.

Settlement or Resettlement as Part of a Family Unit

In many instances in the past, unaccompanied children have left their homelands with extended family members or, while en route or in an asylum camp, have attached themselves to nonrelated families. However, in camp these "informal" placements have often been ended: sometimes because the child, the family, or both, no longer wanted to remain together; more often, because of the pressure of resettlement from these camps, aggravated by immigration laws and resettlement policies that have favored an unaccompanied child over an accompanied child, or a smaller nuclear family over a larger extended family. In such cases the child has been stripped from his principal source of emotional and social support, and an important community resource for taking care of unaccompanied children has been wasted. Furthermore, in many cases, unaccompanied children in asylum situations have been denied the most appropriate interim placement—foster family care—because of the possibiltiy of their being split from that family by resettlement.

To avoid these difficulties and to promote the continuation of ongoing attachments, an unaccompanied child who is part of an extended or foster family unit which is to be settled in a second country or resettled in a third country should remain with that family except when this would obstruct an impending reunion with a natural parent or other significant adult relative whom the child wants to join. Only in cases where the family is unwilling or unable to continue caring for the child, or in the case of older children, when the child does not want to be resettled with the family, should the relationship be dissolved in favor of other options. Furthermore, ongoing relationships between unaccompanied children and extended or foster families would be recognized and supported throughout the entire resettlement process. The persons responsible for unaccompanied children in the asylum situation should provide the necessary documentation of the child's relationship with their adult care-givers and ensure they are resettled as a family unit. Agencies in the receiving nations must, in turn, recognize the importance of these less formal bonds by granting these families the same resettlement consideration afforded natural families. Once resettled, these adult care-givers should also be provided the same rights and financial assistance normally given a national who provides foster care to an unaccompanied child.

The exception to this procedure involves cases in which the resettlement of an unaccompanied child would obstruct an impending reunion with a natural parent or another adult relative. An impending reunion is one for which a claim has been made and which will occur within a reasonable time period. Conversely, this exception does not imply that an unaccompanied child who is still part of an active tracing program, yet for whom no claim has been made, should be removed from a viable family placement because of resettlement of that family. Rather, such a child should normally be resettled with that family, but tracing should continue until either a reunion or an adoption has occurred.

Resettlement to Ensure Adequate Placements

Under the proposed system of care, the principal objective is to ensure that each unaccompanied child receives age-appropriate adult care. In cross-border situations, as in-country emergencies, the principles established in the Interim Placement Section thus apply. This necessitates that all infants and young children, as well as most unaccompanied children under fourteen years of age, be placed in families, while group care and independent living arrangements may be appropriate for certain adolescents over that age. The accomplishment of this principal objective requires that all concerned parties, including agencies and individuals whose traditional concerns have focused solely on resettlement, undertake efforts to first, secure adequate placements for these children in the asylum situation, and when necessary, to then advocate for resettlement opportunities for the entire displaced population and not just its unaccompanied children. Only when age-appropriate adult care cannot be secured in the asylum locale should special resettlement status be granted to unaccompanied children. Resettlement of unaccompanied children from asylum situations to provide placements is thus the last resort after all means to create adequate placements within the local community have been exhausted. As noted above, this approach applies to intercountry adoption as well as resettlement.

The Child's Role in Settlement and Resettlement Decisions

The possibility of repatriation, local settlement, or resettlement is an important issue for most displaced unaccompanied children. By definition, these children, unlike most children, do not have a parent or guardian to make the decision for them. Though it is recommended that all unaccompanied children receive a guardian or representative as soon as possible, even this procedure does not eliminate the question of the child's role in decisions about his repatriation, settlement, and resettlement. For all displaced children these issues involve serious decisions about their future, and whether or not newly appointed guardians or other authorities also participate, the degree of the child's involvement in the decision remains an important question.

Nations vary considerably in the degree of importance they give a child's wishes in decisions which affect him. Their laws concern mainly issues of custody and adoption, however, which may not be comparable to issues such as repatriation, settlement, and resettlement. International legislation bears on the child's role in decisions in two ways: through general provision of participation rights to all people and through special mention of children's rights to participate in decisions about them. The Draft Convention on the Rights of the Child, for example, requires that the views of the child be heard in all matters and his wishes be "given weight in accordance with his age and maturity." Similarly, the Hague Convention on the Civil Aspects of International Child Abduction permits courts to follow the wishes of the child who "has attained an age and degree of maturity at which it is appropriate to take account of its views."

When is it appropriate for the child to decide on his repatriation, settlement, or resettlement to the same extent as similarly situated adults? Studies have shown that children over the age of fourteen do not differ from adults in their capacity to understand available options, make choices, or offer rational reasons for their decisions when they are given full and accurate information and are not subject to duress. UNHCR treats unaccompanied Southeast Asian children of fifteen and older as adults regarding the choice of repatriation, settlement, and resettlement. The Fourth Geneva Convention generally provides special protection only to children under the age of fifteen, treating older children as adults. In sum, since children age fourteen and older can generally understand, make decisions, and offer reasons for their decisions to the same extent as adults, respect for their dignity, freedom, and legal personhood requires that they be allowed to decide on their repatriation, settlement, or resettlement to the same extent as adults. Consistent with the other recommendations made here, the child should be able to consult with his guardian or representative before making this decision.

Children between the ages of nine and fourteen should also be permitted to make these decisions unless their choice is clearly contrary to their best interests. Research shows that children of this age can make rational choices, though they do differ from adults in their capacity to understand the choices and offer rational reasons for their decisions. This ability to make rational decisions, coupled with a preference for respecting their autonomy dictated by international law directs that they, too, be able to participate in decisions about repatriation, settlement, and resettlement. Rejection of choices that are clearly contrary to the child's own welfare by the appropriate authorities serves as a safeguard when needed. The UNHCR practice with unaccompanied Southeast Asian children is somewhat similar, treating those between ages ten and fifteen who are found to be of sufficient maturity as adults.

Participation by the child in decisions concerning repatriation, settlement, and resettlement presupposes several necessary conditions. First, of course, is the absence of duress or coercion, either express or implied. Second, is as full as possible description of the various choices and their consequences in a form that is comprehensible to the child. A third condition, as already mentioned, is a guardian or representative with whom the child can consult. As between the child and his guardian, the child's choices should be taken into account as recommended here, but this suggestion does not imply that the child should be compelled to make these decisions without the assistance of an adult who is serving his interests.

For unaccompanied children under the age of nine, the appropriate authorities must decide questions of repatriation, settlement, and resettlement according to the best interests standard. While such children who are capable of forming their own views should be provided an opportunity to express them, the child's wishes need not be dispositive. The ultimate choice must be made by the appropriate authorities considering all information including the child's desires that may bear on the child's best interests.

Family Reunion after Settlement or Resettlement Outside the Child's Country of Origin

Settlement and resettlement of unaccompanied children outside of their countries of origin can cause obstacles to later family reunions because of national restrictions on departure and entrance in addition to the ordinary logistical and financial difficulties. The child may, for example, be barred from return to his home country, or more commonly, limits on immigration may prevent his adult relatives from joining him in his new country of residence. These potential problems constitute an additional reason to limit second- and third-country settlement and resettlement to those instances where they are clearly necessary to protect the child's best interests, or where they are based on a preexisting legal right. For this reason, too, resettlement should not take place where family reunion is imminent, as recommended above.

Once unaccompanied children have been settled or resettled under the narrowly defined conditions recommended here, however, the receiving countries should be willing to accept the child's close family members (parents, siblings, grandparents, at the least) at a later date when they have been located and have expressed the wish to join the child in order to make family reunion possible. A similar approach was taken by the United Nations High Commissioner for Refugees with unaccompanied Cambodian children resettled from Thailand after 1979 by requiring countries resettling these children to agree in advance to accept their remaining family members. Such a condition puts receiving countries on notice at the time they agree to accept the child that close family members located later must also be allowed to immigrate. This ensures that beneficial family reunions will be able to take place and prevents resettlement from actually acting to perpetuate family separation. By the same token, the child's country of origin should be willing to permit the child to reenter to rejoin his family if he wishes to do so.

UNACCOMPANIED CHILDREN OUTSIDE THEIR COUNTRIES OF ORIGIN: RECOMMENDATION 3

Placements for unaccompanied children settled or resettled outside their countries of origin should ensure the children's best interests.

With resettlement, what changes is not the developmental needs of unaccompanied children but the environment in which these needs must be met. In other words, an unaccompanied child's need for preservation of existing relationships with significant people, continuity of cultural and community ties, and sustained and age-appro-

priate adult care remain paramount, irrespective of whether the child is cared for in his or her homeland or in a different country. To be sure, these needs are more difficult to meet when unaccompanied children are placed in a different sociocultural community. This is why resettlement should only be considered when the better alternative of placement in or repatriation to the native locale is not possible. When, however, resettlement does take place, the best interests principle should continue to determine placement decisions for unaccompanied children. Consequently, the interim placements previously discussed apply in second and third settlement and resettlement countries as well. However, in addition to these recommendations and guidelines, the following are also required.

Location of Settlement and Resettlement Placements

As already discussed in this section, special settlement or resettlement status or quotas for unaccompanied children are generally not recommended. Rather, unaccompanied children are subject to the same constraints and opportunities as are other members of the displaced population. When movements do occur, however, special attention to the needs of unaccompanied children is indeed necessary. This consideration must be oriented towards maximizing the unaccompanied child's opportunity to remain with members of their own cultural and linguistic group. Lacking families of their own, unaccompanied children are dependent upon the continuty of a familiar community to help maintain their language and native customs, and to assist them to find the needed balance between their past and future ways of life.

In practical terms, this requires that the appropriate authority, often the UNHCR, undertake efforts to ensure that unaccompanied children are settled or resettled with other members of the displaced population, giving preference to countries where members of their own cultural group already reside. Only in this way will agencies in receiving countries be able to involve culturally similar adults in all aspects of the children's care and placement. In contrast, unaccompanied children should not be sent to countries which are not offering settlement or resettlement opportunities to families or other adults from the same displaced population.

Providing Information About the Children

For children settled or resettled alone, information about parents, siblings, and extended family members is often crucial for continued tracing efforts as well as when the child becomes an adult, if reunions do not take place. This is especially true for younger children whose capacities for remembering family details are usually quite limited. For settlement and resettlement agencies, full assessments on each and every unaccompanied child they will receive are also essential. Such information allows these agencies to prepare for the reception of the unaccompanied children, and assists in the determination of which placements will most likely meet the best interests of individual children. In the past, the transfer of both kinds of information has seldom occurred, and may be rightly considered a major contributing factor to inadequate preparation, insufficient services, and placement failures in second and third countries.

Acculturation

In the past, settlement and resettlement nations have used different approaches to the care and placement of unaccompanied children. Specific placement alternatives for these children are often influenced by the receiving nation's general resettlement policies for refugees. These policies tend to reflect the host society's or, at minimum, the receiving agency's expectations of the refugees, and place very different adaptational demands on the newcomers themselves. Table 25-1 summarizes the links between a receiving nation's resettlement policy, the adaptational demands this policy places on the refugee, and the placement alternatives most often utilized for unaccompanied children.

At the one end of this spectrum are policies which assimilate the newcomers into the host society. These policies generally disperse refugees throughout the host nation under the assumption that they want and will benefit from becoming members of the majority culture as quickly as possible. Such policies demand that the refugee give up his or her old patterns of life and embrace a new identity as a member of the majority culture. The placement of unaccompanied children in majority foster families is an extension of this policy.

At the other end of this spectrum is the multicultural model. Under this approach, refugees are generally viewed as displaced persons who will return home as soon as possible. For unaccompanied children, this assumption oftens extends to fam-

Table 25–1. National Policy, Cultural Expectations, and Commonly Used Placements for Unaccompanied Children

Resettlement Policy and Expectation	Adaptational Demands on the Refugee	Placement for Unaccompanied Children
Assimilation		
(Policy) Disperse refugees throughout the host society	Giving up of old cultural patterns and embracing a new identity as a member of the majority culture	Majority Family Care
(Assumption) Refugees benefit from quickly becoming members of the majority culture		
Acculturation		
(Policy) Encourage clustering of refugees within the host society	Continuation of personal identity and external adaptation to the new society	Clustered Majority Family Care
(Assumption) Refugees need to maintain their own cultural heritage while learning to function in the host society		Culturally Similar Family Care
		Community-Based Group Care
Multicultural		
(Policy) Separate refugees from the host society	Conservation of native patterns and limited integration into the new society	Institutions
(Assumption) Refugees are displaced people who do not want to become members of the new society		Children's Villages

ily reunion as well. Integration into the host society is kept to a minimum in an effort to promote smoother transition back to native homelands and families. In the case of unaccompanied children, the use of institutions and children's villages accords most with this approach to resettlement.

Between the extremes of the assimilation and the multicultural approaches to resettlement is the acculturation model. By clustering refugees together, acculturation policies attempt to encourage the newcomers to maintain their own cultural heritage while learning to function within the host society. The individual is thus able to maintain a more constant personal identity as he or she adapts externally to new circumstances. Ethnically similar foster families, community-based group homes, and, to a lesser extent, majority foster families clustered together in the same community are placement options which fall within the assimilation model of resettlement.

The middle perspective, acculturation, most accords with the best interests standard and is the recommended approach to the care and placement of resettled unaccompanied children. Unlike the assimilation model and the use of majority family care, unaccompanied children are placed with adults and live among peers who speak the same language, share the same religious beliefs, and practice the same customs and traditions. This familiarity eases the distress that results from drastic cultural change and provides for the smoother continuation of personal identity. In contrast to the multicultural model, which relies upon institutional forms of care that can isolate unaccompanied children from the host society, the acculturation perspective acknowledges the historical fact that the majority of unaccompanied refugee children in the past have not returned to native homelands, and provides them the opportunity to become integrated members of the host society.

Reception Program

A reception program for unaccompanied children and a brief evaluation period that predates the actual resettlement placement has many advantages. It gives the child time to recover from the change of environment and to share initial impressions with peers. The child meets and gets to know the social worker and the staff who will be planning his or her resettlement. Initial information about the host culture, actual placements, and the legal and financial status of the unaccompanied child can be presented in the child's own language. Most importantly, it provides the resettlement agency the opportunity to make its own evaluations of the child's needs and desires and to select an appropriate family or group placement. Similarities and differences in expectations, developmental needs, and personal preferences cannot be fully considered when unaccompanied children are placed directly in families or group homes.

The more comfortable the child feels in the reception program the better use the child will make of his or her time there. The child is likely to feel most welcomed and cared for in a small program where staff members who speak the youngster's own language are readily available. Reception programs should foster regular contact with the host community; if the child remains isolated within a reception center he or she will gain little sense of the new country. If a family placement is selected for the child, he or she may also benefit from several visits with that family before moving into their home.

The reception program should be goal-oriented and time-limited. The purpose is to acquaint the child with the host society, determine the actual placement, and prepare the child for what follows. Conversely, reception care is not a time for major preparation, language training, or educational input. Agency preparation should occur well in advance of the children's arrival, and language and learning will be more readily accomplished after the children are placed in the community. The reception program should not extend beyond the time required to accomplish its main goals.

This guideline does not apply to infants and younger children for whom a reception program would be experienced as a disruption or a break in appropriate adult care. For such children, foregoing a period of reception care in favor of immediate family placements is likely to be a less harmful course of action.

Summary of Placements

Under the rubric of "unaccompanied children" are infants, young and school-aged children, as well as adolescents, who arrive in second and third countries with different needs, experiences, and cultural backgrounds. In some instances, resettled unaccompanied children and adolescents may still be in contact with their parents and retain hopes of rejoining them; in other cases, the parents may have died or been missing for years. It is thus unlikely that the diverse needs of these children will be adequately met when resettlement programs pursue a singular approach to their care and placement. While guidelines need to be established, it is recommended that all three of the placement alternatives that fall within the acculturation perspective—culturally similar families, group homes, and tightly clustered majority family placements—be available for unaccompanied children resettled in second and third countries.

1. Unaccompanied children should, whenever possible, be placed in culturally similar families.

2. Small group homes that are integrated into the host community and staffed by culturally similar adults should be considered when placements in culturally similar families are not feasible. Group homes may be especially suitable for large sibling groups and for certain adolescents who do not desire family placements. Again, it should be the developmental needs and personal preferences of the adolescent that determine whether a family or a group placement is more appropriate (see Interim Placement Section).

3. Majority family care for unaccompanied children under fourteen years of age should be considered when culturally similar family placements are not available. It is best when such placements are tightly clustered together in the same community so that the children involved have daily access to one another. It is also preferable when close friends and other members of their own cultural group also reside in the same community.

4. Semi-independent living arrangements may be suitable for certain older adolescents and for young adults who require only transitional care before accepting adulthood responsibilities (see Interim Placement Section).

5. Follow-up, counseling, and language training services are critical to the children's outcomes and should be considered a mandatory component of any resettlement program.

6. Family tracing and reunification, or conversely, adoption, should proceed according to the guidelines established in the Long-Term Placement Section.

APPENDIX

Basic Information Required
for the Documentation
of Unaccompanied Children

1. Basic personal data
   a. Name (also in original script if different from that which is being recorded)
   b. Other names
   c. Date of birth
   d. Place of birth
   e. Sex
   f. Nationality
   g. Tribe or ethnic origin
   h. Languages spoken
   i. Religion
   j. Education
2. Current address of the child
3. A recent photograph
4. Information about accompanying siblings
   a. Age
   b. Sex
   c. Relationship
   d. Current address
5. Information about present care and placement
   a. Present care-givers
   b. Length of time they have cared for the child
   c. How this association came about
6. Details of previous care arrangements
7. Family tree and related information (including father, mother, siblings, grandparents, aunts, uncles, and other relatives)
   a. Names
   b. Last address
   c. Date and place of birth
   d. Relationship to the child
   e. Profession

8. Information about the family/child separation
    a. Date of separation
    b. Place of separation
    c. Reasons/circumstances of the separation
    d. When the child last saw either parent or other family members
    e. If the death of mother or father is presumed, why the informant believes this to be true.
9. History of the child
    a. Important events in the child's life before and after separation
    b. Information about subsequent placements or places of residence and their importance in the child's life.
10. Reunion information
    a. With whom the child wishes to be reunited if they could be located
    b. Their relationship to the child
11. Psychosocial information
    a. An appraisal of current relationships that are meaningful to the child
    b. Whether age-specific developmental needs of the child are being met
    c. Information about the child's current emotional state
12. Other information relevant to determination of legal status
    a. Legal status of past placements
    b. Information relevant to the determination of refugee status, where applicable
        (1) Reasons for and circumstances of departure from native country or country of habitual residence
        (2) Does child fear persecution for reasons of race, religion, nationality, membership of a particular social group, or political opinion?
        (3) If so, on which ground(s) and why
        (4) Specific past events forming the basis of such fear
        (5) Future expectations if returned to country in which persecution is feared
    c. Information concerning settlement, resettlement, where applicable
        (1) Does child wish to repatriate?
            i. Only for reunion with family
            ii. Yes, even without family reunion
            iii. Not under any circumstances
        (2) Does child have relative or other sponsors in second or third country? If so,
            i. Name
            ii. Relation
            iii. Address
            iv. Last contact
            v. Does relative or sponsor wish child to join him?
        (3) Does child wish to settle in country of asylum?
            i. If so, why?
            ii. Where?
            iii. With whom?
        (4) Does child wish to resettle in another country?
            i. Have resettlement possibilities been explained?
            ii. Where does he wish to resettle?
            iii. Why?
13. Current medical status and past medical record
14. Other information relevant to tracing
    a. Identifying features, birthmarks, or scars
    b. Description of people or places remembered by the child

   c. Names and addresses of other persons who may provide additional information about the child

   d. Any other information that might be helpful in locating family members or understanding more fully the circumstances of the parent/child separation

15. Other information of importance to the daily care of the child

16. The child's own intentions, wishes, and future plans

Notes and References

PART I

1. Dorothy Macardle, *Children of Europe* (London: Victor Gollancz, 1949), pp. 80 and 156. At the end of the war the number of children without homes or parental care was estimated to be 60,000 in Holland and 200,000 in Poland.

2. Roman Hrabar, Zofia Tokarz, and Jacek E. Wilczur, *The Fate of Polish Children During the Last War* (Warsaw: Interpress, 1981), quoting UNESCO, p. 202.

CHAPTER 1

Notes

1. Dorothy Legarreta, *The Guernica Generation* (Reno: University of Nevada Press, 1984), p. 25.

2. Max Huber, "La Croix Rouge au Secours de L'Espagne," 1 *Bulletin International des Sociétés de la Croix Rouge*, 67, no. 410 (1936): 860–861.

3. Patrick Murphy Malin, Report to the Committee on Spain and to the American Friends Service Committee, IUCW Archives, p. 19.

4. Legarreta, op. cit., pp. 1–11.

5. Memorandum regarding the proposed relief work of the International Commission in Republican Spain, n.d.

6. Ibid., p. 19; Malin, op. cit., p. 19.

7. Domingo Ricart, Report on Refugees in North Catalonia, Barcelona, 23 October 1937.

9. Michael Hansson and Howard E. Kershner, International Commission for the Assistance of Child Refugees, News Bulletin, Paris, 5 April 1929.

10. International Commission for the Assistance of Child Refugees, Paris, April 1940, p. 8–9.

11. Ibid., p. 8.

12. Ibid., p. 7.

13. Ibid., p. 10.

14. *The Guernica Generation* by Dorothy Legarreta is a comprehensive and lucid study of the evacuation experiences of these children.

Selected References

Comité d'Accueil aux Enfants d'Espagne. *L'Accueil aux Enfants d'Espagne.* Versaille: La Gutenberg, 1937.

Comité International de Coordination et d'Information pour l'Aide de l'Espagne Républicane. Historique de la Guerre d'Espagne. Paris, 1 March 1938.

Cooper, Kanty. *The Uprooted—Agony and Triumph Among the Debris of War.* London, New York: Quartet Books, 1979.

International Commission for the Assistance of Child Refugees. History. Paris: ICAR, 1940.

Legarreta, Dorothy. *The Guernica Generation: Basque Refugee Children of the Spanish Civil War.* Reno, Nevada: University of Nevada Press, 1984.

Murphy Malin, Patrick. Report to the Committee on Spain and to the American Friends Service Committee. October, 1937.

Office International pour l'Enfance. L'Aide aux Enfants Espagnols Refugiés en France. Paris: Office International pour l'Enfance, 1938.

Republique Espagnole, Ministèr de l'Instruction Publique. Colonies d'Enfants. Paris: Conseil National pour les Enfants Evacue's, 1937.

West, Dan. Needy Spain. Philadelphia: American Friends Service Committee, 1938.

CHAPTER 2

Notes

1. The Facts Speak for Youth Aliyah, p. 5.

2. Society of Friends, Report of the Children's Department of the Friends Centre, Vienna, November 1938–September 1939, p. 1.

3. Ibid., op. cit., p. 2–4.

4. Letter From Castendyck to Lenroot, 1945.

5. Nettie Sutro, "Refugee Children in Transit," *International Child Welfare Review* 5, mo. 2–3 (1951), p. 59.

6. Union International de Secours aux Enfants, Report from the Delegation of the S.C.I.U. at Rome, Rap.Del. W/3/1245, December 1945, p. 3.

7. Dorothy Macardle, *Children of Europe* (London: Victor Gollancz, 1949), p. 13.

8. Denise Grunewald, Etude sur le Probleme des Enfants Deportes en Allemagne, Rastatt, 15 Mai 1948, p. 1.

9. Roman Hrabar, Sofia Tokarz, and Jacek E. Wilcaur, *The Fate of the Polish Children During the Last War* (Warsaw: Interpress, 1981), p. 145.

10. Macardle, op. cit., p. 34.

11. "The Lebensborn," Ibid.

12. Malcolm J. Proudfoot, *European Refugees 1939–1952* (London: Faber and Faber, 1957), p. 268.

13. Macardle, op. cit., p. 292.

14. American Council of Voluntary Agencies for Foreign Service, 35 Years of Cooperation—A History of the American Council of Voluntary Agencies for Foreign Service, n.d., p. 3.

15. Peter Alister Smith, p. 78.

16. Proudfoot, op. cit., p. 278.

17. U.N.R.R.A., European Regional Office, Report on Unaccompanied Children, March 1947, p. 2.

18. U.N.R.R.A., Central Headquarters for Germany, Minutes of Inter-zonal conference on Child Search and Repatriation, n.d., p. 2–3.

19. Enfants Non-Accompagnes, p. 6.

20. Ibid., p. 23.

21. Louise Pinsky, "The Children," in *Flight and Resettlement,* by H. B. H. Murphy (Paris: UNESCO, 1955), p. 49.

22. U.N.R.R.A., European Regional Office, op. cit., Table 1.

23. Proudfoot, op. cit., p. 268.

24. Macardle, op. cit., p. 198.

25. John C. Caldwell, *Children of Calamity* (New York: The John Day Company, n.d.), p. 143.

26. Pertti Olavi Kavén, "Jatkosodan Aikaiset Lastensiirrot Ruotsiin," Diss. Helsingin Yliopisto., 1981, p. 13.

27. Macardle, op. cit., p. 219.

28. Kavén, op. cit., p. 94.

29. Ibid., p. 69.

30. Ibid., p. 86.

31. Birgitta Nylund, "Unaccompanied Minors: Their Legal Status in Sweden, Norway, Denmark," unpublished, August 1983, p. 32.

32. Ibid.

33. Pertti Kavén, "Evacuation of Finnish Children to Sweden During World War II," Presented at Children and War Symposium, Siuntio Baths, Finland 24 March to 27 March 1983, p. 2.

34. Kavén, op. cit., p. 70.

35. Richard Padley and Margaret Cole, *"Evacuation Survey"* (London: George Routhledge & Sons, 1940), p. 42.

36. Henry S. Maas, "The Young Adult Adjustment of Twenty Wartime Residential Nursery Children," *Child Welfare* (February 1963): 59.

37. Susan Isaacs, Sibyl Clement Brown, and Robert H. Thouless, eds., *"The Cambridge Evacuation Survey"* (London: Methuen & Co., n.d.), p. 1.

38. Ibid.

39. Padley and Cole, op. cit., p. 43.

40. Ibid., p. 12.

41. Gillian Wagner, *"Children of the Empire"* (London: Weidenfeld and Nicholson, 1982), p. 248.

42. Ibid.

43. Ibid., p. 250.

44. Ibid., p. 254.

Selected References

American Council of Voluntary Agencies for Foreign Service, Inc. Papers on Voluntary Resettlement Agency with Regard to Adoption of Children, n.d.

Baker, Ron. "The Refugee Experience—Communication and Stress—Past Impressions of a Survivor." Paper delivered at the International Association of Schools of Social Work Research Seminar, University of Sussex. Brighton, August 1982.

Barnett House Study Group. *London Children in War-Time Oxford.* London: Oxford University Press, 1947.

Barocas, Harvey A. Children of Purgatory: Reflections on the Concentration Camp Survival Syndrome. New York, Postgraduate Center for Mental Health, n.d.

Beckh, H. G. "Les Regroupements des Familles en Europe à l'Époque de la Deuxième Guerre Mondiale." *Revue Internationale De La Croix Rouge* no. 714 (1979): 171–184.

———. "Les Regroupements des Familles en Europe à l'Époque de la Deuxième Guerre Mondiale." *Revue Internationale De La Croix Rouge* no. 723 (1980): 115–129.

————. "Les Regroupements des Familles en Europe à l'Époque de la Deuxième Guerre Mondiale." *Revue Internationale De La Croix Rouge* no. 734 (1982): 71–87.

Bentwich, Norman. *They Found Refuge: An Account of British Jewry's Work for Victims of Nazi Oppression.* London: Cresset Press, 1956.

Bernert, Elisabeth, and Fred C. Ikle. "Evacuation and the Cohesion of Urban Groups." *The American Journal of Sociology* 58 (September 1952): 133–138.

Blackey, Eileen. Report on Unaccompanied United Nations Children in Germany. UNRRA internal report, 24 June 1946.

Bodman, Frank. "Child Psychiatry in War-Time Britain." *Journal of Educational Psychology* 35 (1944): 293–301.

Bradbury, Dorothy E. and Katherine B. Oettinger. Five Decades of Action for Children: A History of the Children's Bureau. Washington, D.C., U.S. Department of Health, Education, and Welfare, 1962.

Brauner, Alfred. *Ces Enfants ont Vécu la Guerre.* Paris: Les Editions Sociales Francaises, n.d.

Brody, Sylvia. "The Son of a Refugee." *The Psychological Study of the Child* (1973): 169–191.

Bross, Thérèse. War-Handicapped Children: Report on the European Situation. Paris, UNESCO, 1950.

Caldwell, John C. *Children of Calamity.* New York: John Day, n.d.

Children and Youth Aliyah. Freedom and Work for Jewish Youth: Report for the Third World Youth Aliyah Conference. London: Children and Youth Aliyah, 1939.

Children and Youth Aliyah Committee for Great Britain. 25 Years of Youth Aliyah. London: The Children and Youth Aliyah Committee for Great Britain, 1959.

Children's Bureau. Memorandum. Admission of Refugee Children to the United States. 7 July 1945.

Church World Service. Children in Transit: Final Report of the Children's Division, Immigration Service of the Central Department of Church World Service. New York, 30 June 1953.

Close, Kathryn. *Transplanted Children: A History.* New York: United States Committee for the Care of European Children, Inc., 1953.

Collected Papers on the Admission of European Children to the USA after World War II.

Daumas, Maurice. "Deux Expériences de Centres Éducatifs pour Enfants. Sommaire: Les Répercussions de la Guerre sur les Enfants Français." *Revue de la Fondation Française pour Étude des Problèmes Humains* 4 (Novembre 1945): 75–81.

Davie, Maurice R. *Refugees in America: Report of the Committee for the Study of Recent Immigration from Europe.* Westport, Conn.: Greenwood Press, 1974.

Displaced Persons Commission. Sponsorship of Unaccompanied Children, Orphan and Non-Orphan. Washington, D.C., n.d.

Europaeisches Kinderelend im Spaetherbst. Genf: Vereinigtes Hilfswerk vom Internationalen Roten Kreuz, Dezember 1945.

Federal Security Agency, Social Security Administration, Children's Bureau. Problems in the International Placement of Children. Washington, D.C., July 1948.

Frankenstein, Carl, ed. *Between Past and Future: Essays and Studies on Aspects of Immigrant Absorption in Israel.* Jerusalem: The Henrietta Szold Foundation for Child and Youth Welfare, 1953.

Freud, Anna, and Dorothy Burlingham. *The Writings of Anna Freud. Infants Without Families: Reports on Hampstead Nurseries 1939–1945.* New York: International University Press, 1973.

Friedlander, Walter, and Dewey Myers. *Child Welfare in Germany Before and After Nazism.* Social Service Monographs. Chicago: University of Chicago Press, n.d.

Gershon, Karen. *We Came as Children.* London: Victor Gollancz, 1966.

Grunewald, Denise. Étude sur le Problème des Enfants Déportés en Allemagne. Rastatt, May 1948.

————. Report of International Social Service Work in the French Zone of Occupation of Germany. New York, January 1948.

Hannah, Charles. *A Boy in Your Situation.* London: Andre Deutch, 1977.

Hastings-Hungerford, Joan. "Post-War Search for Missing Children Nearing Its End." *International Child Welfare Review* 5, no. 2–3 (1951): 52–58.

Hicklin, Margot. *War-Damaged Children: Some Aspects of Recovery.* Acton, England: The Association of Psychiatric Social Workers, 1946.

Ilan, Eliezer. "The Treatment of a Refugee Child in a Home for Disturbed Children and a Follow-Up Thirty Years Later." n.d.

Inter-Allied Psychological Study Group. *Psychological Problems of Displaced Persons.* London, UNRRA, 1945.

International Refugee Organization. The International Refugee Organization Provisional Order No. 86. Resettlement of Children. Division of Responsibility Between Child Welfare and Resettlement Division. Geneva, 29 September 1948.

————. Enfants Non-Accompagnés. Geneva, 1951.

International Union for Child Welfare. "The Fate of the Young Refugees Remaining in a Few European Countries." *International Child Welfare Review* 5, no. 2–3 (1951): 94–98.

————. How Best to Promote the Psychological, Educational, and Social Adjustment of Refugee and Displaced Children in Europe. Geneva, UNESCO, May 1952.

Isaacs, Susan, ed. *The Cambridge Evacuation Survey: A War-Time Study in Social Welfare and Education.* London: Methuen, 1941.

Jewish Agency for Palestine. Five Years of Youth Immigration into Palestine, 1934–1939. Jerusalem, Central Bureau for the Settlement of German Jews in Palestine, 1939.

————. Let Facts Speak for Youth Aliyah. Jerusalem: Child and Youth Immigration Bureau, 1946.

Joffo, Joseph. *A Bag of Marbles: The True Story of Two Small Boys on the Run from the Gestapo.* London: Corgi Books, Transworld Publishers, 1976.

Kee, Robert. *Refugee World.* London: Oxford University Press, 1961.

Levy, Edna. Case Work in Reception Centers of Refugee Children. New York, Child Welfare League of America, September 1945.

Macardle, Dorothy. *Children of Europe.* London: Victor Gollancz, 1949.

Markowski, Frank. International Social Services Intercountry Case Work Services in Cases of Unaccompanied Children. Geneva, International Social Services, n.d.

McCahon, William H. Registration for Displaced Persons Orphan Program.

McNeil, Margaret. *By the Rivers of Babylon: A Story Based Upon Actual Experiences of Relief Work Among the Displaced Persons of Europe.* London: Bannisdale Press, 1950.

McTigue, James J. Memo to Harry N. Rosenfield. Interpretation of 'Abandonment' and 'Separation' as used in section 2(e) and 2(f) of the Displaced Persons Act, As Amended. 30 July 1951.

Ministry of Health. *Hostels for Difficult Children: A Survey of Experience Under the Evacuation Scheme.* London: His Majesty's Stationary Office, 1944.

Moskowitz, Sarah. "Roots from Scraps." Paper delivered at the Conference on Genocide, Tel Aviv, Israel, 21 June 1982.

Murphy, H. B. M. *Flight and Resettlement.* Paris: UNESCO, 1955.

Nathan-Chapotot, Roger. *La Qualification Internationale des Réfugiés et Personnes Déplacées dans le Cadre de L'ONU.* Paris: A. Perdone, 1949.

Ockenden Experience. Ockenden's Experience (1952–1980) of Care and Rehabilitation and Integration of Unaccompanied Refugee Minors. n.d.

O'Connor, Edward M. Procedures for Processing Orphans. Displaced Persons Commission, 19 October 1950.

Office Centrale Suisse d'Aide aux Réfugiés, Service de Presse et de Propagande. Histoires de Quelques Enfants d'Aujourd'hui. Zurich, 1947.

Padley, Richard and Margaret Cole. *Evacuation Survey: A Report to the Fabian Society.* London: George Routhledge & Sons, 1940.

Papanek, Ernst, and Edward Linn. "The Boy Who Survived Auschwitz." Reprinted from *The Saturday Evening Post,* Curtis Publishing Company, 1964.

Pearse, Dorothy. Historical Documentation on Child Welfare Services in the British Zone, German Operations. UNRRA, DP Br. No. 21.

Personnes Déplacées. La Collection Chemins du Monde. Paris: Editions de Clermont, 1945–1947.

Pfister-Ammende, Maria. "Displaced Soviet Russians in Switzerland." *Uprooting and After.* New York: Springer-Verlag, 1973. Co-author, Charles Zwingmann. pp. 73–102.

Pluckwell, George. *Children of the War.* London: Regency Press, 1966.

Proudfoot, Malcolm J. *European Refugees, 1939–52: A Study in Forced Population Movement.* London: Faber and Faber, 1957.

Random letters and papers on the situation in Germany before and after World War II. 1944–1950.

Ristelhueber, Rene. *Au Secours des Réfugiés.* Plon: "Présences," n.d.

Save the Children Fund. *Children in Bondage: A Survey of Child Life in the Occupied Countries of Europe and in Finland.* London, New York: Longmans Green, 1942.

Service Social d'Aide aux Emigrants. "Le Problème des Relations Familiales des Circonstances Exceptionnelles." Paper prepared for Child Welfare Congress, Zagreb, 4 August to 4 September 1954. Geneva, 1955.

Social Security Agency. Standards for Care of Displaced Children Coming to the United States Under the Displaced Persons Act of 1948. March 1949.

Society of Friends. Report of the Children's Department of the Friend's Center. Vienna, n.d.

Spender, Stephen. *Learning Laughter.* London: Weidenfeld and Nicholson, 1952.

Stern, Carl S. and Blanche L. Miller. Memorandum Concerning Official Basis of the Program of the United States Committee for the Care of European Children, Inc. n.d.

Strachey, St. Loe. *Borrowed Children: A Popular Account of Some Evacuation Problems and Their Remedies.* New York, The Commonwealth Fund, 1940.

Study Group for Post-War Refugee Problems. Collected Papers. Geneva, 1944.

Thelin, Georges. Le Tragique Destin des Enfants dans une Guerre Moderne. Geneva, International Union for Child Welfare, n.d.

UNESCO. L'Enfance Victime de la Guerre: Problèmes d'Éducation. Paris, 1949.

UNRRA. Collected papers from UNRRA archives. New York.

———. Report on Unaccompanied Children. March 1947. Una. 8369.

———. UNRRA Mission to Austria: Child Welfare in the Displaced Persons Programs. Vienna, 1946.

———. Central Committee of the Council. Proposed ACA Directive on Determination of Nationality for Unaccompanied Children. CC (46) 129. 2 December 1946.

———. Welfare Division. The Registration and Identification of Displaced Unaccompanied Children in Enemy Territory. May 1945.

Wolf, Katherine M. "Evacuation of Children in Wartime: A Survey of Literature with Bibliography," photocopy, n.d.

CHAPTER 3

Notes

1. Margorie M. Whiteman, *Digest of International Law,* Vol. V (n.p., u.p., u.d.), p. 282.

2. United Nations Welfare Mission, Athens, *Child Protection in Greece: Basic Circulars of the Ministry of Welfare, 1945–1948,* UNRRA Archives, 1948, n.p.

3. UNESCO, *"Appeal on Behalf of Greek Children,"* n.d., n.p.

4. Ibid.

5. *New York Times,* 2 April 1947, Letters to the Times.

6. United Nations Special Committee on the Balkans, *Report on Removal of Greek Children to Albania, Bulgaria, Yugoslavia, and other Northern Countries,* 78th Meeting of the Special Committee on 21 May 1948, A/Ac.16/251/Rev. 1, p. 1.

7. "The Abduction of Greek Children" (Athens: Committee of Greek Women, u.d.).

8. United Nations Special Committee on the Balkans, Ibid.

9. "Village D'Enfants en Grece," *Revue Internationale de la Croix Rouge* (Septembre 1951): 760–762.

10. Repatriation of Greek Children, Unpublished Annual Reports from the League of Red Cross Societies and the International Committee of the Red Cross to the United Nations, p. 1.

11. United Nations General Assembly, *Resolutions* 193 c(III), 288 B(IV), and 382 c(V).

12. "Repatriation of Greek Children," photocopy, n.d.

13. Notes es Documents, "Repatriement des Enfants Grecs Déplacés," *Revue Internationale de la Croix Rouge* (Janvier 1953): p. 18.

14. Aileen Fitzpatrick, "Greek Children Join Their Parents in Australia," *International Child Welfare Review* 5, no. 2–3 (1951): 81–84.

15. Comité International de La Croix-Rouge, "Activités Diverses. Regroupement de Familles Grecques." n.d., p. 146.

Selected References

Fitzpatrick, Aileen. "Greek Children Join Their Parents in Australia." *International Child Welfare Review.* 5 (1951): 81–84.

ICRC and League of Red Cross Societies. Repatriation of Greek Children. Report A-85. Presented at the Eighteenth International Red Cross Conference. Toronto, July–August 1952.

Leet, Glen. Child Protection in Greece. Basic Circulars of the Ministry of Welfare, 1945–1948. Athens, United Nations Welfare Mission, 1948.

Perret, Francoise. Letter to author. 20 September 1983.

UNESCO. Appeal on Behalf of Greek Children. June 1949.

United Nations Special Committee on the Balkans. Collected papers. 1948.

————. Report on Removal of Greek Children to Albania, Bulgaria, Yugoslavia, and other Northern Countries. 21 May 1948.

United States Committee for the Care of European Children, Inc. Greek Children Who Arrived on U.S. Committee Assurance. New York, 29 August 1952.

CHAPTER 4

Notes

1. *Encyclopaedia Britannica,* 1965, s.v. "Korean War," L. 473.

2. "Danger of Renewed Conflicts Still Exists," Korea News-Review, 2 July 1983, p. 8.

3. Relief and Welfare. UNKRA. ROAG–2/3.1–20. Refugees. p. 4.

4. "I.U.C.W. and Relief for Korean Children," *News Letter* (Geneva: I.U.C.W. September 1951): p. 1.

5. Dean E. Hess, *"Battle Hymn"* (New York: McGraw-Hill, 1956), p. 156.

6. "The Terrible Plight of Korean Children," *Supplement to the I.U.C.W. News Letter* (May 1953): p. 6.

7. Hess, op. cit., p. 191.

8. Orphans in Korea. 18 April 1955. UNKRA. ROAG–2/3.1–20. Orphanages. p. 3.

9. Korea's Children. UNKRA. ROAG–2/3.1–20. Orphanages, p. 2.

10. Relief and Social Services. UNKRA. ROAG–2/3.1–20. Relief and Social Services, p. 11.

11. Orphans in Korea. UNKRA. ROAG–2/3.1–20. Orphanages, p. 1.

12. John N. Thurston to Gracie Pfost, 23 March 1954. UNKRA. ROAG–2/13.1–20. Orphanages.

13. Helen Miller, "Korea's International Children," *Lutheran Welfare* (Summer 1971): p. 16.

14. David C. Chi, The Institutional Care of Children in Korea. p. 7.

15. Relief and Social Services. UNKRA. ROAG–2/3.1–20. Relief and Social Services, p. 11.

16. David C. Chi, Ibid.

17. Miller, Ibid.

18. Miller, Ibid.

19. Ibid.

20. Yoon Gu Lee, "Socio-Cultural Aspects of Abandoning Children," Paper Delivered at the Kava Conference, 17 June 1964.

21. Miller, op. cit., p. 19.

22. Ibid., p. 18.

23. Ibid., p. 13.

24. Ibid., p. 14.

25. Ibid., p. 16.

26. Foster Parents Plan International, Plan's Involvement with Orphaned Children in the Republic of Korea, n.d., p. 1.

27. Miller, op. cit., p. 21.

Selected References

American-Korean Foundation. Report of the Rusk Mission to Korea, March 11–18, 1953. New York, 8 April 1953.

Byma, Sydney. "Overseas Adoptions Threaten Development of Local Services." *Canadian Welfare* 3 (May/June 1974): 7–11.

Chi, David C. Institutional Care of Children in Korea. Seoul, Chung-Ang University, 1984.

Church World Service Survey Team. Children of Tragedy: Report on Intercountry Adoption, mimeo, n.d.

Hak-Mook, Kim. Opening Horizons for Voluntary Efforts, photocopy, n.d.

Hess, Deane. *Battle Hymn.* New York: McGraw-Hill, 1956.

Miller, Helen. "Korea's International Children." *Lutheran's Social Welfare* (Summer 1971): 12–23.

O'Conner, Louis Jr. "The Adjustment of a Group of Korean and Korean-American Children Adopted by Couples in the United States." Ph.D. dissertation, The University of Tennessee, Knoxville, 1964.

Valk, Margaret A. Korean-American Children in American Adoptive Homes. New York, Child Welfare League of America, 1957.

Winick, Myron, Knarig Katchadurian Meyer, and Ruth C. Harris. "Malnutrition and Environmental Enrichment by Early Adoption." *Science* 190 (December 1975): 1173–1175.

CHAPTER 5

Notes

1. Signe Dreijer, "The Hungarian Refugee Situation in Austria," Xerox, December 1956, p. 1.

2. F. Wurst, "The Educational, Moral, Professional and Social Needs of Young Hungarian Refugees," U.U.C.W. Bulletin, 1957, p. 141.

3. ————. "Refugees from Hungary to Austria," single sheet from I.U.C.W., n.d., n.p.

4. ————. "Aid to Hungarian Children," I.U.C.W. Bulletin, 1957, Pl.d., p. 104.

5. International Social Service, "Report on Hungarian Unattended Youths," ISS Report to UNHCR, October 1957, p. 8.

6. International Social Service, "Summary of ISS Activities Project with the United Nations High Commissioner for Refugees Regarding Unaccompanied Hungarian Youths (14–18)" 8 Oct 1957. Statistic used a summary of figures.

7. N.A., "Work of the ICRC Information Bureau and the Reuniting of Families," n.p., n.d.

8. American Council for Voluntary Agencies for Foreign Service, "Report on the Program for Unaccompanied Hungarian Minors," Report to the Committee on Migration and Refugee Problems, 25 April 1958, p. 6.

Selected References

American Council on Voluntary Agencies for Foreign Service, Inc. Report on Program for Unaccompanied Hungarian Minors: Memorandum on Migration and Refugee Problems. 25 April 1958.

Bursten, Martin A. *Escape from Fear: An Eyewitness Report of the Flight of 200,000 Hungarians and its Aftermath.* Syracuse: Syracuse University Press, n.d.

Close, Kathryn. "Speed in Resettlement—How Has It Worked?" *Children* 4 (July–August 1957): 123–131.

Committee on Migration and Refugee Problems in ACVA. "Final Report of Ad Hoc Committee on Children's Program for Unaccompanied Hungarian Minors," photocopy, n.d.

Hilfskomittee fur die Opfer des Kommunismus. *Ungarns Jugend Klugt An!* Bern, 1958.

International Social Service. Report on Hungarian Unaccompanied Youth for the Office of the UNHCR. 1957.

————. Summary of ISS Activities in Austria. Project with the UNHCR Regarding Unaccompanied Hungarian Youths. Geneva, 8 October 1957, p. 14–18.

————. Report on the Hungarian Unaccompanied Minors for the UNHCR. Geneva, 1957.

International Union for Child Welfare. Help to Hungarian Refugee Children and Young People in Two Years' International Action for Children. Geneva, 1958.

"Those Who Were Left Behind." *Children* 4 (November–December 1957): 229–231.

Weinstock, Alexander S. "Some Factors That Retard or Accelerate the Rate of Acculturation with Specific Reference to Hungarian Immigrants." *Human Relations* 17 (1964): 321–341.

Wurst, F. "The Educational, Moral, Professional and Social Needs of Young Hungarian Refugees." *International Union for Child Welfare Bulletin* (1957): 140–142.

CHAPTER 6

Notes

1. José Llanes, *Cuban Americans: Masters of Survival* (Cambridge: Abt Books, 1982), p. 8.

2. Bryan O. Walsh, "Cuban Refugee Children," *Journal of Inter-American Studies of World Affairs,* 13:378–415 (1971), p. 379.

3. Ibid., p. 388.

4. Katherine Oettinger and John Thomas, "Cuba's Children in Exile," U.S. Department of Health, Education, and Welfare Social and Rehabilitation Service, Children's Bureau (1967), p. 4.

5. Walsh, op. cit., p. 396.

6. Ibid., p. 412.

7. Oettinger and Thomas, op. cit., p. 5.

8. Ibid., p. 6.

9. Kay N. Rogers, Raquel E. Cohen, Jose Szapocznik, *Cuban Entrant Unaccompanied Minors* (July 1980), n.p.

10. Katherine Brownell Oettinger, "Services to Unaccompanied Cuban Refugee Children in the United States," *Cuban Refugee Programs,* Carlos E. Curtis, ed. (New York: Arno Press, 1980), p. 383.

11. Kathryn Close, "Cuban Children Away From Home," *Children,* Vol. 10–No. 1 (1963), p. 7.

12. Oettinger and Thomas, op. cit., p. 8.

Selected References

American Council for Voluntary Agencies. Review of the Cuban Refugee Program. September 1976.

Close, Kathryn. "Cuban Children Away From Home." *Children* 10 (January–February 1963): 3–10.

Edie, David B. Milestone Report: Cuban Unaccompanied Minors Program. Wisconsin Resettlement Assistance Office, 1982.

Llanes, Jose. *Cuban Americans: Masters of Survival.* Cambridge, Mass.: Abt Books, 1982.

Naditch, Murray P. and Richard F. Morrissey. "Role Stress, Personality, and Psychopathology in a Group of Immigrant Adolescents." *Journal of Abnormal Psychology* 85 (1976): 113–118.

Oettinger, Katherine Brownell. "Services to Unaccompanied Cuban Refugee Children in the United States." *Social Service Review* (December 1962).

———, and John Thomas. Cuba's Children in Exile: The Story of the Unaccompanied Cuban Refugee Children's Program. Washington, D.C., Children's Bureau, 1967.

Pierce, John M. and Jeanne M. DeAngelis. "Cuban Youth Resettled in Pennsylvania." *Residential Group Care* (Spring 1981): 11–12.

Rogers, Kay N., Raquel E. Cohen, and Jose Szapocznik. Report: Cuban Entrant Unaccompanied Minors. 9 July 1980.

Santisteban, David and Jose Szapocznik. Cuban Adolescent Management Program: Final Report. Coral Gables, Florida, University of Miami, 6 March 1981.

Skotko, Vince. A Manual for Management of Cuban Unaccompanied Minors. Coral Gables, Florida, University of Miami, 6–7 January 1981.

Walsh, Bryan O. "Cuban Refugee Children." *Journal of Inter-American Studies and World Affairs* 13 (July–October 1971): 378–415.

Watson, Ed. Cuban Entrant Unaccompanied Minors. Memo to CWLA Affiliated Agencies. 9 September 1980.

CHAPTER 7

Notes

1. Senator Charles E. Goodell, "Study Mission to Biafra," presented at the National Press Club, 1969, foreword.

2. Frederick Forsyth, *The Making of an African Legend: The Biafra Story* (Middlesex: Penguin Books, 1977), p. 257.

3. A. E. Ifekwunigwe, "A Memorandum on the Welfare Scheme for Biafran Children," photocopy, January 1969, p. 1.

4. Ifekwunigwe, Ibid., p. 3.

5. Jointchurchaid (International), "World Council of Churches Reports on Nigeria/Biafra Relief," Press Release, 1969, p. 1.

6. Audrey E. Moser, "Final Report on IUCW Action in Nigeria 1970–73," International Union for Child Welfare, June 1973, p. 5.

7. A. J. Zerfas, "Medical Report on the Nigerian Children Repatriated from Overseas to Nigeria November 1970 to February 1971 and Their Progress to end July 1971," International Union for Child Welfare, n.d., p. 21.

8. Alain Borgognon, "Some Reflections Regarding the Rehabilitation Programme for Children in Nigeria," International Union for Child Welfare, April 1971, p. 5.

9. Moser, op. cit., p. 11.

10. "Conclusions made by I.U.C.W. on the Statistical Evidence Collected," 14 March 1972, International Union for Child Welfare archives, p. 1.

Selected References

Agency for International Development, U.S. Department of State. Disaster Emergency Relief Report. Washington, D.C., 15 February 1969.

Akiwowo, Akinsola. "The Importance of Child Welfare in Development Planning." Paper delivered at the National Seminar on Child Welfare, Nigeria, 9–13 April 1973.

Doran, Dermont. "The Children's War." *NC News Service of the United States Catholic Conference* (October 1969).

Forsyth, Frederick. *The Making of an African Legend: The Biafra Story.* Harmondsworth, Middlesex, England: Penguin Books, 1977.

Goodell, Charles E. Total Embargo Until February 19, 1969 at 12 Noon. Study Mission to Biafra.

International Union for Child Welfare. IUCW Rehabilitation Programmes in Nigeria, January 1970–September 1971. Geneva, 1971.

—————. Report on an Orientation Course on The Foster Care Programme for War-Displaced Children. Mgbidi-Orlu Nigeria, Children's Reception Center, 14–26 November 1971.

Knotts, Beryl E. Child Care Services in East Central State: A Manual for Social Workers. East Central State Ministry of Health and Social Welfare, Social Welfare Division, 1971–1973.

Moser, Audrey E. "Cross-Cultural Comparison of Different Methods of Child Care." Seminar on the Follow-up Study on the 1970–1973 Rehabilitation Programme for War-Displaced Children in Nigeria, Geneva, 20–23 September 1978.

—————. Final Report on IUCW Action in Nigeria, 1970–1973. Geneva, June 1973.

—————. "Nigeria: Child Rehabilitation and Family Assistance Programme." *International Child Welfare Review,* 17/18 (June 1973): 9–16.

Obikeze, D. S. "Evacuation as a Child Welfare Intervention Measure: The Case of the Nigerian Civil War." *International Social Work* 22, no. 2 (1979): 2–8.

—————. Follow-up Study of Rehabilitation Programme for War-Displaced Children in Nigeria. Nsukka, University of Nigeria, April 1978.

—————. "Issues in Child Welfare Intervention: A Study of the Rehabilitation Programme for the War-Displaced Children in Nigeria." *Occasional Papers of the IUCW* 2 (1980): 11–28.

—————. "What Treatment Mode? An Assessment of Alternative Child Care Methods for War-Displaced Children in Nigeria." *International Social Work* 23, no. 1 (1980): 2–15.

Save the Children Fund. Collected Tracing Records of Children in Medical Centers in Nigeria. 1970.

Zerfas, A. J. Medical Report on the Nigerian Children Repatriated from Overseas to Nigeria, November 1970 to February 1971 and Their Progress to the End of July 1971. Geneva, International Union for Child Welfare, 1972.

General Sources

Atlas of the 20th Century. London: Hamlyn-Bison, 1982.
The New Columbia Encyclopedia. Ed. by William H. Harris and Judith S. Levey. 4th ed. New York: Columbia University Press, 1975.

CHAPTER 8

Notes

1. John S. Bauman (ed.), *Vietnam War: An Almanac* (New York: Work Almanac Publications, 1985), p. 358.
2. Jean and John Thomas, "Visit to the Republic of Vietnam," Letter printed in *A Special Memorandum on the Findings of LIRS/LCUSA Concerning South Vietnamese Children and Intercountry Adoption* (New York: Lutheran Immigration and Refugee Service, 1974), photocopy.
3. Dr. Duong-Cam Chuong, Director of Public Health in South Vietnam, "Evaluation in Child Welfare," paper presented at a seminar organized by I.U.C.W. in the Hague, 5th–10th September, 1966.
4. Dr. Duong-Cam Chuong, op. cit.
5. Thomas, op. cit., p. 6.
6. Thomas, op. cit., p. 5.
7. Thomas, op. cit.
8. John Califf, Letter to Donald Anderson, printed in *A Special Memorandum on the Findings of LIRS/LCUSA Concerning South Vietnamese Children and Intercountry Adoption* (New York: Lutheran Immigration and Refugee Service, 1974), photocopy.
9. Thomas, op. cit., p. 6.
10. Judith Coburn, "The War of the Babies," *Village Voice,* 14 April, 1975, p. 15.
11. Thomas, op. cit.
12. Susan Abrams, "The Vietnam Babylift," *Commonweal,* September 24, 1976, p. 617–621.
13. Coburn, op. cit., p. 15.
14. Abrams, op. cit.
15. "Operation Babylift" (United States Agency for International Development, 1975), photocopy.
16. Ibid., p. 2.
17. Ibid., p. 23.
18. Ibid.
19. "The Orphan Lift," *Time,* April 21, 1975.
20. Abrams, op. cit., p. 618.
21. Barbara M. Brown, "Operation Babylift and the Exigencies of War—Who Should Have Custody of Orphans," *Northern Kentucky Law Review,* 1981 7(8): 81–91.
22. Richard H. Rahe, John G. Looney, Harold W. Ward, Tranh Minh Tung, and William T. Liv, "Psychiatric Consultation in a Vietnamese Refugee Camp," *American Journal of Psychiatry,* 185 (Feb. 1978), p. 187.
23. Ibid., p. 188.
24. Susan S. Forbes and Patricia Weiss Fagen, "Unaccompanied Refugee Children: The Evolution of U.S. Policies." (Washington, D.C.: Refugee Policy Group, 1984), photocopy.

25. Bruce Grant, *The Boat People* (Harmondsworth, Middlesex, England: Penguin Books, 1979), p. 99.

26. Ingrid Walter, "Resettlement in the United States of Unattached and Unaccompanied Indochinese Refugee Minors, 1975–1978" (Geneva: Lutheran Immigration and Refugee Service, n.d.), TS.

27. Philip A. Holman, Memorandum to State Administrators and Other Interested Organizations and Agencies, "Indochinese Refugees–Unaccompanied Minors–Information," (Washington, D.C.: Department of Health, Education, and Welfare, 1979), photocopy.

28. Audrey Moser, "Unaccompanied Children and Adolescents Among the Indochinese Refugees," Notes on speech presented at ICVA meeting, Geneva, 21 July 1979.

29. J. Morison-Turnbull, "Surveys of Unaccompanied Minors in Malaysia and Hong Kong," (Geneva: United Nations High Commissioner for Refugees, 1979), Memo based on K. Tiborn's "Report on Unaccompanied Minors Among Boat Refugees in West Malaysia," 1979, TS.

30. Patricia Nye, "A Survey on Unaccompanied Minors in the VN Refugee Camps in Hong Kong," TS, Geneva, UNHCR, 1979.

31. Ibid., p. 12.

32. Ibid., p. 15.

33. "Draft Procedural Guidelines Concerning the Registration, Resettlement, and Family Reunion of Unaccompanied Minors Amongst South-East Asian Refugee Groups," Regional Meeting on Resettlement-Kuala Lumpur, UNHCR, 1979.

34. American Council of Voluntary Agencies, 1977.

35. International Council of Voluntary Agencies, "Recommendations Concerning Unaccompanied Refugee Minors from South-East Asia," ICVA: 30 July, 1979, photocopy

36. *Handbook for Social Services* (Geneva: United Nations High Commissioner for Refugees, 1984).

37. "Unaccompanied Children in Emergencies: The Canadian Experience," (Ontario, Canada: York University, Refugee Documentation Project, 1984).

38. Ibid., p. 4.

39. Kent Smith, "In-Camp Care of Vietnamese Unaccompanied Minors in First-Asylum Camps," (UNHCR Discussion Paper, 1985), photocopy.

40. Elizabeth Lloyd, "A Study Including Policy Recommendations of Children Leaving Indochina Without Their Parents," (Canberra, Australia: Department of Immigration and Ethnic Affairs, 1983), photocopy.

41. Ibid., p. 4.

42. Ibid.

43. Lloyd, op. cit., p. 4.

44. E. van der Hoeven and H. de Kort, *Reception and Care of Unaccompanied Vietnamese Minors in the Netherlands,* (The Hague, The Netherlands: Coordinatiecommissie wetenschappelijk onderzoek kinderbescherming, 1984).

45. This estimate is based on the following figures: Australia, 6,500, includes all Indochinese minors not with parents; U.S., 5,000, based on the number of Indochinese children who at the time of admission to the United States were not with parents or relatives; France, 6,500, total unaccompanied Indochinese children to receive services; Germany, 1,300, total of Vietnamese minors not with parents or relatives on admission to Germany; Netherlands, 870, total of all Vietnamese minors under 21 years of age not with parents; Canada, 500, only unaccompanied Indochinese children arriving without relatives and not joining relatives; the number resettled to other countries is estimated at 1,000.

46. Helga Jockenhovel-Schiecke, ed., "Unaccompanied Refugees in European Resettlement Countries," (Conference Report, Frankfurt: 1984).

47. Nic Baker, "U.S. Programs for Indochinese Unaccompanied Minors," (Washington, D.C.: Center for Applied Linguistics, 1981), photocopy.

364 *Notes and References*

48. "Unaccompanied Children in Emergencies: The Canadian Experience," op. cit., p. 11.

49. Lloyd, op. cit., p. 12.

50. H. Brissimi, "Unaccompanied Minors: Requests for Adoption," UNHCR Memo to Regional Representatives, 1979, photocopy.

51. Nancy Schulz, "Project Haven—A Multifunctional Program for Resettlement of Unaccompanied Minors," (Catholic Community Services, n.p., n.d.), TS.

52. American Voluntary Agencies, 1977.

53. International Council of Voluntary Agencies, "Recommendations Concerning Unaccompanied Refugee Minors from South-East Asia," ICVA: 30 July, 1979, photocopy.

54. Diane Zulfacar, "Surviving Without Parents: Indo-Chinese Refugee Minors in NSW," (New South Wales: University of New South Wales, School of Social Work, n.d.), photocopy.

55. "Unaccompanied Children in Emergencies: The Canadian Experience," op. cit.

56. Jockenhovel-Schiecke, op. cit.

57. Do-Lam, Chi La, "Insertion scolaire et sociale d'une population d'adolescents d'origine Vietnamienne de la région parisienne: Étude des processus psychologiques d'adaptation, recherche des critères de prévention de l'inadaptation." Thèse pour le doctorate de 3ème cycle psychologie appliquée, University of Paris V, René Descartes, Sciences Humaines, Sorbonne, 1982.

58. Chris Mougne, *Vietnamese Children's Home: A Special Case for Care?* (London: Save the Children Fund, Mary Datchelor House, 1985).

59. van der Hoeven and de Kort, op. cit.

60. United States Catholic Conference, *Voyagers in the Land: A Report on Unaccompanied South-East Asian Refugee Children* (New York: United States Catholic Conference, 1983).

61. Pennsylvania Department of Public Welfare, "National Mental Health Needs Assessment of Indochinese Refugee Populations," (Pennsylvania: Department of Public Health, Office of Mental Health, Bureau of Research and Training, 1979), p. iv.

62. van der Hoeven and de Kort, op. cit., p. 57.

63. Carolyn L. Williams and Joseph Westemeyer, "Psychiatric Problems Among Adolescent Southeast Asian Refugees: A Descriptive Study" (University of Minnesota: Adolescent Health Program, n.d.), p. 10, photocopy.

64. Saik Lim, "Service Delivery to Indo-Chinese Refugee Children: The South Australian Experience," *Australian Social Work,* 1979, 32(3): n.p.

65. "Unaccompanied Children in Emergencies: The Canadian Experience," op. cit., p. 92.

66. Lloyd, op. cit., p. 12.

67. "Unaccompanied Children in Emergencies: The Canadian Experience," op. cit., p. 98.

Selected References

Baker, M. C. "Family Dislocation Amongst Vietnamese in Australia," *Australia,* n.d., 26: 9–10.

Beach, Hugh and Lars Ragvald. *A New Wave on a Northern Shore: Indochinese Refugees in Sweden* (Statens invandrarverk: Arbetsmarknadsstyrelsen, February 1982).

Boland, Kerry. "Ethnic Background Papers on Isolated Indo-Chinese Refugee Children," (Regional Services Division, n.p., 1980).

Branscombe, Dr. Martha (Chair) and Dr. Gardner Munro (Rapporteur). *"Report of the Meeting on Placement and Adoption of Vietnamese Children in American Homes,"* Washington, D.C., July 25–26, 1973, sponsored by the Agency for International Development.

Brown, Barbara M. "Operation Babylift and the Exigencies of War—Who Should Have Custody of Orphans," *Northern Kentucky Law Review,* 1981, 7(8): 81–91.

Citizens' Committee for Children of New York, Inc. *In Search of Safe Haven: Foster Care Programs for Unaccompanied Indochinese Refugee Minors in New York* (New York: Citizens' Committee for Children of New York Inc., 1980).

Citizens' Committee for Children of New York, Inc. *Unaccompanied Refugee Minors: Policies and Programs* (New York: Citizens' Committee for Children of New York, Inc., 1981).

Ethnic Communities' Council of New South Wales. *A Framework for the Co-operative Management of a Total Care Programme for Khmer Unaccompanied Minors,* (New South Wales, Australia: The Ethnic Communities' Council of New South Wales, Department of Youth and Community Service, 1982).

Harvey, Ian J. "Adoption of Vietnamese Children: An Australian Study," n.s., n.d.

Hole, Arni. *Unaccompanied Young Vietnamese in Norway, How Are They?—A Partial Study in the Ministry of Social Affairs Study: Refugees' Adjustment to the Norwegian Society* (Oslo, Norway: Ministry of Social Affairs, 1983).

Jockenhovel-Schiecke, Helga. *Report: On the Acceptance and Integration of South-East Asian Refugees in France* (Frankfurt: Internationaler Sozialdienst, 1980), photocopy.

Leak, Jenny. *"Smiling on the Outside, Crying on the Inside," The Prevalence and Manifestation of Emotional Stress in Refugee Children from Vietnam* (Adelaide, Australia: South Australian College of Advance Education, Stuart Campus, 1982).

Mathews, Ch. *Report: On the Experience and Work with "Unaccompanied Minor Refugees from Indochina"* (Frankfurt: Internationaler Socialdienst, 1980).

Miller, Barry. "Refugees: Their Mental Health Needs," Testimony Presented to Immigration, Refugees, and International Law Subcommittee of the Judiciary Committee of the House of Representatives, May 10, 1979, photocopy.

Mortland, Carol A. and Maura G. Egan. "Vietnamese Youth in American Foster Families: Differing Worlds," manuscript, 1983, photocopy.

Mulock Houwer, Dan Q. R. *Report on Visit to South Vietnam: From 9th to 17th December, 1966* (Geneva, Switzerland: International Union for Child Welfare, n.d.), Xerox.

Pearce, Joyce. *Refugees as Unaccompanied Minors in the United Kingdom: 1939–1983—With Special Perspective from the Ockenden Venture* (Ockenden Programme for Unaccompanied Refugee and Displaced Children, n.p., 1983), photocopy.

Redick, Liang Tien and Beverly Wook. "Cross-Cultural Problems for Southeast Asian Refugee Minors," *Child Welfare,* 1982, 61(6): 365–373.

Rodier, Claire. "The Care of Unaccompanied Southeast Asian Minors in Belgium, France, and the Netherlands," TS, n.d.

Rudnik, Joan P. and Larry P. Molsad. "Implications of Cross-Cultural Foster Care: Stages of Adjustment," TS, n.d.

Schaffner, Mary E. and Conrad Keelenberg. "Adoption of Vietnamese Children: The Dilemma of the Consent Requirement," TS, n.p., 1977.

Schulz, Nancy. "Project Haven—A Multifunctional Program for Resettlement of Unaccompanied Minors," (Catholic Community Services, n.p., n.d.), TS.

Spence, Susan. "Some Considerations on the Adoption of Vietnamese Children: An International Social Welfare Issue," *International Welfare,* 1975, 18(4): 10–20.

Stalcup, S. Alex, Mark Oscherwitz, Martin S. Cohen, Frank Crast, Dan Broughton, Fred Stark and Robert Goldsmith. "Planning for a Pediatric Disaster—Experience Gained from Caring for 1600 Vietnamese Orphans," *The New England Journal of Medicine,* 1975, 293 (October): 691–695.

Tiborn, Kate. "A Follow-up of the Conditions of Unaccompanied Minors and Handicapped Persons Among the Refugees from Vietnam Resettled in Sweden," (The National Board of Health, n.p., 1981).

Williams, Carolyn L. "The Southeast Asian Refugees and Community Mental Health," (manuscript, University of Minnesota: 1983), photocopy.

Williamson, Jan. "A Compilation of the Available Information Concerning Unaccompanied Minors in Vietnam, Kampuchea, and Laos from the 1970s to the 1980s," (manuscript, n.p., December, 1983).

Zigler, Edward. "A Developmental Psychologist's View of Operation Babylift," *American Psychologist,* 1976, May, 329–340.

Zumbach, Pierre. "The Adoption and Evacuation of Children," (Geneva: International Union for Child Welfare, letter, n.d.).

CHAPTER 9

Notes

1. Ben Kiernan and Chanthou Boua, ed., *Peasants and Politics in Kampuchea 1942–1981* (London: Zed Press, 1982), p. 282, quoting Stanic.

2. *"Human Rights, Wars, and Mass Exodus: Overview of the Decade, 1970–1980,"* drawn from "Human Rights and Massive Exoduses" by Saruddin Aga Khan as Special Rapporteur of the Commission on Human Rights, *Transnational Perspectives,* Vol 7, No 1, 1981, p. 31.

3. J. Millington, "Orphans in the Khmer Republic," USAID/Phnom Phen, 9 August 1974, p. 1, circulated by Dao Spencer to Member of the Children's Committee, American Council of Voluntary Agencies for Foreign Service, Inc., 10 September 1974.

4. Sydney H. Schanberg, "For Hordes of Cambodian Orphans, 3 Small Orphanages—And 2 Toys," *New York Times,* 17 October 1973.

5. J. Millington, op. cit., p. 2.

6. Ibid., p. 3.

7. Ibid., p. 4.

8. Ibid., p. 5.

9. Ibid.

10. Prince Saruddin Aga Khan, op. cit., p. 31.

11. Michael Vickery, *Cambodia 1975–1982* (Boston: South End Press, 1984), p. 177.

12. Barbara Melunsky, UNHCR Bangkok, "Narrative Report on Children's Project for Kampucheans," April 1981.

13. Ibid., p. 4.

14. Lee Rudakewych, memo to CRS-USCC, 6 January 1980, photocopy, p. 2.

15. Office of the High Commissioner for Refugees, "Kampucheans: High Commissioner Sets Up Special Centres for Children," Press release describing letter to countries concerned with assistance to newly arrived refugees from Kampuchea, 5 December 1979.

16. UNHCR, "Discussion paper: Recent UNHCR Involvement with Unaccompanied Refugee Minors," prepared for Seminar organized by Redd Barna on Unaccompanied Children in Emergencies, 24–26 June 1985.

17. Michael Hegenauer, "Current Report on Children's Centres Khao I Dang," Catholic Relief Services, n.d.

18. Marie de la Soudiere and Corrianne Graham, "An Assessment of Facts and Options in the Care of Unaccompanied Kampuchean Children in Thailand" (International Rescue Committee), photocopy, 25 November 1980, p. 3, 4.

19. UNHCR/Radda Barnen Team of Social Workers, "Second Report on Kampuchean Unaccompanied Minors in the UNHCR Holding Centre of Khao I Dang," Part Two of *Second Report of the Radda Barnen Team on Kapuchean Unaccompanied Minors in Two UNHCR Holding Centres in Thailand,* August 31, 1980, photocopy, p. 12.

20. Marie de la Soudiere and Corrianne Graham, op. cit., p. 6.

21. Jan Williamson, "A Compilation of the Available Information Concerning Unaccompanied Minors in Vietnam, Kampuchea, and Laos from 1979 to 1983," unpublished, December 1983, p. 134.

22. UNHCR, "Review of Kampuchean Refugee Care and Maintenance in Thailand," internal document, May 1982, p. 33.

23. Kris Buckner, "The Role of Hospital Records and Personnel in Tracing of Unaccompanied Minors" (International Rescue Committee), photocopy, 11 August 1980.

24. Ibid.

25. Kate Tiborn, Monica Svederoth, Sylvia Karlsson, "Survey on Unaccompanied Children Centres in Khao I Dang UNHCR Holding Center for Kampuchean Refugees in Thailand," unpublished, 20 April 1980, p. 4.

26. UNHCR/Radda Barnen Team of Social Workers, Ibid., op. cit., p. 11.

27. Ibid., p. 7.

28. Kate Tiborn, Monica Svederoth, Sylvia Karlsson, op. cit., p. 5–7.

29. Marie de la Soudiere and Corrianne Graham, op. cit., n.p.

30. Everett M. Ressler, "Analysis and Recommendations for the Care of the Unaccompanied Khmer Children in the Holding Centers in Thailand," *Study of Placement Options for Unaccompanied Kampuchean Children in Thailand,* photocopy, Interagency Study Group in Bangkok, December 1980, p. 12–15.

31. UNHCR, "Options and Timing for Long-Term Solutions," Bangkok, photocopy, n.d.

32. UNHCR/Radda Barnen Team of Social Workers, "General Report on the Radda Barnen Team's Assignment to Do Documentation and Final Recommendations on Kampuchean Unaccompanied Minors in UNHCR Holding Centres in Thailand," Part One of *Second Report of the Radda Barnen Team on Kampuchean Unaccompanied Minors in Two UNHCR Holding Centres in Thailand,* August 31, 1980, photocopy, p. 9.

33. Ibid.

34. Ibid., p. 11.

35. Christine Mougne, "Care Provision and Resettlement of Unaccompanied Minors—A Report of a Private Visit to Phanat Nikhom Refugee Camp, Thailand, July–August 1984," photocopy, March 1985, p. 21.

Selected References

Helene Blanc, "Abandoned Babies in a Thai-Kampuchean Border Camp—A Six-month Study: November 80–April 81, High-Risk Children," Medecins Sans Frontieres, photocopy, n.d.

Neil Boothby, "Khmer Children: Alone at the Border," *Indochina Issues,* December 1982.

The Committee for the Coordination of Services to Displaced Persons in Thailand, "Care of Unaccompanied Children in Holding Centers for Kampucheans," Guidelines approved by the CCSDPT ad hoc committee 4 December 1980.

International Committee of the Red Cross, "Rules of Conduct as Regards the Problem of Unaccompanied Children," memo, 2 December 1981.

John Farvolden, "Tracing and Reunification for Unaccompanied Kampuchean Children on the Thai/Kampuchea Border," photocopy, 12 August 1981.

Charlotte Herrman, "The Children's Centres Relative to the Camp as a Whole," *Study of Placement Options for Unaccompanied Kampuchean Children in Thailand,* Xerox copies distributed by an Interagency Study Group in Bangkok, December 1980.

Helga Jockenhovel-Schiecke, *The Unaccompanied Minor Refugees from Kampuchea in the Camps in Thailand,* International Social Service–German Branch, November 1981.

"Deathwatch: Cambodia," *Time,* 12 November 1979.

UNHCR, *UNHCR Kampuchean Operation Manual No. 15—The Care of Unaccompanied Minors,* Bangkok, March 1980.

UNHCR, "The Promotion of Durable Solutions for Unaccompanied Minor Children in South-East Asia," 7 March 1983.

Jan Williamson, "Centers for Unaccompanied Children Khao I Dang Holding Center," *Disasters,* Vol. 5, No. 2, pp. 100–104, 1981.

CHAPTER 10

1. n.a., "Final Statistics for Returned Children Until March, 1972," Follow-up/Family Assistance Program for the East Central State, IUCW archives, n.d., n.p. The number 327 is given as "number of orphans reunited."

2. Everett M. Ressler, "Analysis and Recommendations for the Care of the Unaccompanied Khmer Children in the Holding Centers in Thailand," Redd Barna, November 1980.

3. Helene Blanc, "Abandoned Babies in a Thai-Kampuchean Border Camp: High-Risk Children," Medecins sans Frontieres, A six-month study: November 80–April 81, p. 7.

4. Ibid., p. 6.

5. Dorothy Lagarreta, *The Guernica Generation* (Reno: University of Nevada Press, 1984), p. 326.

6. Ibid., p. 304.

7. Everett M. Ressler, *Widows and Unaccompanied Children in Guatemala,* unpublished, Redd Barna, 1985.

CHAPTER 11

1. Freud, A. *Normality and Pathology in Childhood,* New York: International Universities Press, Inc., 1965.
Erikson, E. *Childhood and Society,* New York: Norton, 1950.
Piaget, J. *The Construction of Reality in the Child,* New York: Ballantine Books, 1964.

2. Freud, A. *op. cit.,* 1965.

3. See Part III.

4. Whiting, B. B. and Whiting, J. W. M. *Children in Six Cultures: A Psychological-Cultural Analysis,* Cambridge, Mass.: Harvard University Press, 1975.

5. *Ibid.*

6. Winnicott, D. W. "The Theory of the Parent-Infant Relationship," *International Journal of Psychoanalysis,* 41, 585–595, 1960.

7. Erikson, E. *Childhood and Society,* New York: Norton, 1950.

8. Rohner, R. P. *They Love Me, They Love Me Not: A World-Wide Study of the Effects of Parental Acceptance and Rejection,* New Haven: HRAF Press, 1975.

9. Fraiberg, S. *Every Child's Birthright: In Defense of Mothering,* New York: Basic Books, 1977.

10. Ainsworth, M. D. S. and Bell, S. M. "Attachment, Exploration and Separation: Illustrated by the Behavior of One-Year-Olds in a Strange Situation," *Child Development,* 41, 49–67, 1970.
Bowlby, J. *Attachment and Loss* (Volumes 1 and 2), New York: Basic Books, 1969.

11. Fadouti-Milenkovic, M. and Uzgiris, I. C. "The Mother-Infant Communication System," in: Uzgiris, I. C. (ed.), *Social Interaction and Communication During Infancy* (New Directions for Child Development, No. 4), San Francisco: Josey-Bass, 1979.

12. See for example:
Bowlby, J. *Maternal Care and Mental Health,* Geneva: World Health Organization, 1951.
Bowlby, J. *op. cit.,* 1969.
Provence, S. and Lipton, R. C. *Infants in Institutions,* New York: International Universities Press, 1962.
Provence, S. "A Clinician's View of Affect Development," in: Lewis, M. and Rosenblum, L. A. (eds.) *The Development of Affect,* New York: Plenum Publishing Corporation, 293–307, 1978.

Robertson, J. and Robertson, J. "Young Children in Brief Separations: A Fresh Look," *The Psychoanalytic Study of the Child*, 26, 264–315, 1971.

Rutter, M. *Maternal Deprivation Reassessed*, Middlesex: Penguin, 1972.

Solnit, A. J. "Developmental Perspective of Self and Object Constancy," *The Psychoanalytic Study of the Child*, 37, 201–218, 1982.

Spitz, R. A. *The First Year of Life*, New York: International Universities Press, 1965.

Yarrow, L. J. "Separation from Parents During Early Childhood," in: Hoffman, M. L. and Hoffman, L. W. (eds.), *Review of Child Development Research*, New York: Russell Sage Foundation, Vol. 1, 1964.

13. Mead, M. *Coming of Age in Samoa*, New York: Morrow, 1928.

14. Ainsworth, M. D. S. *Infancy in Uganda: Infant Care and the Growth of Love*, Baltimore: The Johns Hopkins Press, 1967.

15. Whiting, B. B. "Wisdom and Child-Rearing," *Merrill-Palmer Quarterly of Behavior and Development*, 20, 9–19, 1974.

Whiting, J. W. M. "Effects of Climate on Certain Cultural Practices," in: Goodenough, W. H. (ed.), *Explorations in Cultural Anthropology*, New York: McGraw-Hill, 1964.

16. Erikson, E. *op. cit.*, 1950.

17. Schafer, R. "The Loving and Beloved Superego in Freud's Structural Theory," *The Psychoanalytic Study of the Child*, 15, 163–188, 1960.

18. Werner, E. E. *Cross-Cultural Child Development: A View from Planet Earth*, Monterey, California: Books/Cloe Pub. Co., 1979.

19. Levine, R. A. *Parental Goals: A Cross-Cultural View*, Bureau of Educational Research, University of Nairobi, 1974.

20. Barry, H. A., Child, I. L., and Bacon, M. K. "Relation of Child Training to Subsistence Economy," *American Anthropologist*, 6, 51–63, 1959.

Berry, J. W. and Annis, R. C. "Ecology, Culture and Psychological Differentiation," *International Journal of Psychology*, 1, 207–229, 1966.

21. Cole, M. and Scribner, S. *Culture and Thought: A Psychological Introduction*, New York: Wiley, 1974.

22. Piaget, J. "The Mental Development of the Child," in: Elkind, D. (ed.), *Six Psychological Studies of Jean Piaget*, New York: Random House, 4–70, 1967.

Piaget, J. "How Children Form Mathematical Concepts," *Scientific American*, 189, 20, 74–79, 1953.

Piaget, J. *The Origins of Intelligence in Children*, New York: International Universities Press, 1953.

23. Erikson, E. *op. cit.*, 1950.

24. *Ibid.*

25. Whiting, B. B. and Whiting, J. W. M. *Children of Six Cultures: A Psychocultural Analysis*, Cambridge, Mass.: Harvard University Press, 1975. 26.

26. Werner, E. E. *op. cit.*, 1979, 211–234.

27. Erikson, E., *op. cit.*, 1950.

Erikson, E. *Identity: Youth and Crisis*, New York: Norton, 1968.

28. Erikson, E., *op. cit.*, 1950, 261–262.

29. Erikson, E. "The Problem of Ego Identity," *Journal of the American Psychoanalytic Association*, 4 (1), 56–121, 1956.

30. Douvan, E. and Adelson, J. *The Adolescent Experience*, New York: Wiley, 1966.

Larson, L. E. "Influence of Parents and Peers During Adolescence: The Solution Hypothesis Revisited," *Journal of Marriage and the Family*, 34, 67–74, 1972.

Meissner, W. W. "Parental Interaction of the Adolescent Boy," *Journal of Genetic Psychology*, 107, 225–233, 1965.

31. Werner, E. E., *op. cit.*, 1979.

32. *Ibid.*

33. Erikson, E. "Identity and Uprootedness in Our Time," *Insight and Responsibility,* New York: Norton, 1964.

CHAPTER 12

1. Blaufarb, H., and Levine, J. "Crisis Intervention in an Earthquake," *Social Work,* 17 (4), 16–19, 1972.

Dunlap et al. "Young Children and the Watts Revolt," Southern California Permanent Medical Group, 1966.

Newman, C. J. "Children of Disaster: Clinical Observations at Buffalo Creek," *American Journal of Psychiatry,* 133 (3), 306–312, 1976.

Perry, H. S., and Perry, S. E. "The Schoolhouse Disaster: Family and Community as Determinants of the Child's Response to Disaster," *Disaster Study 11,* Washington, D.C.: National Academy of Science.

2. Fraser, M. *Children in Conflict,* London: Secker and Warburg, 73, 1973.

3. Crawshaw, R. "Reactions to a Disaster," *Archives of General Psychiatry,* 9, 157, 162, 1963.

Elder, J. H. "A Summary of Research on Reactions of Children to Nuclear War," *American Journal of Orthopsychiatry,* 35, 12–123, 1965.

Silber, E.; Perry, S.; and Blocj, D. "Patterns of Parent-Child Interaction in a Disaster," *Psychiatry,* 21, 159–167, 1958.

4. Farberow, N. L. and Gordon, N. S. *Manual for Child Health Workers in Major Disasters,* Washington, D.C.: U.S. Department of Health and Human Services, 1981.

5. Ciuca, R.; Downie, C.; and Morris, M. "When a Disaster Happens: How Do We Meet Emotional Needs?" *Americal Journal of Nursing,* 77, 3, 454–456, 1977.

Howard, J. et. al. (eds.) *Emergency and Disaster Management: A Mental Health Sourcebook,* Bowie, Md.: The Charles Press, 1976.

Raphael, P. "Crisis and Loss: Counseling Following a Disaster," *Mental Health in Australia,* 1, 4, 118–122, 1975.

6. International Union for Child Welfare, "Post Disaster Programmes," *International Child Welfare Review,* 17–18, June, 1973.

7. Schelsky, H. "Die Fluchtlingsfamilie," *Christ Unterwegs,* (Munchen), 7, 9–11, Juli 1951, and 8, 3–8, Aug. 1951.

8. Fajrajzen, S. "A Report Prepared for the Italian O.S.E.," as cited in: *The Psychological, Educational and Social Adjustment of Refugee and Displaced Children in Europe,* UNESCO, 1952.

9. Pfister, M. "Erfajrungen Bei Kriegsgeschadigten Jugendilichen Fluchtligeng Und Ruchwanderern," *Sonderadruck Aus Gesundheit Und Wohlfahrt,* 2, 51–54, Jarhrg. 1948.

10. Burbury, W. M. "Effects of Evacuation and of Air Raids on City Children," *British Medical Journal,* 2, 640, 1941.

11. Bodman, F. "War Conditions on the Mental Health of the Child," *British Medical Journal,* II, 486–488, 1941.

12. Carey-Tretzer, C. J. "The Results of a Clinical Study of War-Damaged Children Who Attended the Child Guidance Clinic, The Hospital for Sick Children, Great Ormond Street, London," *Journal of Mental Science,* 95, 535–599, 1949.

13. Freud, A. and Burlingham, D. *War and Children,* New York: Medical War Books, 37, 1943.

14. Pfister, Maria, *op. cit.,* 1948.

Meierhof, M. "First Experience in Medical-Psychological Work at the Pestalozzi Children's Village at Trogen," *Mimeograph,* UNESCO, 1949.

15. Sutter, J. M. "Conséquences Éloignees des Émotions de Guerre Chez les Enfants," *Pédiatre*, 41, 345, 1952.

16. Swedish Save the Children, "A Report on Finnish Children Transferred to Sweden and Subsequently Repatriated," cited in *The Psychological, Educational and Social Adjustment of Refugee and Displaced Children in Europe*, UNESCO, 23–24, 1952.

17. See, for example:

Addesa, D. J. "Refugee Cuban Children: The Role of the Catholic Welfare Bureau of the Diocese of Miami, Florida, in Receiving, Caring for, and Placing Unaccompanied Cuban Refugee Children," *M.S.W. Thesis*, Fordham University, 1964.

Garzon, C. G. "A Study of the Adjustment of 34 Cuban Boys in Exile," *D.S.W. Thesis*, School of Social Service, Florida State University, 1965.

Gil, R. M. "The Assimiliation of Problems of Adjustment to the American Culture of One Hundred Cuban Refugee Adolescents, Attending Catholic and Public High Schools, in Union City and West New York, New Jersey, 1959–1966," *M.S.W. Thesis*, Fordham University, 1968. Also see:

Almeida, J. "The Massive Cuban Immigration: A Psychological View, Comments on the Group's Dynamics," paper presented at the Annual Meeting of the American Psychiatric Association, Miami, May, 1969.

Copey-Blanco, M.; Montia, P.; and Suarez, L. "A Study of Attitudes of Cuban Refugees Towards Assimilation," *M.S.W. Thesis*, Barry College, Miami, 1968.

Morrissey, R. and Naditch, M. "Role Stress, Personality and Psychopathology in a Group of Immigrant Adolescents," *Journal of Abnormal Psychology*, 85, 113–118, Fall, 1976.

Szapocznik, J. et al. "Cuban Value Structure: Treatment Implications," *Journal of Consulting Clinical Psychology*, 46, 961–970, October, 1978.

Walsh, B. "Cuban Refugee Children," *Journal of Inter-American Studies and World Affairs*, 13, 378–415, July–October, 1971.

18. See Part I.

19. Obideze, D. S. "Issues in Child Welfare Intervention: A Study of the Rehabilitation Programmed for War-Displaced Children in Nigeria," in *Occasional Papers of the IUCW: The Protection of Children in Armed Conflicts (Two)*, International Union for Child Welfare, 11–27.

20. *Ibid.*, 15–16.

21. Bolin, R. C. "Family Recovery from Natural Disaster: The Case of Rapid City," *Dissertation Abstracts International*, 37, 8–A, 5385, 1977.

Quarantelli, E. L. and Dynes, Russell R. "Images of Disaster Behavior: Myths and Consequences, Preliminary Paper 5," *Disaster Research Center*, The Ohio State University.

Quarantelli, E. L. "Real and Mythological Problems in Community Disasters, Preliminary Paper 72," *Disaster Research Center*, The Ohio State University.

Young, M. "The Role of the Extended Family in Disaster," *Human Relations*, 7, 383–391, 1954.

22. Burbury, W. M.; Freud, A.; Burlingham, D., *op. cit.*, 1943.

Langmeier, J. and Matejcek, Z. *Psychological Deprivation in Childhood*, New York: Halsted Press Books, 1963.

23. Ziv, A. and Israel, R. "Effects of Bombardment on the Manifest Anxiety Level of Children Living in Kibbutzim," *Journal of Consulting and Clinical Psychology*, 40, 287–291, 1973.

Ziv, A.; Kruglanski, A.; and Schulman, S. "Children's Psychological Reactions to Wartime Stress," *Journal of Personality and Social Psychology*, 30, 24–30, 1974.

Zuckerman-Bareli, "The Effects of Border Tension on the Adjustment of Kibbutzim and Moshavim on the Northern Border of Israel," in Spielberger, C. and Sarason, I. (eds.), *Stress and Anxiety (Volume 8)*, Washington, D.C.: Hemisphere Publishing Corporation, 1982.

24. Pinsky, L. "The Effects of War on Displaced Children," (unpublished Materials in the World Federation For Mental Health Files, and summarized in the booklet): *Basic Materials*

Used by Working Groups, Second World Mental Health Assembly, New York: National Committee for Mental Hygiene, 1949.

Pinsky, L. "The Children," in Murphy, H. B. M. (ed.), *Flight and Resettlement,* London: UNESCO, 1955.

25. Markowitz, A. (personal communication), Bangkok, Thailand, December 5, 1982.

26. Boothby, N. *Cambodia's Unaccompanied Refugee Children: Loss, Recovery and Renewal of Hope* (in progress).

CHAPTER 13

1. Burt, C. "War Neurosis in British Children," *Nervous Child,* 2, 324, 1943.

2. Freud, A. and Burlingham, D., *op. cit.,* 1943.

Freud, A. and Burlingham, D. *Infants Without Families: Reports on the Hampstead Nurseries (The Writings of Anna Freud)* Volume III, New York: International Universities Press, 1973.

3. Issacs, S. *The Cambridge Evacuation Survey,* London: Methuen, 1941.

Carey-Tretzer, C. J., *op. cit.,* 1949.

Freud, A. and Burlingham, D., *op. cit.,* 1973.

Pfister, Maria, *op. cit.,* 1940.

4. Van Grevel, N. "Het Joodse Orlgos-Pleegkind," *School Voor Sociale Wetenschappen En Maatschappelikj Werk,* Dactyl., (Rotterdam) (no date).

Close, K. "Transplanted Children," report prepared for *The U.S. Committee for the Care of European Children,* Inc., 1953.

Issacs, S., *op. cit.,* 1941.

5. Van Grevel, N., *op. cit.,* (no date).

Close, K., *op. cit.,* 1953.

Langmeier, L. and Matejcek, Z., *op. cit.,* 1963.

6. Van Grevel, N., *op. cit.,* (no date).

Close, K., *op. cit.,* 1953.

Langmeier, L. and Matejcek, Z., *op. cit.,* 1963.

7. See, for example, summary reviews presented in:

Wolf, K. M. "Evacuation of Children in Wartime," *Psychoanalytic Study of the Child,* 1, 389–404, 1945.

Langmeier, L. and Matejcek, Z., *op. cit.,* 1963.

8. Bodman, F., *op. cit.,* 1941.

9. Freud, A. and Burlingham, D., *op. cit.,* 1943.

10. Issacs, S., *op. cit.,* 1941.

11. *Ibid.*

12. Fifty-four of those who did not become part of a family unit came from an orphanage in England and were cared for in a group setting in New York, and 15 others in a boarding school arranged by British and American citizens in advance of their arrival. Thus only sixteen children were unable to adjust to family life,Close,K. *op. cit.,* 1953.

13. Wall, W. D. "Education et Santé Mentale, Problémes d'écuation," No. 11, Paris: UNESCO, 1955.

14. Rasanen, E. "On the Experience of Evacuated Children" in *Children and War,* (Proceeding of Symposium at Siuntio Baths, Finland, March 24–27, 1983), Peace Union of Finland.

15. *Ibid.* p. 71.

16. Huyck, E. and Fields, R. "Impact of Resettlement on Refugee Children," in: *International Migration Review,* Volume 15, New York: Center for Migration Studies, Summer/Spring, 1981, p. 252.

17. Mierhoff, M. cited in: *The Psychological, Educational and Social Adjustment of Refugee and Displaced Children in Europe,* Geneva: UNESCO, 1952, p. 39.

18. Sprengel, R. cited in: *The Psychological, Educational and Social Adjustment of Refugee and Displaced Children in Europe,* Geneva: UNESCO, 1952, p. 37, 38.

19. Freud, A. and Dann, S. "An Experiment in Group Upbringing," *Psychoanalytic Study of the Child,* 1951, 6, 127–68.

20. Solnit, A. J. "Some Adaptive Function of Aggressive Behavior," in: *Psychoanalysis— General Psychology,* (eds.) R. M. Loewenstein; L. M. Newman; M. Schur; and A. J. Solnit, New York: International Universities Press, 1966, 169–89.

21. Boothby, N. (unpublished field notes) Thailand, 1981–1982.

22. Bowlby, J. *Maternal Care and Mental Health,* Geneva: World Health Organization, 1951.

23. Rutter, M. *Maternal Deprivation Reassessed,* Middlesex, England: Penguin, 1972.

24. *Ibid.*

25. Tizard, B. and Joseph, A. "Cognitive Development of Young Children in Residential Care: A Study of Children Aged 24 Months," *Journal of Child Psychology and Psychiatry,* 1970, 11, 177–86.

Tizard, B. and Rees, J. "The Effects of Early Institutional Rearing on the Behavioral Problems and Affectional Relationships of Four-Year-Old Children," *Journal of Child Psychology and Psychiatry,* 1975, 16, 61–74.

Tizard, B. and Hodges, J. "The Effects of Early Institutional Rearing on the Development of Eight-Year-Old Children," *Journal of Child Psychology,* 1978, 19, 99–119.

26. Rutter, M., *op. cit.,* 1972.

27. Huh, S.; Cheol-Wha, C.; and Kim, K. M. (personal communications) Seoul, Korea, July, 1983.

28. Zigler, E. "America's Babylift of Vietnamese Children: What Is to Be Learned for a Psychological Change," in: *The Child and His Family,* volume 5, (eds.) E. J. Anthony, T. C. Chiland, New York: Wiley and Son, 1978, 107–116.

29. Brazelton, T. B. "Restoring Kampuchea's Children," *Indochina Issues,* 20, 1981, 4–5.

30. Langmeier, J. and Matejcek, Z., *op. cit.,* 1963.

31. Foster Parents Plan International, Plan's Involvement with Orphaned Children in the Republic of Korea, (unpublished paper), (undated).

Anderson, J. Plan/ROK Director (personal communications), Providence, R.I., June, 1983.

Kim, K. M. (personal communications), Seoul, Korea, July, 1983.

32. Bowlby, J., *op. cit.,* 1951.

33. Rey, D., cited in: *The Psychological, Educational and Social Adjustment of Refugee and Displaced Children in Europe,* op. cit., 1952, p. 27.

34. Collins, R., cited in: *The Psychological, Educational and Social Adjustment of Refugee and Displaced Children,* op. cit., 1952, 27–28.

35. Boothby, N. *Cambodia's Unaccompanied Refugee Children: Loss, Recovery and Renewal of Hope,* op. cit., in progress.

Boothby, N. "The Horror—The Hope," *Natural History Magazine,* January 1983.

Boothby, N. "Khmer Children: Alone at the Border," *Indochina Issues,* 32, December 1982.

36. Meerloo, J. *Psychological Problems of Displaced Persons,* London: UMRRA, 1945.

Wain, B. *The Refused,* New York: Simon and Schuster, 1981.

Boothby, N. (unpublished field notes), op. cit., 1981–1982.

Eddie, D. "Milestone Report: Unaccompanied Minors Program," *Wisconsin Resettlement Assistance Office,* February, 1982.

37. Meerloo, J., *op. cit.,* p. 5.

38. Markowitz, A. (personal communications) Bangkok, Thailand, December 1981, Cambridge, Mass., February 1983.

39. Boothby, N. *Cambodia's Unaccompanied Refugee Children: Loss, Recovery and Renewal of Hope,* op. cit., in progress.

40. Eddie, D., op. cit., 1982.

41. Szapocnik, J. *Cuban Adolescent Management Program: Final Report,* Spanish Family Guidance Center, Department of Psychiatry, University of Miami, March, 1981.

42. UNESCO, The Psychological, Educational and Social Adjustment of Refugee and Displaced Children in Europe, op. cit., 1952.

43. Kim, K. M. (personal communication) Seoul, Korea, July 1983.

44. Felsman, J. K. "Abandoned Children: A Reconsideration," *Children Today,* May–June 1984, 13–19.

Felsman, J. K. "Street Urchins of Cali: On Risk, Resiliency and Adaptation in Childhood," *Doctoral Thesis,* Harvard Graduate School of Education, 1981.

Felsman, J. K. "Street Urchins of Colombia," *Natural History,* April 1981.

45. Felsman, J. K. "Street Urchins of Colombia," op. cit., 1981, p. 43.

46. Langmeier, J. and Matejcek, z., op. cit., 1963.

47. *Ibid.*

48. UNESCO, op. cit., 66–67.

49. Pinsky, L. op. cit., 1949.

Pinsky, L. op. cit., 1955.

50. Boothby, N. (unpublished field notes), op. cit., 1981–1982. *Indochina Issues,* op. cit., December 1982.

51. Eddie, D., op. cit., 1982.

52. Hardy, R. K. and Looney, J. G. "Problems of Southeast Asian Children in a Refugee Camp," *American Journal of Psychiatry,* 134, April 1977.

Liu, W. T. and Murauta, A. K. "Mental Health of Vietnamese Refugees Part IV," *Bridge,* Summer 1978.

53. Liu, W. T. and Murauta, A. K., op. cit., 1978, p. 2.

54. Murphy, H. B. M. "The Camps," in *Flight and Resettlement,* H. B. M. Murphy (ed.), Paris: UNESCO, 1955.

55. Bakis, E. "D. P. Apathy," in: *Flight and Resettlement,* H. B. M. Murphy (ed.), Paris: UNESCO, 1955.

56. Mamdani, M. *From Citizens to Refugees: Uganda Asians Come to Britain,* London: F. Pinter Ltd., 1973.

Hardy, R. K. and Leoney, J. G., op. cit., 1977.

Liu, W. T. and Murata, A. K., op. cit., 1978.

Starr, P. D. "Stressful Life Events and Mental Health Among Vietnamese Refugees: Inoculation and Synchronization," paper presented at the Southern Sociological Society Annual Meeting, Knoxville, TN., March 1980.

Stein, B. N. "The Refugee Experience: Defining the Parameters of a Field of Study," *International Migration Review,* New York: Center for Migration Studies, Spring/Summer, 1981.

Boothby, N. *Cambodia's Unaccompanied Refugee Children: Loss, Recovery and Renewal of Hope,* op. cit., in progress.

57. Boothby, N., *Ibid.*

58. Ressler, E., et al. "Study of Placement Options For Unaccompanied Kampuchean Children in Thailand," Interagency Study Group, (unpublished report), Thailand, December, 1980.

59. Williamson, J. An Experiment in the Care of Unaccompanied Minors in Refugee Camps, (unpublished report UNHCR), 1980.

60. Boothby, N., (unpublished field notes), op. cit., 1981–1982.

61. *Ibid.*

62. *Redd Barna,* (personal communications), December 1981, July 1983.

63. *Ibid.,* July 1983.

CHAPTER 14

1. Goldstein, J.; Freud, A.; and Solnit, A. J. *Beyond the Best Interest of the Child,* New York: The Free Press, 1973.
2. Winnicott, D. W. *The Child and the Outside World,* London: Tavistock Publications, Ltd., 86, 1962.
3. Freud, A. and Burlingham, D., op. cit., 1943.
4. Issacs, S., op. cit., 1941.
5. *Ibid.*
6. Wolf, K., op. cit., 1945.
7. Close, K., op. cit., 1953.
8. Goldstein, J.; Freud, A.; and Solnit, A. J., op. cit., 1973.
9. Boothby, N., (unpublished field notes), op. cit., 1981–1982.
10. Rasanen, Eila, op. cit., 1983.
Wall, W. D., op. cit., 1955.
Winnicott, D. W., op. cit., 1962.
11. Winnicott, D. W., op. cit., 1962.
12. Johnson, B. S. *The Evacuees,* London: Victor Gollancz, Ltd. 1968.
13. Close, K., op. cit., 1953, 20–21.
14. Gordon, M. *Assimilation in American Life.* New York: Oxford University Press, 1964.
15. Addesa, D. J., op. cit., 1964.
Garzon, C. G., op. cit., 1965.
Gil, R. M., op. cit., 1968.
16. Rasanen, E., op. cit., 1983.
17. *Ibid.,* 71–72.
18. International Union for Child Welfare; "Post-Disaster Programmes," op. cit., 1975.
19. Van Grevel, N., op. cit., 1952.
20. *Ibid.,* p. 101.
21. Maas, H. S. "The Young Adult Adjustment of Twenty Wartime Residential Nursery Children," *Child Welfare,* 57–72, February 1963.
22. *Ibid.,* 66–67.
23. Rasanen, E., op. cit., 1983.

CHAPTER 15

1. Barinbaum, L. "Identity Crisis in Adolescence: The Problem of an Adopted Child," *Adolescence,* 9, 547–554, 1974.
Brinich, P. M. "Some Potential Effects of Adoption on Self and Object Representations," in: *Psychoanalytic Study of the Child,* A. J. Solnit (ed.), 35, 1980.
Kadushin, A. *Child Welfare Services,* New York: Macmillan, 1974.
Krementz, J. *How It Feels to Be Adopted,* New York: Knopf, 1982.
Sants, H. J. "Genealogical Bewilderment in Children with Substitute Parents," *British Journal of Medical Psychology,* 37, 133–141, 1964.
Schwartz, E. M. "The Family Romance Fantasy in Children Adopted in Infancy," *Child Welfare,* 49, 386–391, 1970.
2. Mech, E. V. "Adoption: A Policy Perspective," in: *Review of Child Development Welfare Review,* 29, 20–25, 1976.
3. Goldstein, J.; Freud, A.; Solnit, A. J., op. cit., 1973, p. 54.
4. Joe, B. "In Defense of Intercountry Adoption," *Social Service Review,* 52, 1978, p. 6.

5. Kadushin, A. "Problems Relating to the Formation of Principles for Adoption and Foster Placement of Children," in: *Adoption and Foster Placement of Children,* U.N. (ed.), New York: U.N., 1980.

6. Reid, J. "Rethinking Intercountry Adoption," *Child Welfare League of America Newsletter,* Spring–Summer 1975, 7–8.

7. Melone, T. "Adoption and Crisis in the Third World: Thoughts on the Future," *International Child Welfare Review,* 29, 1976, 24–25.

8. Berlin, I. N. "Anglo Adoption of Native Americans: Repercussions in Adolescence," *Journal of the American Academy of Child Psychiatry,* 17, 387–388, 1979.

9. Simon, R. J. and Alstein, H. *Transracial Adoption,* New York: Wiley and Sons, 1977.

10. Graham, L. B. "Children from Japan in American Adoptive Homes," in: *Casework Papers,* National Conference on Social Welfare, (ed.), New York: Family Services Association of America, 1957.

Kim, D. S. "How They Fared in American Homes: A Follow-up Study of Adopted Korean Children in the U.S.," *Children Today,* March–April, 1977, 3–6.

Kim, S. P. "Behavior Symptoms in Three Transracially Adopted Asian Children: Diagnosis Dilemma," *Child Welfare,* 59, 1980, 213–224.

Rathbun, C. and Kolody, R. L. "A Groupwork Approach in Cross-Cultural Adoption," *Children,* 14, 1967, 117–21.

11. di Virgilio, L. "Initial Adjustment of Adopted Children from Other Countries," in: *Casework Papers,* National Conference of Social Work, (ed.), New York: Family Service Association of America, 1956.

Gardell, I. *A Swedish Study of Intercountry Adoption,* Stockholm, Sweden: Libertryck, 1980.

Hochfeld, E. "The Alien Child in the Adoptive Family," *Social Casework,* 41, 1960, 123–127.

Hosbergen, R. A. C. "Adoption of Foreign Children in the Netherlands," *International Child Welfare Review,* 49, 1981, 28–37.

Pruzan, V. *Født I Udlandet-Adopteret I Danmark* [Born in a Foreign Country—Adopted in Denmark], Copenhagen, Denmark: Danish National Institute of Social Research, 1977.

Rathbun, C.; di Virgilio, L.; and Waldfogel, S. "The Restitutive Process in Childhood Following Radical Separation from Family and Culture," *American Journal of Orthopsychiatry,* 28, 1958, 408–415.

Rathbun, C.; McLaughlin, H.; Bennett, C.; and Garland, J. "Later Adjustments of Children Following Radical Separation from Family and Culture," *American Journal of Orthopsychiatry,* 35, 1965, 604–609.

Wolters, W. H. G. "Psychosocial Problems in Young Foreign Adopted Children," *Acta Paedopsychiatrica,* 46, 1980, 67–81.

12. di Virgilio, L., op. cit., 1956.

Gardell, I., op. cit., 1980.

Hoksbergen, R. A. C., op. cit., 1981.

McDermott, R. E., "Oriental Adoptive Placements," *Catholic Charities Review,* 49, 1965, 24–25.

Rathbun, C.; di Virgilio, L.; and Wadfogel, S., op. cit., 1980.

13. Hockfeld, E., op. cit., 1960.

Hoksbergen, R. A. C., op. cit., 1980.

Kim, S. P., op. cit., 1980.

McDermott, R. E., op. cit., 1965.

Rathbun, C.; di Virgilio, L.; and Wadfogel, S., op. cit., 1958.

Valk, M. A. "Adjustment of Korean-American Children in their American Adoptive Homes," in *Casework Papers,* National Conference on Social Welfare, (ed.), New York: Family Services Association of America, 1957.

Wolters, W. H. G., op. cit., 1980.

14. di Virgilio, L., op. cit., 1956.

Gardell, I., op. cit., 1980.

Kim, S. P., op. cit., 1980.

Valk, M. A., op. cit., 1957.

Wolters, W. H. G., op. cit., 1980.

15. Hachfeld, E., op. cit., 1960.

16. Rathburn, C.; McLaughlin, H.; Bennett, C.; and Garland, J., op. cit., 1965.

17. Welter, M. "Comparison of Adopted Older Foreign and American Children," Ph.D. diss., Western Reserve University, 1965.

18. Pruzan, U., op. cit., 1977.

19. Hoksbergen, R. A. C., op. cit., 1981.

Gardel, I., op. cit., 1980.

20. Gardel, I., op. cit., 1980, p. 16.

21. Kim, D. S., op. cit., 1977.

22. Gardel, I., op. cit., 1980.

Hochfeld, E., op. cit., 1960.

Rathburn, C.; di Virgilio, L.; and Waldfogel, S., op. cit., 1958.

23. Gardel, I., op. cit., 1980.

24. Brazelton, T. B., "Instant Adoption—A Comment," *Children Today,* November–December, 1980.

Kim, S. P., op. cit., 1980.

Kim, S. P.; Hong, S.; and Kim, B. S. "Adoption of Korean Children by New York Area Couples: A Preliminary Study," *Child Welfare,* 58, 1979, 422–426.

Rathburn, C. and Kolodny, R. L., op. cit., 1967.

Wolters, W. H. G., op. cit., 1980.

Zigler, E., "A Developmental Psychologist's View of Operation Babylift," *American Psychologist,* 31, 1976, 329–340.

25. Brazelton, T. B., op. cit., 1980, p. 6.

26. "Foreign Adoption," *Washington Post,* February 17, 1983, p. 18.

27. Kim, S. P., op. cit., 1980.

Wolters, W. H. G., op. cit., 1980.

28. Wolters, W. H. G., op. cit., 1980.

29. Goldstein, J., Freud, A., Solnit, A. J. op. cit., 1973, p. 4.

30. See Part I.

31. Hardy, R. K. and Lonney, J. G., op. cit., 1977.

Jones, C. E. and Else, J. F. "Racial and Cultural Issues in Adoption," *Child Welfare,* 58, 1979, 373–382.

Liu, W. T. and Murata, A. K., op. cit., 1978.

Report of the Joint Committee for Refugees from Vietnam, London: Home Office, 1982.

32. Hardy, R. K. and Looney, J. G., op. cit., 1977.

Liu, W. T. and Murata, A. K., op. cit., 1978.

van der Hoeven, E. and de Kort, H., (personal communications), Holland, May 12, 1983.

33. Jones, C. E. *Vietnamese Refugees* (Research and Planning Unit Paper No. 13), London: Home Office, 1982.

Mathews, C. *Report on the Experience and Work with Unaccompanied Minor Refugees from Indochina in Australia,* I.S.S. Germany, 1980.

van der Hoeven, E. and de Kort, H., op. cit., 1983.

34. Jones, C. E., op. cit., 1982.

Mathews, C., op. cit., 1980.

35. van der Hoeven, E. and de Kort, H., op. cit., 1983.

36. Liu, W. T. and Maurata, A. K., op. cit., 1978.

37. Hardy, R. K. and Looney, J. G., op. cit., 1977, p. 49.

38. Mathews, C., op. cit., 1980.

Adler, P. (personal communications) New York, April 20, 1983.

39. Baker, N., in: "U.S. Programs for Indochinese Unaccompanied Minors," *Compiled Proceedings: Helping Indochinese Families in Transition,* W. H. Meredith and B. J. Twiten (eds.), Lincoln, NE: Nebraska University, (ERIC Document Reproduction Service No. ED 206-768), 1981.

Rudnik, J. P. and Molstall, L. P. *Implications of Cross-Cultural Foster Care: Stages of Adjustment,* Lutheran Social Services of Minnesota Unaccompanied Minors Program, 1982.

Walter, I. "Resettlement in the United States of Unattached and Unaccompanied Indochinese Refugee Minors (1975–1978) by Lutheran Immigration and Refugee Service," *International Migration,* 17, 1979, 139–161.

40. Barker, N., op. cit., 1981.

Citizen's Committee for Children of New York Inc. *In Search of Safe Haven: Foster Care Programs for Unaccompanied Minors in New York,* 1980.

Citizen's Committee for Children of New York Inc. *Unaccompanied Refugee Minors: Policies and Programs,* (ERIC Document Reproduction Service No. ED 211-216), 1981.

A Framework for the Cooperation Management of a Total Care Programme for Khmer Unaccompanied Minors, Sidney, Australia: Austcare, 1982.

Jockenhovel-Schiecke, H. and Mathews, C. *First Assessment of the Acceptance and Placement of South-East Asian Unaccompanied Minors in the Federal Republic of Germany,* I.S.S. Germany, 1980.

Schulz, N. *Project Haven—A Multifunctional Program for Resettlement of Unaccompanied Minors,* Catholic Community Services, New Jersey, 1982.

Hammarberg, C., (personal communications), Lutheran Child and Family Service, Southeast, PA, April 22, 1983.

41. Citizen's Committee, op. cit., 1980.

42. Jockenhovel-Schiecke, N. and Mathews C., op. cit., 1980, p. 6.

43. Gogler, F. "Letter to J. Croissandean," in: *Le Monde de'Education,* April 14, Paris, France: Terre D'Asile, 1983.

Labe, Y. M. and Croissanclean, J. "Coupables d'etre en Vie (Guilty of Being Alive)," *Le Monde de L'Education,* December 1982, 63–66.

44. Quito and Mennesohn, (personal communications), Croix Rouge, May 17, 1983.

45. Labe, Y. M. and Crossendeau, J., op. cit., 1982.

46. Walter, I., op. cit., 1979.

47. Jockenhovel-Schiecke, H. "Unaccompanied Indochinese Minors in West Germany," in: *The Unaccompanied Minor: Global View—National Response,* New York: I.S.S., 1983.

Jockenhovel-Schiecke, H. and Mathews, C., op. cit., 1980.

Duche, D. J.; Bathien, N.; Basquin, M.; Frechette, D.; and Schmit, G. "Limites de la Prise en Charge en Placement Familial des Enfants du Sud Est Asiatique" (Limitations in the Practice of Family Sponsoring of Southeast Asian Refugee Children), unpublished manuscript, 1983.

van der Hoeven, E. and de Kort, H., op. cit., 1983.

Mathews, C., op. cit., 1980.

48. Schulz, N. and Sontz, A. *Voyagers in the Land: A Report on Unaccompanied Southeast Asian Refugee Children,* Washington, D.C.: U.S. Catholic Conference Migration and Refugee Services, 1983.

49. Boothby, N. Cambodia's Unaccompanied Refugee Children: Loss, Recovery and Renewal of Hope, op. cit., in progress.

50. Duche, D. J. et al., op. cit., 1983.

Jockenhovel-Schiecke, H. and Mathews, C., op. cit., 1980.

Hos, J. (personal communications), Foundation Opbouw, May 13, 1983.

51. Duche, D. J. et. al., op. cit., 1983.

52. Walter, I., op. cit., 1981.

53. Baker, N., op. cit., 1981.

54. Ellis, A. A. *The Assimilation and Acculturation of Indochinese Children into American Culture,* Sacramento, CA: Department of Social Services, 1980.

Clarens, Y. (personal communication), L'enfant Refuge, May 19, 1983.

55. Schulz, N., op. cit., 1982.

Walter, I., op. cit., 1979.

Hammarberg, C., op. cit., 1983.

Schulz, N. (personal communications) U.S.C.C., April 22, 1983.

56. Citizen's Committee, op. cit., 1981.

Hammarberg, C., op. cit., 1983.

57. Wolkind, S. "Fostering the Disturbed Child," *Journal of Child Psychology, Psychiatry, and Allied Disciplines,* 19, 1978, 393–397.

58. Duche, D. J. et al., op. cit., 1983.

Rudnik, J. P. and Mostead, L. P., op. cit., 1982.

Walter, I., op. cit., 1979.

Hos, J., op. cit., 1983.

van der Hoeven, E. and Kort, H., op. cit., 1983.

59. Boothby, N. Cambodia's Unaccompanied Refugee Children: Loss, Recovery and Renewal of Hope, op. cit., in progress.

60. Jockenhovel-Schiecke, H., op. cit., 1983.

61. Do-Lam, C. L. "Insertion Scholaire et Sociale de'une Population d'adolescents d'origine Vietnameine de la Region Parisienne (Academic and Social Adjustment of a Group of Vietnamese Adolescents in the Paris Area)," Ph.D. diss., University of Paris, Paris, France, 1982.

62. Bennoun, R. and Kelly, P. *Indo-Chinese Youth,* Indochina Refugee Association of Victoria, Fitzroy, Australia, 1981.

Carlin, J. "The Catastrophically Uprooted Child: Southeast Asian Refugee Children," in: *Basic Handbook of Child Psychiatry:* Volume I, J. D. Noshpits (ed.), 1979, 290–300.

Pearce, J. *Ockenden (UK) Programme for Unaccompanied Children,* Ockenden, England, 1981.

Walter, I., op. cit., 1979.

Brand, R. (personal communications), Save the Children, May 23, 1983.

Hammarberg, C., op. cit., 1983.

van der Hoeven, E. and de Kort, H., op. cit., 1983.

van Westerlo (personal communications), Foundation Opouw, May 16, 1983.

63. Rudnik, J. P. and Molstad, L. P., op. cit., 1982, p. 11.

64. Okura, K. P. "Indochina Refugees: Families in Turmoil," Paper Presented at the Meeting of American Orthopsychiatric Association, New York [ERIC Document Reproduction Services No. ED 206-790] April, 1981.

65. Rudnik, J. P. and Molstad, L. P., op. cit., 1982.

66. Brown, M. (personal communications), Lutheran Family Services of New England, February 2, 1983.

67. Jockenhovel-Schiecke, H., op. cit., 1983.

Walter, I., op. cit., 1979.

68. Mathews, C., op. cit., 1980.

69. *Ibid.,* p. 5.

70. Bennoun, R. and Kelly, P., op. cit., 1981.

Walter, I., op. cit., 1979.

Hammarberg, C., op. cit., 1983.

van Harten, J., op. cit., 1983.

71. Mathews, C., op. cit., 1980, p. 7.

72. Adler, P. (personal communications), Louise Wise Foundation, April 21, 1983.

73. Shultz, N. and Sontz, A., op. cit., 1983.

74. *Ibid.,* pp. 51–52.

75. *Ibid.,* p. 52.

76. Jones, C. E. and Else, J. F. "Racial and Cultural Issues in Adoption," *Child Welfare,* 58, 1979, 373–382.

77. Billingsley, A. and Giovanonni, J. "Research Perspective on Interracial Adoption," in: *Race Research and Reason: Social Work Perspectives,* R. Miller (ed.), New York: National Association of Social Workers, 1969.

Jones, E. D. "On Transracial Adoption of Black Children," *Child Welfare,* 51, 1972, 156–164.

Ladner, J. A. *Mixed Families,* Garden City, N.Y.: Anchor/Doubleday, 1977.

78. Ladner, J. A., op. cit., 1977.

79. Brand, R. "A Happy Place to Be," *Community Care,* October 22, 1981.

Pearce, J., op. cit., 1981.

Pearce, J. *Unaccompanied Minors Amongst Refugees Groups,* Ockenden, England, 1981.

van der Hoeven, E. and de Kort, H.; op. cit., 1983.

Jockenhovel-Schiecke, H. and Mathews, C.; op. cit., 1980.

Boland, K. and Bodna, K.; *Group Living Program for Isolated Refugee Children,* Victoria, Australia: Department of Social Welfare, 1980.

80. Save the Children, *Vietnamese Children's Centre,* England: Save the Children, 1983.

81. Brand, R., op. cit., 1981.

Save the Children (personal communication), England, May 23, 1983.

82. van Harten, J., op. cit., 1983.

van Westerlo, L., op. cit., 1983.

83. van der Hoeven, E. and de Kort, H., op. cit., 1983.

van Harten, J., op. cit., 1983.

Hos, J., op. cit., 1983.

84. van Westerlo, L., op. cit., 1983.

85. Wolins, M. and Piliavin, I. *Institution or Foster Family: A Century of Debate,* New York: Child Welfare League of America, 1964.

86. Council of Europe, *Resolution (77) 33 on Placement of Children,* Strasbourg, France: Council of Europe, 1977.

David, M. and Lezine, I. *Early Child Care in France,* London: Gordon and Breach, 1975.

Luscher, K. K.; Ritter, V.; and Gross, P. *Early Childhood Care in Switzerland,* London: Gordon and Breach, 1973.

Mayer, M. F.; Richman, L. H.; and Balcerzak, E. A. *Group Care of Children: Crossroads and Transitions,* New York: Child Welfare League of America, 1977.

Wolins, M. and Piliavin I., op. cit., 1964.

87. David, M. and Lezine, I., op. cit., 1975.

Luscher, K. K. et al., op. cit., 1973.

Wolins, M. and Piliavin, I., op. cit., 1964.

88. Council of Europe, *Social Measures Regarding the Placement of Children in Community Homes or Foster Families,* Strasbourg, France: Council of Europe, 1971.

Wolins, M. and Piliavin, I., op. cit., 1964.

89. Mayer, M. F. et al., op. cit., 1977.

90. Wolins, M. *Successful Group Care,* Chicago: Aldine, 1974.

91. Luscher, K. K. et al., op. cit., 1973.

Mayer, M. F. et al., op. cit., 1977.

Wolins, M., op. cit., 1974.

Wolins, M. and Piliavin, I., op. cit., 1964.

92. Meisel, J. F. and Loeb, M. B. Unanswered Questions about Foster Care, *Social Service Review,* 30, 1956, p. 246.

93. Mayer, M. F. et al., op. cit., 1977, p. 72.

94. Robertson, J. and Robertson, J. "Young Children in Brief Separations: A Fresh Look," *The Psychoanalytic Study of the Child,* 26, 1971, 264–315.

Rutter, M. *Maternal Deprivation Reassessed,* Middlesex, England: Penguin, 1972.

Rutter, M. "Maternal Deprivation 1972–1978: New Findings, New Concepts, New Approaches," *Child Development,* 50, 1979, 283–305.

Bowlby, J., op. cit., 1951.

95. Robertson, J. and Robertson, J., op. cit., 1971.

Rutter, M., op. cit., 1972, 1981.

96. Council of Europe, op. cit., 1977.

Kadushin, A. *Child Welfare Services,* New York: Macmillan, 1974.

Mayer, M. F. et al., op. cit., 1977.

Rutter, M., op. cit., 1981.

97. Bennoun, R. and Kelley, P., op. cit., 1981, pp. 6–7.

98. Blacher, T., op. cit., 1980, p. 22.

99. Jockenhovel-Schiecke, H. *Unaccompanied Minor Refugees and Juveniles in the Federal Republic of West Germany (1979–1983),* prepared for: Seminar on Unaccompanied Refugee Minors in European Resettlement Countries, I.S.S. Germany, 1984.

100. Mennesohn and Quito (personal communications), Croix Rouge Francais, May 17, 1983.

101. Jockenhovel-Schiecke, H., op. cit., 1983.

Leer, A. M. *Is It Possible for Groups of Orphaned Children and Young People from a Foreign Culture to Adjust for example to Danish Society?,* Copenhagen, Denmark: Danish Refugee Council, 1979–1980.

Christen, B. (personal communications), Pestalozzi Children's Village, May 4, 1983.

Citizen's Committee, op. cit., 1980, 1981.

Schulz, N., op. cit., 1982.

102. Baker, N., op. cit., 1981.

103. Jockenhovel-Schiecke, H., op. cit., 1982.

Christen, B., op. cit., 1983.

104. Christen, B., op. cit., 1983.

105. Labe, Y. M. and Croissandeau, J., op. cit., 1982.

106. Baker, N., op. cit., 1981.

Labe, Y. M. and Croissandeau, J., op. cit., 1982.

107. Labe, Y. M. and Croissandeau, J., op. cit., p. 64.

108. Schulz, N. (personal communications), 1983.

109. Council of Europe, op. cit., 1971, p. 24.

110. Spitz, R. A. "Anaclitic Depression," *The Psychoanalytic Study of the Child,* 2, 1946, 313–342.

Bowlby, J., op. cit., 1951.

Yarrow, L. J. Separation from Parents During Early Childhood, in: *Review of Child Development Research: Volume I,* M. L. Hoffman and L. W. Hoffman (eds.), New York: Russell Sage Foundation, 1964, 89–136.

Mayer, M. F. et al., op. cit., 1977.

Rutter, M., op. cit., 1972, 1981.

111. Rutter, M., op. cit., 1981, p. 45.

112. Council of Europe, op. cit., 1977.

Kadushin, A., op. cit., 1974.

Mayer, M. F. et al., op. cit., 1977.

Rutter, M., op. cit., 1981.

113. UNESCO, op. cit., 1952, p. 50.

114. Baker, N., op. cit., 1981.

Citizen's Committee, op. cit., 1980.

Do-Lam, C. L., op. cit., 1982.

Jockenhovel-Schiecke, H., op. cit., 1983.

115. Leer, A. M., op. cit., 1979–1980.

116. Jockenhovel-Schiecke, H., op. cit., 1983, 47–48.

117. UNESCO, op. cit., 1952, p. 50.

118. Bennoun, R. and Kelley, P., op. cit., 1981.

Mathews, C., op. cit., 1980.

van Harten, J., op. cit., 1983.

Schulz, N., op. cit., 1982.

Walter, I., op. cit., 1979.

119. Bennoun, R. and Kelley, P., op. cit., 1981.

Mathews, C., op. cit., 1980.

van der Hoeven, E. and de Kort, H., op. cit., 1983.

120. Bennoun, R. and Kelley, P., op. cit., 1981.

121. van Harten, J., op. cit., 1983.

122. Schulz, N., op. cit., 1983.

123. Citizen's Committee, op. cit., 1982.

Walter, I., op. cit., 1979.

124. Durker, E. J. "Special Problems of Custody for Unaccompanied Refugee Children in the United States," in: *1982 Michigan Yearbook of International Legal Services: Transnational Legal Problems of Refugees,* New York: Clark Boardman, 1982, 197–225.

125. Children's Defense Fund, op. cit., 1978.

Fanshel, D. "Status Change of Children in Foster Care: Final Results of the Columbia University Longitudinal Study," *Child Welfare,* 55, 1976, 143–171.

Golstein, J.; Freud, A.; and Solnit, A. J., op. cit., 1973.

Rutter, M. "Parent-Child Separation: Psychological Effects on the Children," *Journal of Child Psychology, Psychiatry, and Allied Disciplines,* 12, 1971, 233–260.

Rutter, M., op. cit., 1981.

Vasaly, S., *Foster Care in Five States,* Washington, D.C.: Government Printing Office, 1976.

126. Baker, N., op. cit., 1981.

127. Fanshel, D., op. cit., 1976.

128. Palmer, S. E., *Children in Long-Term Care—Their Experiences and Progress,* London, Ontario, Canada: Family and Children's Services, 1976.

Vasaly, S., op. cit., 1976.

129. Children's Defense Fund, op. cit., 1978.

Fanshel, D., op. cit., 1976.

Geiser, R., *The Illusion of Caring: Children in Foster Care,* Boston: Beacon Press, 1973.

Goldstein, J.; Freud, A.; Solnit, A. J., op. cit., 1973.

Palmer, S. E., op. cit., 1976.

Vasaly, S., op. cit., 1976.

130. Vietnamese Children's Resettlement Advisory Group, "Recommendations of the Vietnamese Children's Resettlement Advisory Group," unpublished manuscript, 1975.

131. Report on the Conditions, op. cit., 1982.

132. Dominguez, V., prepared statement for: *Hearings Before the Subcommittee on Immigration, Refugee and International Law, Committee on the Judiciary, U.S. Senate, 96th Congress, 1st Session,* Washington, D.C.: Government Printing Office, May 9, 1979, 325–329.

133. Bogardus, E. S. "Cultural Pluralism and Acculturation," *Sociology and Social Research,* 34, 1949, 125–129.

Domiguez, V., op. cit., 1979.

Simpson, G. E. "Assimilation," in: *International Encyclopedia of the Social Sciences: Volume I,* D. Sills (ed.), New York: Macmillan, 1968.

Spicer, E. N. "Acculturation," in: *International Encyclopedia of the Social Sciences: Volume I,* D. Sills (ed.), New York: Macmillan.

Teske, R. N. C. and Nelson, B. H. "Acculturation and Assimilation: A Classification," *American Ethnologist,* 1, 1974, 351–367.

134. Teske, R. H. C. and Nelson, B. H., op. cit., 1974.

135. C.B.S. Evening News, November 29, 1983.

136. Hodgetts, C. "What Future for the Vietnamese in Britain?" *Exile,* 1 (1), 1983, p. 3.

137. *Ibid.,* p. 3.

138. Baker, N., op. cit., 1981.

Carlin, J., op. cit., 1979.

Citizen's Committee, op. cit., 1981.

Pearce, J., op. cit., 1981.

Schjoth, T. C. *Adaptation and Integration of Refugee Children in Norway with Particular Notice to Unaccompanied Refugee Children,* Geneva, Switzerland: Intergovernmental Committee for European Migration (Fourth Seminar on Adaptation and Integration of Permanent Immigrants Paper No. INF/21), 1979.

139. Duche, D. J. et al., op. cit., 1983.

140. Domiguez, V., op. cit., 1979, p. 327.

141. Baker, N., op. cit., 1981.

Citizen's Committee, op. cit., 1980.

Do-Lam, C. L., op. cit., 1982.

Jockenhovel-Schiecke, H., op. cit., 1983.

142. Shulz, N. and Sontz, A., op. cit., 1983.

143. Adler, P., op. cit., 1983.

144. Jockenhovel-Schiecke, H., op. cit., 1983, p. 54.

CHAPTER 16

1. This chapter draws on the essays on the national laws of thirteen countries collected in Law and the Status of the Child, a project of the UNITAR Research Division, edited by Anna Mamalakis Pappas, and published in 13 *Colum. Human Rights L. Rev.* vol. 1, 2 (1981) (hereinafter cited as UNITAR Report). The thirteen countries included in that report are: Australia, China, Colombia, the Congo, Cuba, Czechoslovakia, Egypt, England, Greece, Israel, Kenya, Norway, and the United States. Other major sources for this section are the International Encyclopedia of Comparative Law, vol. 4, *Persons and Family* (J. C. B. Mohr) (hereinafter cited as IECL) and Lawasia Family Law Series (J. E. Sihambing and H. A. Finlay ed.), (hereinafter cited as Lawasia).

2. Pappas, *supra* n., 1 at xxxiii.

3. J. Foyer, *The Reform of Family Law in France,* in: The Reform of Family Law in Europe 75 (A. G. Chloros, ed. 1978).

4. *Cf.* J. Goldstein; A. Freud; & A. Solnit, Before the Best Interests of the Child 17–18, (1979). There are "strong presumptions in law (a) that parents have the right, the capacity, and the obligation to care for their children in accord with their own notions of childrearing; and (b) that children have the right to uninterrupted and permanent membership in a family with such parents."

5. See "Removed Children," *infra* at 219.

6. DRC Principle 2. The obligation of parents to provide material and emotional support is found primarily in national legislation; see *e.g.,* UNITAR Report. Parental obligations may be enforced by a civil suit against the parents [*e.g.,* USA: Wash. Rev. Code Sec. 26.09.100

(1979)] or by criminal prosecution [*e.g.,* Kenya: Penal Code, Laws of Kenya Ch. 63, sec. 216, 217]. The Draft Convention on the Rights of the Child places "primary responsibility for the upbringing and development of the child" on parents and guardians, DCRC Art. 8(1). The American Declaration of the Rights and Duties of Man, Art. 30, squarely states: "It is the duty of every person to aid, support, educate and protect his minor children." Parental support duties are the subject of two international private law conventions concerning the law and enforcement of support judgments. Hague Convention Concerning the Recognition and Enforcement of Judgments Regarding Child Support Obligations, April 15, 1958, and the Hague Convention on the Law Applicable to Maintenance Obligations, October 2, 1973.

7. S. J. STOLJAR, CHILDREN, PARENTS AND GUARDIANS, 4 IECL Chap. 7, at 41.

8. *See e.g.,* French Civil Code Art. 371-2.

9. Havelka and Raduanova, *Czechoslovak Law and the Status of the Child,* in UNITAR REPORT, *supra* n. 1, at 274–5; S. J. STOLJAR, *supra* n. 7 at 43–46.

10. ICCPR Art. 18(4); ICESCR Art. 13(3); ACHR Art. 12 (4); *cf.* DRC Principle 9.

11. DRC Principle 6.

12. DCRC Art. 8(1).

13. DCRC Art. 6(1).

14. DDFPA Art. 3.

15. UDHR Art. 16(3); ACHR Art. 17; ADRM Art. 8; ECHR Art. 8; ICESCR Art. 10(1) (this convention adds "particularly while it is responsible for the care and education of dependent children"). *See also,* DCRC Preamble ¶ 5.

16. The Draft Convention on the Rights of the Child proposes to reinforce the integrity of the family in several ways. First, under the Convention "applications by a child or parents to enter or leave a State Party for the purpose of family reunification shall be dealt with by States Parties in a positive, humane and expeditious manner." Second, children separated from one or both parents would have the right to maintain "personal relations and direct contact" on a regular basis. Third, "where such separation results from any action initiated by a State Party," the State is obligated to provide the child or parent "with essential information concerning the whereabouts of the absent member(s) of the family unless the provision of the information would be detrimental to the well-being of the child."

17. See UNITAR REPORT.

18. ICESCR Art. 10.

19. DRC Principle 6; *see also* Principles 2, 4.

20. DCRC Preamble ¶ 5.

21. *Id.* Art. 8(2).

22. *See generally,* UNITED NATIONS COMMISSION ON THE STATUS OF WOMEN, PARENTAL RIGHTS AND DUTIES, INCLUDING GUARDIANSHIP, E/CN.6/474/Rev. 1 (1968).

23. DCRC Art. 6(3).

24. E.g., USSR: Soviet Fundamental Principles of 1968, Art. 18(5); Greece: Code of Civil Procedure of Greece, Art. 681B.

25. Preamble ¶ 3.

26. DCRC Art. 6 ter.

27. G.A. Res. 260 (III)A. December 9, 1948, 78 U.N.T.S. 277.

28. Australia: *e.g.,* Children's Services Act 1965 sec. 46 (Queens.).

29. People's Republic of the Congo: Act of July 24, 1889.

30. Norway: The Child Welfare Act, Act of July 17, 1953, No. 14, Sec. 20.

31. United States: UNITAR Report at 694.

32. DRC Principle 9.

33. DDFPA Art. 5.

34. DCRC Art. 6(2).

35. DCRC Art. 8(1).

36. J. GOLDSTEIN, A. FREUD & A. SOLNIT, BEYOND THE BEST INTERESTS OF THE CHILD, *supra* n. 4, at 3–14.

37. *See* pp. 88–105 *infra.*

38. D. LEGARRETA, THE GUERNICA GENERATION (1984).

39. A. DE ONAINDIA, EXPERIENCIAS DEL EXILIO: HOMBRE DE PAZ EN LA GUERRA (1974) cited in D. Legarreta, *supra* n. 38.

40. Law 5, 1975, Art. 1, *amending* Civil Code, Art. 282.

41. H. D. KRAUSE, THE CREATION OF RELATIONSHIPS OF KINSHIP, 4 IECL. 6 Chap. 6, p. 81 (1976).

42. European Treaty Series No. 58, Art. 5(1)(a). Civil law countries typically look to the parties' personal (i.e., national) law to determine the validity of a consent to adoption, while common law countries apply their own domestic law. 2 A. EHRENZWEIG & E. JAYNE, PRIVATE INTERNATIONAL LAW 235 (1973). The Hague Convention on Adoption takes the position that the court or agency granting the adoption "shall apply the national law of the child relating to consents and consultations," Art. 5.

43. *Nguyen Da Yen v. Kissinger,* 528 F.2d 1194 (9th Cir. 1975); Carbonneau, *'Operation Babylift'—The Dilemma Surrounding Child Custody Controversies,* in THE FAMILY IN INTERNATIONAL LAW: SOME EMERGING PROBLEMS 87 (R. Lillich ed. 1981).

44. *Huynh Thi Anh v. Levi,* 586 F.2d 625 (6th Cir. 1978).

45. In the *Huynh Thi Anh* case *supra* n. 44, a settlement was reached in which the American foster parents retained physical custody of the children until they reached the age of 16, no adoption would take place, the plaintiff uncle and grandmother would receive liberal visitation rights, and at age 16 each child would be allowed to choose where he wanted to live. Interview with Martin Guggenheim, counsel for plaintiffs, in New York City (August 1983). *In Hao Thi Popp v. Lucas,* no. 1627-75 (Conn. Superior Ct. April 22, 1977) the Court held that the plaintiff mother's relinquishment of parental rights in Vietnam was voluntary, and that it was in the best interest of her children who wanted to remain with the foster parents to do so.

46. *Doan Thi Hoang Anh v. Nelson,* 245 N.W. 2d. 511 (Iowa 1976) (mother refused to sign release or consent to adoption in Vietnam). *Duong Bich Van v. Dempsey,* No. 76-140499 (Mich. Cir. Ct., June 21, 1976), cited in Carbonneau, *supra* n. 43, at 102 (mother refused four times to sign release for adoption in Vietnam).

47. *Le Thi Sang v. Knight,* No. 125898 (Cal. Superior Ct., April 25 and September 22, 1977) cited in Carbonneau *supra* n. 43, at 103.

48. DCRC Art. 11(1).

49. DDFPA Art. 23.

50. Capron, *The Competence of Children as Self-Deciders in Biomedical Interventions,* in WHO SPEAKS FOR THE CHILD 65 (W. Gaylin and R. Macklin, eds. 1982).

51. I. GOLDSCHMIDT, *The Adoption of Children from Latin American Countries,* in PROBLEMS CONCERNING THE ADOPTION OF CHILDREN FROM COUNTRIES OF THE "THIRD WORLD" (International Social Service—German Branch: 1982).

52. STOLJAR, *supra* n. 7, at 79–80.

53. *See e.g.,* Australia: Children's Services Act 1965 sec. 46 (Queens.). Israel: 14 L.S.I. sec. 2 (a minor is "in need of care and protection" if "there is no person responsible for the minor").

54. Czechoslovakia: Family Code, No. 94 (1963) sec. 44(1)(2).

55. Declaration of the Rights of the Child ("Declaration of Geneva") (1924) Art. 1.

56. UDHR Art. 25.

57. ICESCR Art. 10(3).

58. DRC Preamble ¶ 3.

59. DCRC Preamble ¶ 6.

60. Declaration of Geneva, *supra n. 55, at Art. II.*

61. DRC Principle 6.

62. DCRC Art. 10(1).

63. DDFPA Art. 5.

64. Convention Concerning the Powers of Authorities and the Law Applicable in Respect of the Protection of Infants. This convention is discussed *infra,* in Chapter Eighteen.

65. Pappas, *Introduction,* UNITAR REPORT x1.

66. DCRC Art. 1.

67. Fourth Convention, Arts. 23, 24, 38 [(*see also* ICCRP Art. 6(5)], and Protocol I, Art. 78.

68. Art. 4. For discussion of the substance of this convention see pp. 234–35 *infra.*

69. Art. 1.

70. Hague Conference on Private International Law, concluded 24 October 1956, Art. 1.

71. Hague Conference on Private International Law, concluded 15 April 1958, Art. 1.

72. Arts. 24, 50.

73. Principle 6.

74. DDFPA, Arts. 3–7.

75. UNITED NATIONS HIGH COMMISSIONER FOR REFUGEES, HANDBOOK FOR EMERGENCIES 161 (1983).

76. See Chapter Seventeen *infra.*

77. Art. 12.

78. S. STOLJAR, *supra* n. 7, at 78.

79. United Nations Commission on the Status of Women, Parental Rights and Duties, Including Guardianship, E/CN 6/474 Rev. 1 (1968).

80. See *infra* at 259–60.

81. DDFPA Art. 25.

82. S. STOLJAR, *supra* n. 7 at 85; *e.g.,* United States: Uniform Probate Code Sec. 5-209, 8 Uniform Laws Annotated 518 (1975).

83. N. ANDERSON, ISLAMIC FAMILY LAW, 4 IECL Chap. 11, p. 76. *See also* Republic of China: 2 LAWASIA 278–79 (grandparents living in the same family, the head of house, grandparents not living in the same family, paternal uncle, in that order); Spain: Civil Code Art. 211 (paternal grandfather, maternal grandfather, oldest uncle, oldest aunt, in that order).

84. France: Civil Code Art. 404.

85. *E.g.,* Colombia: Civil Code Art. 151.

86. DDFPA Art. 5.

87. *See e.g.,* TenBroeck, *California's Dual System of Family Law: Its Origin, Development, and Present Status, Part I,* 16 STAN. L. REV. 257 (1964); STOLJAR, *supra* n. 7 at 90–91.

88. Civil Code Art. 354.

89. RSFSR FC (1969) Art. 127(1).

90. Germany: Youth Welfare Law sec. 69(4); Colombia: 1 UNITAR Report 131.

91. E. Lloyd, The Children of Indochina, (Dept. of Immigration and Ethnic Affairs: unpublished 1983).

92. C. Rodier, The Care of Unaccompanied Southeast Asian Minors in France and Belgium (unpublished 1983).

93. D. Rouleau, The Unaccompanied Minor Refugee Program in Quebec (unpublished 1984); REFUGEE DOCUMENTATION PROJECT, UNACCOMPANIED CHILDREN IN EMERGENCIES: THE CANADIAN EXPERIENCE 170 (1984).

94. REFUGEE DOCUMENTATION PROJECT, *supra* n. 93, at 154–160.

95. C. Rodier, *supra* n. 92, at 26–28.

96. I. Baer and H. Jockenhovel-Schiecke, Problems of Guardianship Regulations and Possible Adoptions of Unaccompanied Cambodian Refugee Children and Adolescents in German Foster-Families (unpublished 1983).

97. C. Rodier, The Care of Unaccompanied Southeast Asian Minors in the Netherlands (unpublished 1983).

98. B. Nylund, Unaccompanied Minors: Their Legal Status in Sweden, Norway, Denmark (unpublished 1983).

99. *Id.* at 19–20, 25–26.

100. British Refugee Council et al. Unaccompanied Refugee Minors in the United Kingdom, A Report Prepared for Seminar on Unaccompanied Minor Refugees in European Resettlement Countries (unpublished February 1984).

101. 8 U.S.C. § 1522 (d)(2)(B).

102. Durkee, *Special Problems of Custody for Unaccompanied Refugee Children in the United States,* in MICHIGAN YEARBOOK OF INTERNATIONAL LEGAL STUDIES, TRANSNATIONAL LEGAL PROBLEMS OF REFUGEES 203 (1982); J. Paul, Entrance of Unaccompanied Children to the United States (unpublished 1983).

103. M. A. GLENDON, STATE LAW AND THE FAMILY 272 (1977). *See also* Luderitz, *The Legal Position of Children after Divorce in Germany,* in THE CHILD AND THE LAW, (F. Bates ed., Oceana: 1976) at 195: "It is an internationally accepted rule that the welfare of the child determines which parent will be given the right to custody on the separation of the parents."

104. UNITAR REPORT, *supra* n. 1, *passim. See also* LAWASIA Family Law Series, *supra* n. 1.

105. DRC Principle 2.

106. *Id.* Principle 7.

107. DCRC Art. 3.

108. DDFPA Art. 5.

109. Hague Adoption Convention Art. 6.

110. *See* discussion at pp. 234–35 *infra.*

111. Executive Committee of the High Commissioner's Program, Report of the Meeting of the Sub-Committee of the Whole on International Protection, A/AC. 96/599, 12 October 1981, ¶ 27.

112. *See* pp. 231–35 *infra.*

113. DRC Principle 3.

114. *See e.g.,* ICCPR Art. 2; ICESCR Art. 2.

115. BEYOND THE BEST INTERESTS OF THE CHILD (1979).

116. The Draft Declaration on Foster Placement and Adoption, for example, states that "the children's right to security, affection and continuing care should be of greatest importance," and that "it is in the child's best interest to reach [a decision on his future] as quickly as possible." DDFRA Article 5. Preference for continuity of relationships in placement is also becoming more common in national law. *See e.g.,* Luderitz, *The Legal Position of Children after Divorce in Germany,* in THE CHILD AND THE LAW, *supra* n. 103, at 200–01.

117. Mnookin, *Child-Custody Adjudication: Judicial Functions in the Face of Indeterminancy,* 39 LAW AND SOCIAL PROBLEMS No. 3, 221, 264–265 (1975).

118. Goldstein, *The Rights of the Child in Israel,* in 2 UNITAR REPORT 434.

119. *E.g.,* Norway: Social Welfare Act (Lov om sosial omsort), 1964-06-05, no. 2, sec. 3(1),(c) and (d).

120. See UNITAR Report; STOLJAR, *supra* n. 7, at 95: "This approach is now very widespread"; UNITED NATIONS ECONOMIC AND SOCIAL COUNCIL, PROTECTION AND WELFARE OF CHILDREN, 5, E/CN. 5 (504) (1974).

121. DCRC Preamble para. 7.

122. *See* ECOSOC COMMISSION ON HUMAN RIGHTS, REPORT OF THE WORKING GROUP ON A DRAFT CONVENTION ON THE RIGHTS OF THE CHILD, E/CN.4/1984/71/ (February 23, 1984).

123. DDFPA Arts. 7 and 8.

124. DDFPA Art. 10.

125. DRC Principle 6.

126. The discussion here concerns these issues in a general sense. The particular law that

would actually govern these issues would be determined by the jurisdictional and choice of law principles discussed in Chapter Eighteen, *infra.*

127. Variable competence based on the risks and benefits of the decision at hand and the nature of the decision is proposed in Gaylin, *Competence: No Longer All or None,* in WHO SPEAKS FOR THE CHILD 27–54 (W. Gaylin and R. Macklin, eds. 1982).

128. The evolution of legal and philosophical views of the child, in the West at least, has been fairly marked. Hobbes, Locke, and Mills, for example, regarded the child as a creature to be molded according to adult preconceptions. "None of these philosophers would have considered seriously the perspective of children themselves in determining their own best interest," Worsfold, *A Philosophical Justification for Children's Rights* in THE RIGHTS OF CHILDREN 29, 33 (1974).

129. Family Code Arts. 137(1) and 151(1975).

130. 2 UNITAR Report 431.

131. Greece: 2 UNITAR Report 375; People's Republic of the Congo: 1 UNITAR Report 191.

132. 1 UNITAR Report 80.

133. 1 UNITAR Report 15–16.

134. Mnookin, *Child-Custody Adjudication: Judicial Functions in the Face of Indeterminacy,* 39 LAW & CONTEMP. PROBS. 226 (Summer 1975).

135. H. D. KRAUSE, CREATION OF RELATIONSHIPS OF KINSHIP, 4 IECL Chap. 6.

136. DCRC Art. 3(2).

137. DCRC Art. 7.

138. Art. 13.

139. United Nations High Commissioner for Refugees, December 21, 1982.

140. *See also* ICCPR Art. 16.

141. *See also* ACHR Art. 8; ECHR Art. 6.

142. McDougal and Lasswell, *The Identification and Appraisal of Diverse Systems of Public Order,* 53 AM. J. INTL. L. 1, 24 (1959).

143. *Cf.* Tribe, *Childhood, Suspect Classifications and Conclusive Presumptions: Three Linked Riddles,* 39 LAW & CONTEMP. PROBS., No. 3, 8–37 (1975), at 32: "[T]here must be an opportunity, absent strong justification for denying it, for a child to rebut any implied or asserted age-based incapacity." *See also* Worsfold, *supra,* n. 128, at 38–44.

144. Weithorn, *Children's Capacities in Legal Contexts,* in CHILDREN, MENTAL HEALTH AND LAW (N. D. Reppucci et al. ed. 1983); Weithorn, *Involving Children in Decisions Affecting Their Own Welfare,* in CHILDREN'S COMPETENCE TO CONSENT (G. B. Melton et al. ed. Plenum: 1983).

145. J. WALLERSTEIN & J. KELLY, SURVIVING THE BREAKUP, 314–15 (1980).

146. *See* p. 219 *supra.*

147. *See* Chapter Seventeen.

148. *See* Chapter Seventeen.

149. UDHR Art. 12.

150. Concluded October 25, 1980. As of December 7, 1983, this convention had been ratified by Canada, Germany (Fed. Republic), Portugal, and Switzerland.

151. The European Convention on Recognition and Enforcement of Decisions Concerning Custody of Children and on Restoration of Custody of Children, European Treaty Series 105, concerns the narrower problem of *enforcement of judicial decisions* (decisions of an authority) relating to child custody among the members of the Council of Europe.

152. Art. 13.

153. Art. 20.

154. Art. 12.

155. The same result would not necessarily follow under the earlier 1961 Hague Convention on the Protection of Infants, discussed at length at pp. 234–35 *infra.* That convention obligates the State where the child is found to recognize "relationships subjecting the infant to

authority" of the State of the child's nationality and, in certain circumstances, to respect measures of protection taken by the authorities of that State. The law of the child's national State would thus effectively decide any conflict between the claim of the natural parents and the child's best interest. Given the widespread, if not global, acceptance of the best interests principle, however, a State would, under the Convention's public policy (ordre public) exception, be justified in refusing to respect any measures which did not adequately ensure the child's best interest (Art. 16). In practice, the Hague Convention on the Protection of Infants has been interpreted to respect the child's psychological and emotional ties to current caretakers. Wahler, *The Convention on the Protection of Infants and the Judicial Practice in West German Courts,* in 2 THE CHILD AND THE LAW, *supra* n. 103, at 507.

156. UNHCR, HANDBOOK FOR EMERGENCIES 161 (1982).

157. INTERNATIONAL UNION FOR CHILD WELFARE, THE PSYCHOLOGICAL, EDUCATIONAL AND SOCIAL ADJUSTMENTS OF REFUGEE AND DISPLACED CHILDREN IN EUROPE 46 (1952), J. GOLDSTEIN, A. FREUD & A. SOLNIT, BEYOND THE BEST INTERESTS 107-8.

158. STOLJAR, *supra* n. 7, at 81-82, 87.

159. UNRRA, "Unaccompanied Displaced Children Found in Enemy Territory" (unpublished June 19, 1945).

160. Chapter Nineteen *infra.*

161. A. Brownlee, "UNRRA Mission to Austria: Child Welfare in the Displaced Persons Programme" 6 (unpublished 1947).

162. UNRRA *supra* n. 158, at p. 6; Preparatory Commission International Refugee Organization "Exhibit #24" (unpublished October 1, 1947).

163. IRO Operational Manual, Appendix XVI.

164. A. Brownlee, supra n. 161, at 7-12.

165. *Id.* at 15.

166. ECOSOC Res. VII/157, August 24, 1948.

167. PCIRO, Provisional Order No. 75, July 26, 1948.

168. L. HOLBORN, THE INTERNATIONAL REFUGEE ORGANIZATION 501 (1956).

169. See Chapter Eighteen *infra.*

170. Hague Conference on Private International Law, Convention Concerning the Powers of Authorities and the Law Applicable in Respect of the Protection of Infants, Art. 16.

171. Case concerning the Application of the Convention of 1902 Governing the Guardianship of Infants (Netherlands v. Sweden), 1958 I.C.J. 55, 90 (separate opinion of Judge Lauterpacht). This case is also known as the *Boll* case and is discussed in Chapter Eighteen *infra.*

172. Decree Law XVIII/1957, amending sec. 76 of the Act concerning Family and Guardianship.

173. *See generally,* A. GRAHL-MADSEN, TERRITORIAL ASYLUM (1980).

174. UDHR Art. 13(2); ICCPR Art. 12(2). There are also, of course, the rights to seek and enjoy asylum, UDHR Art. 14(1), and rights under refugee law, Chapter 17, *infra.*

175. Cuban Family Code Art. 99. The European Convention on the Adoption of Children, European Treaty Series No. 58, defines the effects of adoption as follows in Article 10, (1):

> Adoption confers on the adoptor in respect of the adopted person the rights and obligations of every kind that a father or mother has in respect of a child born in lawful wedlock.
> Adoption confers on the adopted person in respect of the adoptor the rights and obligations of every kind that a child born in lawful wedlock has in respect of his father or mother.

176. H. D. KRAUSE, CREATION OF RELATIONSHIPS OF KINSHIP, 4 IECL Chap. 6.

177. DDFPA Art. 12.

178. ECOSOC Commission for Social Development, Protection and Welfare of Children: Report of the Secretary General, E/CN.5/504 (1974).

179. P. H. Neuhaus, The Family in Religious and Customary Laws 74, 4 IECL, Chap. 11 (1973).

180. U.N. Department of Economic and Social Affairs, Adoption and Foster Placement of Children 22, ST/ESA/99 (1980).

181. P. H. Neuhaus, *supra* n. 179, at 88; T. P. Gopalakrishnan, Hindu Law 34–91 (1968).

182. Art. 11.

183. *See supra* n. 178.

184. H. D. Krause, *supra* n. 176, European Convention on the Adoption of Children, Arts. 6–8.

185. In addition, the effects of simple adoption may not extend to relatives of the adopter's family for inheritance or support purposes, and the child's natural family may continue to have a subsidiary duty of support.

186. Countries permitting only full adoption include Cuba, Czechoslovakia, England, Norway, Switzerland, the Soviet Union, and the United States (most states). Those permitting both simple and full adoption include Belgium, Colombia, France, Italy, and Spain. Countries permitting simple adoption only include Israel and West Germany.

187. Code Civil Art. 370.

188. Art. 13, sec. 1.

189. H. D. Krause, *supra* n. 176, at 95.

190. *E.g.,* European Convention on the Adoption of Children, Arts. 4, 8; Hague Convention on Adoption, Art. 6; Draft Convention on the Rights of the Child, Art. 11(2).

191. K. H. Neumayer, *General Introduction,* in The Reform of Family Law in Europe, *supra* n. 2, at 3; *see also* ECOSOC Commission for Social Development, ECOSOC, *supra* n. 178, para. 28.

192. *See generally,* 2 A. A. Ehrenzweig & E. Jayme, Private International Law 235 (A. W. Sijthoff: 1973); Chao Shou-po, Comparative Aspects of Conflict of Laws in Domestic Relations 320–366 (1983); I. di Delupis, International Adoptions and the Conflict of Laws (1975).

193. Council of Europe, European Treaty Series No. 58. As of December 5, 1981, the following States had acceded to this Convention: Austria, Denmark, Germany (Fed. Rep.), Greece, Ireland, Italy, Lichtenstein, Malta, Norway, Portugal, Sweden, Switzerland, United Kingdom. In addition, the Nordic Convention (Sweden, Denmark, Finland, Iceland, and Norway) provides that any application for adoption must be made in the country of domicile or permanent residence of the adopter. There are also bilateral agreements concerning adoption among Eastern European States. Adoption and Foster Placement of Children, *supra* n. 180 at 34. The United Nations sponsored a European Seminar on Intercountry Adoption in 1960 which produced twelve fundamental principles for intercountry adoption. The recommendations for legal measures include: the necessary consents must be legally valid in both countries, the child must be able to immigrate into the country of the prospective adopters and obtain their nationality, the legal responsibility for the child should be established promptly in the new country, and steps should be taken to assure that the adoption is legally valid in both countries. European Seminar on Intercountry Adoption, *Report,* UN/TAO/SEM/1960/Rep. 2 (Leysin, Switzerland 22–31 May 1960). For an early international effort at regulating choice of law for adoption, see the "Bustamante Code," Am. J. Intl. L. 273 (1928). This Code provides that the conditions for adoption are governed by each party's personal law.

194. Hague Conference on Private International Law, concluded November 15, 1965.

195. Art. 6.

196. Art. 8.

197. Austria, Switzerland, United Kingdom.

198. Principles 21–25.

199. Art. 11(2).

200. *Supra,* pp. 220–21.

201. Civil law countries typically look to the parties' personal law to determine the validity of a consent to adoption, while common law courts apply their own law. 2 A. A. EHRENZWEIG & E. JAYNE, PRIVATE INTERNATIONAL LAW 235 (A. W. Sijthoff: 1973). The Hague Convention on Adoption takes the position that the court or agency granting the adoption "shall apply the national law of the child relating to consents and consultations," other than those with respect to an adoptor, his family, or his spouse." Art. 5.

202. European Seminar on Intercountry Adoption *supra* n. 193, Principle 5. The European Convention on Adoption Art. 5(4), limits the mother's right of consent following birth:

> A mother's consent to the adoption of her child shall not be accepted unless it is given at such time after the birth of the child, not being less than six weeks, as may be prescribed by law, or, if no such time has been prescribed, at such time as, in the opinion of the competent authority, will have enabled her to recover sufficiently from the effects of giving birth to the child.

203. A. E. VON OVERBECK, PERSONS, 3 IECL Chap. 15, pp. 8–12; Note, *Declarations of Death—A New International Convention,* 25 ST. JOHN'S L. REV. 18, 23 (1950).

204. Note, *supra* n. 203, at 23–25.

205. Civil Code sec. 1681 (1).

206. United States: 37 U.S.C. sec. 555.

207. Szabad and Blum, *Proving Death of Victims of Nazi Oppression,* 24 N.Y.U.L.O. 577, 584–86 (1949).

208. *Id.* at 585.

209. *Id.*

210. 119 United Nations Treaty Series 122, concluded April 6, 1950.

211. Art. 1.

212. Art. 3(1)(iv).

213. Art. 3(2)(iii) and (iv).

214. Sussman, *Declaration of Death in the Israeli Courts,* 2 I.C.L.Q. 614 (1953).

215. The States Party were Israel, Belgium, Pakistan, Germany (Fed. Rep.), Italy, Cambodia, and Guatemala. The Protocols extending the Convention's validity appear at 258 U.N.T.S. 392 and 588 U.N.T.S. 290. The Convention expired on January 24, 1972.

216. European Convention on the Adoption of Children, Art. 5(3):

> If the father or mother is deprived of his or her parental rights in respect of the child, or at least the right to consent to an adoption, the law may provide that it shall not be necessary to obtain his or her consent.

217. *See* section on Parent-Child Separations, *supra,* p. 218-21.

218. *Id.*

219. Civil Code sec. 1747(3).

220. Adoption Act 1976, sec. 16(2)(c) and (e).

221. 1 THE CHILD AND THE LAW, 289 (The Proceedings of the First World Conference of the International Society on Family Law, Berlin, April 1975) (F. Bates ed.: 1976).

222. Adoption Act 1976, sec. 16(2)(a).

223. 1 UNITAR REPORT 437.

224. Greece: Decree Law 610, Art. 11.

225. *E.g.,* England, Adoption Act 1976, sec. 16(2)(a).

226. Adoption Act 1976, sec. 16 (2)(b).

227. 3 All E.R. 613 1966.

228. *Id.* at 617.

229. See section on Parent-Child Separations, *supra,* pp. 26–29.

230. Art. 5(2).

CHAPTER 17

1. Principle 8.
2. *E.g.,* Cuba: 1 UNITAR Report 257.
3. Pappas, *Introduction,* 1 UNITAR REPORT xxix–xxx.
4. Article III.
5. Principle 6.
6. *See generally,* P. MACALISTER-SMITH, INTERNATIONAL HUMANITARIAN ASSISTANCE (1985).
7. There are four Geneva Conventions of 12 August 1949:
   1. Convention for the Amelioration of the Condition of the Wounded and Sick in Armed Forces in the Field;
   2. Convention for the Amelioration of the Condition of Wounded, Sick, and Shipwrecked Members of Armed Forces at Sea;
   3. Convention Relative to the Treatment of Prisoners of War;
   4. Convention Relative to the Protection of Civilian Persons in Time of War.

Protocol I (Protocol Additional to the Geneva Conventions of 12 August 1949, and Relating to the Protection of Victims of International Armed Conflicts) was completed on June 10, 1977 and is designed to "develop the provisions protecting the victims of armed conflicts" and to "supplement" the Geneva Conventions of 12 August 1949. Like its predecessors, it concerns international armed conflicts. Protocol II (Protocol Additional to the Geneva Conventions of 12 August 1949, and Relating to the Protection of Victims of Non-International Armed Conflicts) applies to conflicts "which take place in the territory of a High Contracting Party between its armed forces and dissident armed forces or other organized groups which, under responsible command, exercise such control over a part of its territory as to enable them to carry out sustained and concerted military operations and to implement this Protocol" [Art. 1(1)].

The United Nations General Assembly has passed several resolutions on protection of civilians in war, including the Declaration on the Protection of Women and Children in Emergency and Armed Conflict, G.A. Res. 3318 (XXIX), Dec. 14, 1974. This declaration concerns "periods of emergency and armed conflicts in the struggle for peace, self-determination, national liberation and independence." It adds little to the law of either armed conflict or children, merely stating that women and children "in circumstances of emergency and armed conflict" of the type described above "shall not be deprived of shelter, foods, medical aid or other inalienable rights," in accordance with the provisions of the basic international instruments (para. 6). See also G.A. Res. 2444 (XXIII), Dec. 19, 1968; G. A. Res. 2674 and 2675 (XXV), Dec. 9, 1970.

There has been some discussion of including an article dealing with children in armed conflicts in the Draft Convention on the Rights of the Child. The Informal NGO Ad Hoc Group on the Drafting of the Convention on the Rights of the Child has proposed an article requiring the Parties "in . . . internal and international armed conflicts, [to] take special measures to prevent all children from being subjected to any form of physical or psychological violation and to ensure that they are always among the first to receive protection and care." "Report of Informal Consultations among International Non-Governmental Organizations," 1 INTERNATIONAL CHILDREN's RIGHTS MONITOR No. 2 (1983). In 1985, the Netherlands, Belgium, Sweden, Finland, Peru, and Senegal proposed that the Convention on the Rights of the Child include provisions requiring States Parties to "respect and to ensure respect for rules of humanitarian law applicable in armed conflicts which are relevant to children," and to "refrain in particular from recruiting children into the armed forces . . . and . . . take all feasible measures to ensure that children do not take part in hostilities." ESOC, Commission on Human Rights, Report of the Working Group on a Draft Convention on the Rights of the Child, E/CN. 4/1985/L.1, Annex, p. 1 (1985). These provisions were not considered by the working group at its 1985 session, however.

8. Plattner, *Protection of Children in International Humanitarian Law,* in *Children and War* 199 (M. Kahnert, D. Pitt, I. Taipale eds. 1983); J. PICTET, HUMANITARIAN LAW AND THE

PROTECTION OF WAR VICTIMS 120 (1975). Singer, *The Protection of Children During Armed Conflicts,* 1986 INTL. REV. OF THE RED CROSS 133 (May–June 1986).

9. Protocol I Art. 50; Third Convention Art. 4(A)(1)(2)(3) and (6).

10. Protocol I, Arts. 8(a), 10.

11. Protocol I, Art. 77; Protocol II, Art. 4(3).

12. Fourth Convention, Art. 23.

13. Protocol I, Art. 70(1).

14. Protocol I, Art. 77(2).

15. Protocol I, Art. 77(4).

16. Fourth Convention, Art. 68(4); Protocol I, Art. 77(5).

17. Protocol I, Art. 76(3).

18. Fourth Convention, Art. 38(5).

19. Fourth Convention, Art. 50.

20. Fourth Convention, Art. 81.

21. *Id.,* Art. 89.

22. *Id.,* Art. 94. In noninternational conflicts, States Party to Protocol II undertake to assure all children an education, Art. 4(3)(a).

23. Fourth Convention Art. 24. This provision derives from Article 46 of the Hague Regulations Respecting the Laws and Customs of War on Land, 28 October 1907: "Family honour and rights, the lives of persons and private property, as well as religious convictions and practice, must be respected." *See* J. PICTET, *supra* n. 8, at 122: "Respect for family rights does not merely imply that ties must be preserved but further that, where those ties have been broken by war, they shall be restored by means of correspondence and the reuniting of dispersed families, as the Convention lays down elsewhere."

24. Fourth Convention, Art. 50.

25. *Id.,* Art. 49.

26. *Id.,* Protocol II, Art. 17.

27. G.A. Res. 260 (III)A, 9 December 1948, 78 U.N.T.S. 277.

28. Fourth Convention, Art. 49.

29. *Id.,* Art. 82.

30. Protocol I, Arts. 75(5), 77(4).

31. Protocol II, Art. 4(3)(e). This provision and its equivalent in Art. 17 of the Fourth Convention, *infra* n. 34, have their origin in Article 19 of the "Lieber Instructions":

> Commanders, whenever admissible, inform the enemy of their intention to bombard a place, so that the noncombatants, and especially the women and children, may be removed before the bombardment commences. But it is no infraction of the common law of war to omit thus to inform the enemy. Surprise may be a necessity.

Instructions for the Government of Armies of the United States in the Field, prepared by Francis Lieber, promulgated as General Orders No. 100 by President Lincoln, 24 April 1863, *reprinted in* D. SCHINDLER and J. TOMAN, THE LAWS OF ARMED CONFLICTS 6 (1973).

32. Fourth Convention, Art. 15.

33. *Id.,* Art. 14.

34. *Id.,* Annex I: Draft Agreement Relating to Hospital and Safety Zones and Localities.

35. *Id.,* Art. 17.

36. INTERNATIONAL COMMITTEE OF THE RED CROSS, DRAFT ADDITIONAL PROTOCOLS TO THE GENEVA CONVENTIONS OF AUGUST 12, 1949: COMMENTARY 87 (1973).

37. Report of Committee III, Fourth Session (CDDH/407/Rev. 1; XV, 445) *quoted in* 4 H. LEVIE, PROTECTION OF WAR VICTIMS: PROTOCOL I TO THE 1949 GENEVA CONVENTIONS 113 (1981).

38. H. LEVIE, *supra* n. 37, at 112–13.

39. Remarks of Mr. Ajayi at Plenary Meeting of May 27, 1977 (CDDH/SR. 43; VI 243).

40. *Id.* at 116.

41. Protocol I, Art. 78(1) and (2).

42. *Id.* Art. 78(3).

43. It is significant that the original draft of this Article, which appeared as Article 69 of the ICRC Draft Additional Protocols to the Geneva Conventions of August 12, 1949 (1973) applied to all evacuations of children to foreign countries, whether the children were nationals of the Party arranging the evacuation or not. The final version narrowed its coverage to evacuation of children who are not the nationals of the State arranging the evacuation.

44. Fourth Convention, Art. 24.

45. *Id.* Art. 50.

46. *Id.* Arts. 25, 107.

47. *Id.* Art. 26.

48. *Id.* Arts. 136–141.

49. *Id.* Art. 134.

50. *Id.* Art. 132. Such women and children should be repatriated, returned to their places of residence or accommodated in a neutral country.

51. Protocol II, Art. 4(3)(a). *Cf.* Fourth Convention, Art. 26.

52. Fourth Convention, Art. 50.

53. Art. 24, para. 2.

54. INTERNATIONAL COMMITTEE OF THE RED CROSS, DRAFT ADDITIONAL PROTOCOLS TO THE GENEVA CONVENTIONS OF AUGUST 12, 1949: COMMENTARY, Art. 69(1) (1973).

55. *Id.* at 88.

56. Remarks of Mr. Surbeck, Meeting of Committee III, 5 May 1976 (CDDH/III/SR. 45; XV, 63), in 4 H. LEVIE, *supra* n. 37, at 108.

57. *See* pp. 248–50 *supra.*

58. S. Forbes & P. Fagan, Unaccompanied Refugee Children: The Evolution of U.S. Policies: 1939, 77 (unpublished draft manuscript of the Refugee Policy Group: 1984).

59. EUROPEAN CONSULTATION ON REFUGEES AND EXILES, ASYLUM IN EUROPE 14–15 (3d ed. 1983).

60. Federal Republic of Germany: H. Jockenhovel-Schiecke, "Problems of Guardianship Regulations and Possible Adoptions of Unaccompanied Cambodian Refugee Children and Adolescents in German Foster Families (1983) (unpublished). Holland: C. Rodier, "The Care of Unaccompanied Southeast Asian Minors in the Netherlands" (1983) (unpublished). Sweden: B. Nystrom, "Unaccompanied Minors: Their Legal Status in Sweden, Norway, Denmark" (1983) (unpublished). U.S.: J. Paul, "Entrance of Unaccompanied Children to the United States" (1983) (unpublished).

61. Pub. L. 96–212, March 17, 1980, 94 Stat. 109, section 412 (d), 207.

62. National Security Decision Directive No. 93 (May 1983).

63. Act of June 25, 1948, 62 Stat. 1009.

64. Orphan Act of July 29, 1953, 67 Stat. 229; Refugee Relief Act of Aug. 7, 1953, 71 Stat. 639; Refugee-Escapee Act of Sept. 11, 1957, 71 Stat. 639; Fair Share Refugee Act of July 14, 1960, 74 Stat. 504.

65. Immigration and Nationality Act, as amended, § 101(b)(1)(F), 8 U.S.C. § 1101(b)(1)(F).

66. *Id.* § 204e, 8 U.S.C. § 1154(e).

67. *E.g.,* U.S.: Immigration and Naturalization Act §§ 201(b), 203(a)(2)(3); 8 U.S.C. §§ 1151(b), 1153(a)(2)(3). United Kingdom: Immigration Act 1971; H.C. 80 ¶ 43.

68. Section 2, AuslG.

69. E. VAN DER HOEVEN AND H. DE KORT, RECEPTION AND CARE OF UNACCOMPANIED VIETNAMESE MINORS IN THE NETHERLANDS *1974–1984* 22 (Coordination Commission Scientific Research Child Protection: 1984) (seventy-five percent of Vietnamese unaccompanied minors arriving in the Netherlands had at least three "direct family members" in Vietnam or outside the Netherlands).

70. British Refugee Council et al., Unaccompanied Refugee Minors in the U.K.: A Report Prepared for Seminar on Unaccompanied Minor Refugees in European Resettlement Countries, § 5.1 (1984) (unpublished).

71. United Nations High Commission for Refugees, Statement of December 5, 1979.

72. The Convention is contained in 189 United Nations Treaty Series (U.N.T.S.) 137; the Protocol in 606 U.N.T.S. 267. As of February 1, 1984, ninety-one States were parties to the Convention and Protocol, three to only the Convention and two to only the Protocol.

73. Convention, Art. 1(A)(2); Protocol Art. 1(2).

74. Art. 1(2). The Convention entered into force 20 June 1974, U.N.T.S. No. 14691.

75. OFFICE OF THE UNITED NATIONS HIGH COMMISSIONER FOR REFUGEES, HANDBOOK ON PROCEDURES AND CRITERIA FOR DETERMINING REFUGEE STATUS, HCR/PRO/4 (1979) ¶ 184 [hereafter cited as UNHCR HANDBOOK].

76. *Id.,* ¶¶ 185–6. The "principal of family unity" in the administration of refugee law derives specifically from The Final Act of the 1951 United Nations Conference of Plenipotentiaries on the Status of Refugees and Stateless Persons, ¶ IV/B, 189 U.N.T.S. 37:

> Considering that the unity of the family, the natural and fundamental group unit of society, is an essential right of the refugee. . . .
> Recommends Governments to take the necessary measures for the protection of the refugee's family, especially with a view to:
> (1) Ensuring that the unity of the refugee's family is maintained particularly in cases where the head of the family has fulfilled the necessary conditions for admission to a particular country. . . .

77. 1A GRAHL-MADSEN, THE STATUS OF REFUGEES IN INTERNATIONAL LAW 1, 413 (1966).

78. *Cf.* Pearl, *Dependency in the Immigration Law of the United Kingdom,* in THE CHILD AND THE LAW (The Proceedings of the First World Conference of the International Society on Family Law, Berlin, April 1975) (F. Bates ed. 1976).

79. *E.g.,* UNHCR certificates of foster parent/child relationship Chapter Nineteen, *infra.*

80. IRO Constitution, Annex I, Part I, Section B. ¶ 4.

81. UNHCR HANDBOOK, *supra* n. 75, ¶ 213.

82. UNHCR HANDBOOK, *supra* n. 75, ¶¶ 217–219.

83. *Id.* at ¶¶ 213–219.

84. G. GOODWIN-GILL, THE REFUGEE IN INTERNATIONAL LAW 40 (1983).

85. 1 A. GRAHL-MADSEN, *supra* n. 77, at 212, 220–24, 231–48.

86. If the penalty for unauthorized absence (or presumably, draft evasion) is outrageously severe this may suffice. *Id.,* at 241.

87. Grahl-Madsen himself favors this approach, *id.,* at 250–51. UNHCR also urges refugee status for politically (religious, etc.) motivated unlawful departure, but only, seemingly, if the person is exposed to "severe penalties" for his absence. UNHCR HANDBOOK ¶ 61.

88. This differs from Grahl-Madsen's concept of "indirect persecution," where family members may be seriously affected if one member, particularly the breadwinner, is killed, imprisoned, or prevented from earning a living. 1A. GRAHL-MADSEN, *supra* n. 77, at 423–4.

89. *Cf.* UNHCR HANDBOOK, ¶ 217.

90. Executive Committee of the UNHCR Programme, *Conclusions on the International Protection of Refugees* No. 8 (XXVIII) ¶ (iv), HCR/PRO/2/Rev. 2. (1980); UNHCR HANDBOOK ¶ 192.

91. UNHCR HANDBOOK, ¶ 195.

92. *Id., ¶ 190.*

93. For a discussion of the role of a guardian see pp. 225–27 *infra.*

94. In *Perez-Funez v. District Director, Immigration and Naturalization Service,* 619 F.Supp. 656 (C.D. Cal. 1985) and 611 F.Supp. 990 (C.D. Cal. 1984), an American court

addressed the question of what procedural protections should be provided unaccompanied minors apprehended in the United States who are potentially eligible for asylum. Prior to this decision, the immigration authorities had encouraged most of these children to "voluntarily depart" from the United States and thereby waive the right to apply for political asylum and other forms of discretionary relief. The court noted that with this practice

> unaccompanied children of tender years encounter a stressful situation in which they are forced to make critical decisions. Their interrogators are foreign and authoritarian. The environment is new and the culture completely different. The law is complex. The children generally are questioned separately. In short, it is obvious to the Court that the situation faced by unaccompanied minor aliens is inherently coercive. (619 F.Supp. at 662)

The court ordered that unaccompanied alien minors be given oral and written notice of their rights to retain an attorney and to apply for political asylum. In addition, the minors must be given a current list of free legal services providers and a telephone call to a parent, close relative or friend, or legal services organization before being offered voluntary departure. The court declined, however, to direct that representation by free counsel or review of the waiver of rights by an immigration judge precede voluntary departure for unaccompanied minors.

95. *Supra* n. 76.
96. UNHCR Handbook, ¶¶ 217–219.

CHAPTER 18

1. Subject to whatever international obligations, conventional or otherwise, to which it is bound.
2. A/RES/38/142 (December 19, 1983).
3. The Charter of the United Nations, Art. 2, Section 7, states:

> Nothing contained in the present charter shall authorize the United Nations to intervene in matters which are essentially within the domestic jurisdiction of any State or shall require the members to submit such matters to settlement under the present Charter.

One arguable exception to territorial sovereignty is the doctrine of humanitarian intervention: intervention in the internal affairs of one State by another (or others) when the first State is treating its people in such a way as to deny their most fundamental human rights and to shock the conscience of mankind. G. von Glahn, Law Among Nations 165 (4th ed. 1981) (*citing* Lauberpacht's Oppenheim). The acceptance, definition, and wisdom of the doctrine of human intervention in international law are all widely debated, *see e.g.,* Humanitarian Intervention and the United Nations (R. B. Lillich ed. 1973).

4. *See e.g.,* United Nations Economic and Social Council, *Study on Human Rights and Massive Exoduses,* E/CN.4/1503 (1981).
5. P. M. Cheshire and G. C. North, Private International Law 3 (10th ed. 1979).
6. Case Concerning the Application of the Convention of 1902 Governing the Guardianship of Infants (Netherlands v. Sweden), 1958, I.C.J. Rep. 55.
7. Convention Pour Regler la Tutelle des Mineurs, 12 June 1902, Third Session, Hague Conference on Private International Law.
8. 1958 I.C.J. Rep. 69.
9. 1958 I.C.J. Rep. 71.
10. Art. 9, Convention Concernent la Competence des Autorites et la Loi Applicable en Matiere de Protection des Mineurs. (Convention Concerning the Powers of Authorities and the

Law Applicable in Respect of the Protection of Infants), concluded October 5, 1961 at the Hague Conference on Private International Law. There is no official English text; the English translation relied on here is the one settled at a meeting of members of the Danish, Finnish, Norwegian, Japanese, United Kingdom, and United States delegations on October 25, 1960 at the Hague, and printed in 9 Am. J. Comp. Law 708 (1960). As of Dec. 7, 1983, the Convention had been acceded to or ratified by Austria, France, Germany (Fed. Rep.), Luxembourg, Netherlands, Portugal, Switzerland, and Turkey.

11. A. E. von Overbeck, Persons, 3 I.E.C.L., Chap. 15, pp. 20–21; Stoljar, Children, Parents, and Guardians, 4 I.E.C.L., Chap. 7.

12. *Id.*

13. Convention Relating to the Status of Refugees, Art. 12, 28 July 1951, 189 U.N.T.S. 137; made applicable to States Party to the Protocol Relating to the Status of Refugees, 31 January 1967, by Art. 1 (1) of that Protocol.

14. *Id.* Art. 12(2).

15. *Supra* n. 10.

16. *Id.* Art. 1.

17. *Id.* Art. 2.

18. *Id.* Art. 3.

19. *Id.* Art. 4.

20. *Id.* Art. 7.

21. *Id.* Art. 8 (emphasis added).

22. *Id.* Art. 16.

23. Cheshire and North, *supra* n. 5, at 187.

24. Blom, *The Adoption Act of 1968 and Conflict of Laws,* 22 I.C.L.Q. 109 (1973).

25. *Id.* at 136.

26. Wahler, *The Convention on the Protection of Infants and the Judicial Practice in West German Courts,* in 2 The Child and the Law 507 (The Proceedings of the First World Conference of the International Society on Family Law, Berlin 1975) (F. Bates ed. 1976).

27. The public order exception contained in the Hague Convention on the Protection of Infants, Art. 16, referred to above, would certainly permit such a result, especially in view of the *Boll* decision. The policy grounds for such an approach have been summarized in A. E. von Overbeck, Persons, 3 IECL Chap. 15, p. 20:

> At a time when institutions such as guardianship were regarded essentially from the point of view of legal representation enabling the incapable person to carry out legal transactions, it was only natural to deal with these matters in accordance with the law which governs capacity. This is the attitude of the Convention for the Regulation of the Guardianship for Minors of 12 June 1902 which is based on the principle of nationality. However, since this Convention was signed, new ideas were developed in this field and it has become clear that guardianship and other measures of protection of minors do not serve primarily the purpose of permitting legal transactions to be concluded in the name of an infant, but rather of ensuring his well-being and education. This task can only be carried out conveniently by the authorities of the country where the infant happens to be. Thus the authorities of the infant's domicile or habitual residence appear in practice to assume a role which was not envisaged by the conventions based on nationality.

CHAPTER 19

1. As used here, "international organizations" are those such as UNHCR, UNICEF, and the International Committee of the Red Cross which are internationally chartered. Voluntary organizations are usually chartered as nonprofit charitable corporations in one State, though they may operate internationally.

2. Statutes of the International Committee of the Red Cross, (adopted September 25, 1952, amended January 8, 1964 and May 6, 1971) Art. 4(d).

3. Statutes of the International Red Cross (adopted 1928, revised 1952), Art. VI (5) and (6).

4. Fourth Geneva Convention, Art. 10.

5. *Id.* Arts. 63, 142.

6. *Id.* Art. 3.

7. Protocol I, Art. 5.

8. Protocol I, Art. 74.

9. Fourth Geneva Convention Arts. 136–141.

10. General Assembly Res. 428 (V), 14 December 1950. *See generally,* G. GOODWIN-GILL, THE REFUGEE IN INTERNATIONAL LAW 5–12, 129–136 (1983).

11. *See* United Nations Resolutions and Decisions Relating to the Office of the United Nations High Commissioner for Refugees (3d ed.), HCR/INF/48 Rev. 2.

12. *See generally,* Fowler, *The Developing Jurisdiction of the United Nations High Commissioner for Refugees,* 7 REVUE DES DROITS DE L'HOMME 119 (1974).

13. Statute of the Office of the United Nations High Commissioner for Refugees, ¶ 8(b)(c) and (d).

14. *Id.* at ¶ 2.

15. Final Act of the United Nations Conference of Plenipotentiaries on the Status of Refugees and Stateless Persons, ¶ IV.B., HCR/INF/29 Rev. 2, p. 9 (1951).

16. Pask and Jayne, *Refugee Camps and Legal Problems: Vietnamese Refugee Children,* 22 J. FAM. L. 537, 543 (1983).

17. G.A. Res. 57(I), December 11, 1946, secs. 1 and 2.

18. United Nations, Economic and Social Council, An Overview of UNICEF Policies, Organization and Working Methods, E/ICEF/670/Rev. 2, p. 1.

19. G.A. Res. 417(V), December 1, 1950. *See also* G. A. Resolution 318 (IV) in which the General Assembly *"Notes with concern* the existence of children's emergency needs arising out of war and other calamities as well as the great needs which the Fund's experience has demonstrated as existing in underdeveloped countries."

20. *See generally,* M. EL BARADEI, MODEL RULES FOR DISASTER RELIEF OPERATIONS (1982); P. MACALISTER-SMITH, INTERNATIONAL HUMANITARIAN ASSISTANCE (1985).

21. Except in those rare circumstances where the local order has broken down and any group with the means to enter may do so.

22. *See e.g.,* G.A. Res. 57(I), section 2(d): "The Fund [UNICEF] shall not engage in activity in any country except in consultation with, and with the consent of, the Government concerned."

23. Protocol I, Art. 81(1).

24. *Id.* at Art. 81 (2), (3).

25. *See e.g.,* Protocol I, Art. 81(4).

26. Agreement Relating to UNHCR Assistance with Cuban Refugees (May 16, 1980), 19 INTERNATIONAL LEGAL MATERIALS 1296 (1980).

27. *See* Regulations Relating to the Guardianship or Adoption of Orphans in Centres for Displaced Persons in Thailand, (Royal Thai Government: undated).

28. UNRRA TWE (45) 30.

29. UNRRA Welfare Division, 25 III. 45.

30. AG 383, 7-1, GE AGGM, 6 March 1945.

31. *Id.*

32. Agreement Between the U.S. High Commissioner for Austria and the Preparatory Commission for the International Refugee Organization, PC/LEO/31, Section 5, December 7, 1947.

33. *Id.* at Art. IV (10).

34. *Id.* at Section 8.

35. Agreement Between the Preparatory Commission for the International Refugee Organization and the Allied Commission for Austria (British Element), PC/LEG/5,7, December 1947, Art. III.

36. *Id.* at Art. IV(10).

37. ICRC, "Rules to be Observed Regarding the Problem of Unaccompanied Children," (undated).

38. Statement of the United Nations High Commissioner for Refugees, 5 December 1979.

39. Interview with Simon Cornwell, former Director, Tracing Program, International Rescue Committee, July 18, 1983, at Bangkok, Thailand.

40. Executive Committee of the High Commissioner's Programme, Report on the Meeting of the Subcommittee of the Whole on International Protection A/AC. 96/599, 12 October 1981, ¶ 27.

41. *Id.* at ¶ 2–6.

42. *Id.* at ¶ 27.

43. UNHCR, December 21, 1982.

44. By way of comparison, the International Court of Justice "whose function is to decide in accordance with international law such disputes as are submitted to it," shall apply:

> a) international conventions, whether general or particular, establishing rules expressly recognized by the contesting States;
> b) international custom, as evidence of a general practice accepted as law;
> c) the general principles of law recognized by civilized nations;
> d) . . . judicial decisions and the teachings of the most highly qualified publicists of the various nations, as subsidiary means for the determination of rules of law.

Statute of the International Court of Justice Art. 38(1).

45. L.F.L. OPPENHEIM, INTERNATIONAL LAW: A TREATISE (8th ed. H. Lauterpacht) 877 (1955).

46. C. PARRY, THE SOURCES AND EVIDENCE OF INTERNATIONAL LAW 34 (1965).

47. Vienna Convention on the Law of Treaties, U.N. Doc. A/CONF 39/27, Art. 34; North Sea Continental Shelf Cases, 1969 I.C.J. 25–26.

48. Art. 4.

49. Art. 2.

50. Advisory Opinion on the Jurisdiction of the Courts of Danzig, PCIJ Publ. 3. 15 (1928).

51. Vienna Convention, *supra* n. 48, Art. 26.

52. *Id.* Art. 60(2). This right of suspension or termination does not apply to "provisions relating to the protection of the human person contained in treaties of a humanitarian character, in particular to provisions prohibiting any form of reprisals against persons protected by such treaties," Art. 60(5).

53. *E.g.,* The 1951 Convention Relating to the Status of Refugees Art. 38.

54. Arts. 146, 147.

55. G. VON GLAHN, LAW AMONG NATIONS 13 (2d ed. 1970).

56. I. BROWNLIE, PRINCIPLES OF PUBLIC INTERNATIONAL LAW 14, 695 (3d ed. 1979).

57. *Cf.* C. PARRY, *supra* n. 46, at 114.

58. *See* Art. 38(1)(b) of the Statute of the I.C.J. *supra* n.44. For a discussion of the circumstances in which general practice is recognized as customary international law, *see* I. BROWNLIE, PRINCIPLES OF PUBLIC INTERNATIONAL LAW 4–12 (3d ed. 1979).

59. J. F. STARKE, AN INTRODUCTION TO INTERNATIONAL LAW (7th ed. 1972) 56–88.

60. See Art. 38(1)(d) of the Statute of the I.C.J. *supra* n. 44.

61. A discussion of whether or not the principles of the Declaration of the Rights of the Child have become part of customary international law appears in A. M. Pappas, *Introduction,* UNITAR REPORT, at xxix–xxxvii.

Index

Cuba (*continued*)
  evacuation of unaccompanied children from,
    124, 150, 157
  guardianship in, 232
  lessons from the refugee situation of, 56–57
  Mariel boatlift from, 54, 120
  and the missile crisis, 51, 52, 54
  programs for unaccompanied Cuban children,
    51–57
  refugees from, to the U.S., 51–56, 120, 150, 167,
    170, 171, 177
  repatriation of unaccompanied Cuban children
    to, 54
  resettlement of unaccompanied children from,
    53, 85, 118
  reunion of parents and unaccompanied children,
    of, 54–56, 177
  revolution in, 51–52, 118
  role of the child from, in settlement choice, 232
  unaccompanied children from, as refugees to the
    U.S., 51–57, 150, 157, 166
  and UNHCR, 271
  voluntary agencies and unaccompanied children
    from, 51, 53, 55
Cultural heritage, treatment and placement of
    unaccompanied children with regard to, 174,
    176–80, 187, 190, 191, 194–96, 206–7, 285,
    297, 316, 319, 334, 343–45. *See also* Ethnically
    similar associations and placement;
    Traditional heritage, of unaccompanied
    children, preservation of
Cultural identity, 205–6, 312
Culturally appropriate assistance, 328–29
Customary law
  international, 210
  national, 210
Cyclones, 12
Cyprus, refugee camps in, 24
Czechoslovakia
  and abduction of children, 21, 32, 34–36
  and Jewish children, 20, 237
  and the Red Cross, 23
  and unaccompanied children, 158, 159
  and UNRRA, 22

Dann, Sophie, 158–60
Death, of one or both parents
  as a cause of unaccompanied children, 12, 14, 16,
    25, 38, 46, 59, 69, 89, 91, 104, 106, 107, 116,
    121, 187, 198, 211, 295, 308
  declaration of, 243–44
  effect of, on adolescents, 164–67
  effect of, on unaccompanied children, 12, 148,
    149, 164–66, 170, 171, 178, 181, 331
Declaration of Geneva of 1924 (also known as
    Declaration of the Rights of the Child of
    1924), 212, 217–19, 222–24, 246. *See also*
    Human rights
Declaration of the Rights of the Child of 1924. *See*
    Declaration of Geneva of 1924
Declaration of the Rights of the Child (DRC;
    1959), 212, 217–19, 222–31, 232, 246, 282,
    283, 300. *See also* Human rights

Declarations, 211–12, 279
  definition of, 210
Delegation, to international and voluntary
    organizations, 268, 270–72, 275
  definition of, 268
Denmark
  adoption of foreign unaccompanied children in,
    42, 184
  evacuation of foreign unaccompanied children
    to, 16, 20, 74, 83
  and institutional care, 200
  and UNRRA, 22
Department of Defense (U.S.), and Operation
    Babylift, 72
Dependency
  on adults, 160, 319
  of children, 136–39, 142, 218, 313
  status of, 294
Deportation, 18, 22, 23, 44
Deprivation. *See* Poverty
Destitution. *See* Poverty
Development, of children, in normal patterns, 135–
    46
  adolescence and, 136, 144–46, 314, 315
  age-related, 135, 136
  capacities and, 134–36
  climate and, 138–39
  early childhood and, 136, 139–42
  family and, 135, 136, 312–16
  infancy and, 314–15
  later or middle children and, 136, 142–44, 315–16
  limitations and, 134–39, 151
  needs for, 134–36, 229, 282, 284–85, 313, 320
  neutral, 139
  physical, 135
  psychological, 135–41, 151
  similarities in, 135, 144
  social, 135, 136, 138, 140, 199
  stages in, 135, 136
  studies (often cross-cultural) regarding, 134, 137–
    39, 141, 144, 145
Direct service organizations
  and care of unaccompanied children, 38–39, 49,
    55, 59–66, 76–78, 80–81, 86, 91–97, 99, 100,
    102–5, 110–11, 119–21, 127, 128, 131, 132,
    152, 161–63, 172, 196, 230, 231, 267–69, 271,
    301, 303, 304, 332
  definition of, 291
Displacement. *See also* Refugees
  and camps, 169–73, 188
  causes of, 173
  and dependency, 130
  in Europe after World War II, 236–40
  and guardianship, 237
  law on, 236–40, 253–58
  and refugees, 205–6, 256–57, 275
  role of the child in resolution of, 231–34
  and unaccompanied children, 14–15, 18, 22, 59,
    62, 67, 89, 100, 104, 116, 118, 122, 127, 128,
    173, 231–34, 256–59, 285, 336, 343
  of unaccompanied children outside country of
    origin, 336–46
  and UNHCR, 269, 311, 323
  and UNRRA, 22

overcoming effects of, 163–69
physical and mental development owing to, 161–62, 173, 179
prevention of, 286, 293–97
psychological effects of, during refugee situations, 153, 158–63, 289
psychological effects of, during war, 153–63, 172–80
and psychosocial development, 163–64, 173
signs of distress in unaccompanied children owing to, 154–63, 172–73, 289
studies of, 158–64, 168–72, 174–79, 289
willful or voluntary, 16, 26–27, 52, 78, 115–18, 173, 220–21, 294, 313, 334
Settlement
child's role in decisions on, 340–42
definition of, 265
and guardianship, 311
law on, 263–66, 283–84
national policy on, 344–46
as part of a family unit, 338–40
of unaccompanied children outside their country of origin, 263–65, 336, 338–42, 344–46
Sexual abuse
in camps for refugees during and after World War II, 166
at Fort McCoy, 170
Short-term care. *See* Interim care
Siblings (and other relatives)
in early childhood, 140
family care for, 315–16
group care for, 318, 319
immigration of, as a group, 294, 297
in later or middle childhood, 143
as security, 151, 203
separation of, 125, 284
as sources of information, 324
Singapore, refugees from Indochina to, 77–78
Slave laborers, 18, 21–22
Slavs, 19–21
Social welfare. *See* Programs of assistance, for unaccompanied children
Social welfare services, 119–20, 124, 127, 128, 130–32, 291, 295, 332. *See also* Programs of assistance, for unaccompanied children
Society of Friends, 19
Solnit, Albert J., 160, 174, 182, 186, 229
Sontz, Ann, 195
South Africa, 3, 30
South America, 228
South Korea. *See also* Korea; North Korea
adoption programs in, 42–43, 184
illegitimate children in, 40–42, 118
orphanages in, 38–40, 42, 43, 132
programs for unaccompanied children in, 38–43, 162, 163, 167
refugees in, 38
unaccompanied children of, 37–43, 116, 120, 263
and the U.S., 37
South Vietnam. *See also* North Vietnam; Vietnam
boat people, 76–77, 81, 83, 166
and Cambodia, 89
characteristics of unaccompanied children of, 120, 152, 173

evacuation of children from, 124
extended family in, 68, 71, 75
family tracing for unaccompanied children of, 74, 75
foster care and, 70, 71, 79, 192, 194
group care and, 196–97, 200
immediate postwar exodus from, 75–76, 173
and independent children, 117
intercountry adoption and, 70–75, 181
and Operation Babylift, 71–75, 124, 221, 296
orphanages in, 67–71, 124, 127
refugee camps for unaccompanied children of, 77–80, 152, 166, 171, 173
and the Vietnam war, 67, 71, 75–77, 127, 170
Southeast Asia. *See also* Cambodia; Laos; Vietnam
boat people from, 76–77, 83
camps with unaccompanied children in, 3, 76–80, 83, 152
holding centers for unaccompanied children in, 3
resettlement and, 78–87, 137, 187–98, 200, 203–7, 256
unaccompanied children from, 76–80, 152, 256, 273, 274
Soviet Union. *See* U.S.S.R.
Spain. *See also* Spanish Civil War
and adoption without parental consent, 245
and Cuban refugees, 51
and the U.S.S.R., 16, 126
Spanish Civil War, 5, 6, 12–17
evacuation of children to other countries during, 13–16, 24–26, 220
family tracing services in, 19
governments and unaccompanied children, 13–17
Moroccan troops in, 13, 15
programs for unaccompanied children in, 13, 14, 16–17
and the Red Cross, 129
separation of children from their families in, 9, 13–17, 113
social welfare in, 127
Sprengel, Renate, 158
Staff recruitment and training for action on unaccompanied children, 290, 314
and documentation, 306
State
and best interests of the child, 234–40, 282, 283
and delegation of authority to international and voluntary organizations, 267–72, 275
and the Geneva Conventions of 1949, 248–52
and guardianship, 234, 240
as intervenor, 222, 300
and jurisdiction and choice of law regarding unaccompanied children, 211, 262–67
laws and practices of, 253, 273, 279, 283
and repatriation, 237–39, 262–66
and reunion, 237–39
State Department (U.S.), 50, 70, 72, 78
Steinbock, Daniel J., 5, 7
Street children, 9, 39, 79, 92, 113, 118, 120, 121, 129, 301, 303
causes of, 167–68
and children's villages, 167
in Colombia, 167–68
programs for, 168–69